What is Calculus?

From Simple Algebra to Deep Analysis

What is Calculus?
From Simple Algebra to Deep Analysis

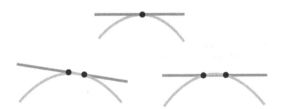

$$\frac{df}{dx}(a) = \lim_{h \to 0} \frac{f(a+h) - f(a)}{h}$$

R. Michael Range

State University of New York at Albany, USA

 **World Scientific**

NEW JERSEY · LONDON · SINGAPORE · BEIJING · SHANGHAI · HONG KONG · TAIPEI · CHENNAI · TOKYO

Published by

World Scientific Publishing Co. Pte. Ltd.
5 Toh Tuck Link, Singapore 596224
USA office: 27 Warren Street, Suite 401-402, Hackensack, NJ 07601
UK office: 57 Shelton Street, Covent Garden, London WC2H 9HE

Library of Congress Cataloging-in-Publication Data
Range, R. Michael.
  What is calculus? : from simple algebra to deep analysis / by R. Michael Range (State University of New York at Albany, USA).
    pages cm
  Includes index.
    ISBN 978-9814644471 (hardcover : alk. paper) -- ISBN 978-9814644488 (pbk. : alk. paper)
  1. Calculus--Textbooks. I. Title.
  QA303.2.R36 2015
  515--dc23
                          2015019347

British Library Cataloguing-in-Publication Data
A catalogue record for this book is available from the British Library.

Printed in Singapore

To my grandchildren
Kareem, Kayan, Joshua, and Alexander
and all other students
who want to learn and understand calculus

Contents

Preface

Calculus is one of the great creations of the human mind. The mathematical ideas, concepts, and tools of calculus have played a major role in the physical sciences since the 17th century, when Isaac Newton (1642-1727) and Gottfried Wilhelm Leibniz (1646-1716) laid the foundations. Among the early successes of calculus was a thorough understanding of the motion of planets and stars, a complicated phenomenon that had intrigued mankind for thousands of years, and still continues to fascinate anyone who ever looks up into a star studded sky. In modern times, these applications evolved, for example, into one of the pillars that support the launching and tracking of communication satellites and that provide the theoretical foundation for space travel. The basic ideas of calculus have branched out and matured into *Analysis,* which for centuries has been viewed—next to *Algebra* and *Geometry*—as one of the three major areas of mathematics.

In essence, calculus allows a precise formulation of rates of change in very general and abstract settings, and it provides the tools to reconstruct, analyze, and make predictions about the process under consideration from information about the relevant rates of change. Historically, the development of calculus has been intimately intertwined with the physical sciences. However, in the last few decades the concepts and tools of calculus have been applied successfully in many other areas of human endeavor, reaching well beyond the classical applications. As areas such as biology, chemistry, economics, finance, and psychology, to name just a few, have become more quantitative, calculus has featured prominently among the mathematical tools used in these disciplines. Consequently numerous academic programs beyond mathematics, physics, and engineering encourage or require their students to learn the fundamentals of calculus, and many high schools, too, are offering introductions to calculus. Clearly there is much interest in calculus today.

Unfortunately, the transition from high school mathematics to calculus is not easy. Students are usually exposed to deep new concepts right at the beginning. In particular, important central applications such as variable velocity, slopes of tangents, and more general rates of change and derivatives are introduced by an approximation process that involves "limits" of certain expressions that formally approach the meaningless quotient 0/0. Therefore it becomes necessary to investigate and understand such "limits" in order to proceed. Algebraic examples involving polynomials, rational functions, roots, and so on, often tend to confuse matters: The limit as the input x approaches the value a, where x must be assumed $\neq a$, is ultimately found—after algebraic manipulations to remove the troublesome zero from the denominator—by what is de facto *evaluation* of an algebraic expression by setting $x = a$. Thus limits tend to get mixed up with evaluation, often leaving one wondering about what seem unnecessary complications. The confusing relationship between limits and evaluation had surfaced already at the origins of calculus in the 17th century, but that did not stop the pioneers from moving forward. The difficulties were only resolved in the 19th century, when mathematicians introduced precise—and necessarily complicated—technical descriptions of *limits*. Since then, these new abstract concepts—in varying degrees of technical detail—have become a major component of any introduction to calculus. Even when discussed in intuitive non-technical language, they present quite a challenge right at the beginning for anyone who wants to learn and understand calculus.

In this book we present a more elementary approach to derivatives for *algebraic* functions that completely avoids limits. More advanced concepts are only introduced later, when algebraic methods no longer work, for example while studying exponential functions. The heart of the matter is an up-to-date version of a fundamental idea that goes back to René Descartes (1596—1650), one of the intellectual giants of his time, and that has remained on the sidelines for centuries.

In more detail, we begin with a *Prelude to Calculus*, in which the ancient tangent problem and some of its variations are introduced and solved for polynomials and other algebraic functions—which are built up by *finite* processes—by using only elementary concepts familiar from high school algebra and geometry. In particular, no mysterious quotients 0/0 appear, and no limits whatsoever are needed at this stage. Basic rules and formulas are established in a direct and most natural way. The reader thus begins to learn about tangents, derivatives, and all the mechanical rules of calculus in a familiar setting, without getting burdened by investigations of more advanced concepts based on limits and infinite processes. At the end of the Prelude we turn our attention to the concrete *exponential function* $f(x) = 2^x$, perhaps the simplest and most familiar type of *non*-algebraic function. Such functions arise naturally in important applications involving, for example, compound interest, population growth, or radioactive decay. It quickly becomes clear that the algebraic tools and finite processes used up to this moment are no longer sufficient. In particular, no elementary techniques allow us to reduce the calculation of the slope of the tangent to simple evaluation. Instead, building upon the algebraic approach, we recognize that as a consequence of an elementary estimate the explicit algebraic derivative introduced earlier can also be captured by an *approximation process*. This new *non*-algebraic idea opens the door to solving the tangent problem for the exponential function. Proceeding along this way, numerical evidence reveals surprising new phenomena, and it becomes apparent that new and more advanced tools are needed.

We are thus ready to move on to the main topic of this book, that is, the "analysis" part of calculus. In Chapter I we review some necessary background material, with particular emphasis on those aspects that are important for our purposes. Much of this should be familiar to the reader from high school mathematics. The one exception is most likely the "completeness" of the *real numbers*, that is, that fundamental property that ensures the *existence* of those numbers that arise as "limits" in the study of non-algebraic functions. Motivated by results in the Prelude, in Chapter II we begin to develop the concepts of limits and tangents, i.e., derivatives, in the setting of exponential functions, so as to keep the discussion of the new and more complicated ideas as concrete as possible. Once this case is understood, it is then an easy step to extend the new concepts, as well as all the rules of differentiation already discovered in the Prelude in the algebraic setting, to the level of generality usually considered in calculus and in mathematical analysis. Other concrete examples studied in detail include logarithms (the inverses of exponential functions) and the trigonometric functions. The latter ones are essential for modeling periodic phenomena such as sound waves, or the motion of the planets around the sun. In Chapter III we discuss some important applications, focusing on simple models involving the basic transcendental functions, so as not to distract the reader with complications that would obscure the simplicity of the basic ideas. Finally, in Chapter IV, we consider the fundamental process of *reversing* differentiation that arises, for example, when one tries to solve the concrete problem of reconstructing a motion from its known velocity and/or acceleration. This requires a new type of approximation procedure that leads us to the *definite integral*, the other central concept in calculus, whose roots, in fact, go back to Archimedes (287 - 212 B.C.) and other Greek mathematicians over 2000 years ago. It is easily seen that integrals are closely related to the geometric notion of area, but their importance goes far beyond that aspect. To further motivate this fundamental idea, we discuss several other applications to concepts such as length of curves, income streams, probability distributions, and work of variable forces. The connection between definite integrals and the reversal of the differentiation process is captured by the so-called *Fundamental Theorem of Calculus*. Beyond its central role, this result also provides an important computational technique to evaluate integrals. We conclude with some significant applications of these results. For example, we use the fundamental theorem to show that another type of limit process, known as the *Taylor series*, provides an approximation of elementary transcendental functions such as exponential and trigonometric functions to any desired level of accuracy by certain explicit polynomials (the Taylor polynomials) of sufficiently high degree. Aside from its theoretical importance, this latter result has great practical applications, as it allows us to find explicit numerical approximations for the *values* of these functions that in most cases cannot be obtained by finite procedures. The highlight of the discussion of Taylor series reveals a deep and surprising connection between exponential functions and trigonometric functions, i.e., between growth and periodicity, that only becomes visible as we expand our horizon to include so-called *complex* numbers, an amazing extension of the familiar (real) numbers.

This introduction to calculus aims to carefully motivate the new ideas that are central to the subject and to discuss them in the proper context, so that the reader will be able to understand them better and also recognize why they are necessary. The topics are developed in a well-ordered sequence that progresses from familiar elementary algebra to the important new concepts that distinguish calculus, culminating with some remarkable deep results in analysis. Rather than developing a large number of formulas and computational techniques—which too often are quickly forgotten—our main goal is to deepen and enhance the understanding of the fundamental concepts and ideas of calculus. We hope that this may be of more lasting value for the reader when she/he applies these ideas and tools in the chosen discipline. Altogether, this book should give the reader a solid foundation in the ideas, main techniques, and classical applications of calculus without getting overwhelmed by distracting technical details. Beyond that, we hope that the reader will also gain some lasting appreciation for the amazing ideas and concepts that have become indispensable for an understanding of the physical world around us.

Suggestions for the Reader

This book has been written for a reader who wants to learn about calculus and understand *why* and *how* deep new mathematical concepts arise naturally as we study the world around us. The presentation is broad enough to suit readers at different levels and with different backgrounds. Many will have been exposed to calculus, either at the level of a first introduction, or perhaps by completion of standard college calculus courses. Others still may never have had the opportunity to explore mathematics beyond high school material, but are interested in learning about a central and historically significant part of mathematics. The Prelude should be studied carefully—with help from sections in Chapter I, as needed—before proceeding with the main part of the book. This applies, in particular, to readers who have already been introduced to calculus. They may be tempted to ignore this part (why do it differently?) and jump ahead to Chapter I, or even to Chapter II. They would thus miss out on much of the motivation for the need for limits and the fundamental property of completeness of the real numbers. Also, in later sections we often refer back to the Prelude for context and motivation. I therefore urge such readers to try to put aside what they learned in an earlier introduction to calculus and to approach the Prelude with an open mind. When moving on to the main part of the book, some material may appear quite difficult on first reading. This is to be expected. The new concepts are not easy, and precise mathematical notation and technical language cannot be completely avoided, even though we try to minimize these formal aspects. The reader should feel free to skip some of the technical details and explanations; one can always come back later to fill some gaps as needed. While moving forward, the reader should at least try to understand the context and the question that is considered at that moment, and keep track of the "big picture". If one feels lost, it might help to go back to the beginning of a section or chapter in order to gain a better perspective.

As for prerequisites, there are two basic requirements. For one, the reader should have the mathematical skills that are usually acquired through completion of high school algebra and geometry courses, including quadratic equations, polynomials, and the algebraic operations such as multiplication and factoring performed with them, and basic geometric concepts such as lines and circles in the plane, and their representations by algebraic equations in a coordinate system. While the essentials will be reviewed at appropriate places, it will help if the reader is already familiar with these topics. On the other hand, no prior knowledge of trigonometry is needed. While derivatives and applications of *sine* and *cosine* functions are studied thoroughly, the necessary background is carefully reviewed in the text. The other requirement is more general: the reader should be able to think clearly and be willing to put forth the effort required to learn and understand some deep and at times abstract concepts that are at the heart of important and central mathematical topics that are widely used in many disciplines.

As for classroom use, this book could be used as a text for an honors calculus class with well motivated students, where the instructor has quite a bit more flexibility in adjusting the course content. Most of the material could be covered in one semester; students would acquire a solid foundation and would then be ready to proceed with multivariable calculus. It could also be used as a text for a first course in "Analysis", to be followed by an "Advanced Calculus" course that would cover in detail the technical aspects and move on to analysis in *several* variables. More broadly, this book should be of interest and helpful as supplementary reading for students as well as instructors of calculus and/or analysis. Perhaps one or the other of the novel ideas found in this book (see the Notes for Instructors for more details) might eventually be adopted by some instructors and authors.

Finally, the material in the Prelude should be of interest to high school teachers. Polynomials and their zeroes and multiplicities are standard topics in high school algebra and/or precalculus courses. Consequently, the approach to finding tangents for polynomials discussed in the Prelude should fit right in. If desired, it could easily be extended to other functions, such as rational or root functions, as well as to all the standard rules for differentiation. In any case, this material would provide a highly non-trivial, yet simple application of standard algebra tools to the solution of a historically central problem.

To assist the reader, key formulas are numbered in sequence within each chapter, for example (I.1), (I.2), ..., (II.1), ... Similarly, statements such as definitions, theorems, lemmas, etc., are numbered in a single sequence within each section. Theorem 2.3 thus identifies the third such statement within Section 2 of the current chapter, and could be followed by Definition 2.4, and so on. A reference such as Section 2.5 identifies subsection 5 in Section 2 in the current chapter. A reference to an item in a different chapter is augmented by the appropriate Roman numeral, e.g., Theorem III.2.3, or Section II.2.5. Exercises are found in the last subsection of each section.

Notes for Instructors

As outlined in the Preface, this book differs significantly from most of the existing calculus and/or analysis texts either designed for the typical calculus sequence with a standard syllabus, or targeted at more special audiences, such as business, biology, or more advanced mathematics students. Its main goals are described in the Preface. The emphasis is on motivating and explaining the relevant concepts so that the reader will be able to understand how the various pieces fit together and recognize the need for the new and at times abstract fundamental ideas that distinguish calculus from high school algebra. The majority of the exercises are chosen so as to reinforce such understanding. The focus on fundamental concepts emphasized in this book should be valuable for all students in disciplines that require a knowledge of calculus, whether or not such students will take any more advanced courses in analysis or mathematics. This applies, in particular, also to students in the mathematical and physical sciences. I believe such students, too, would benefit from an introduction to calculus following the approach developed in this book. It would equip them with a solid foundation and understanding, so that they may then profitably pursue more advanced and technical courses as appropriate.

The Table of Contents provides a detailed outline of the topics covered in this book. In this section I explain the main differences to the more traditional approaches and highlight a few other distinguishing features.

As stated in the Preface, a major new feature is a "Prelude to Calculus". Tangents and derivatives for polynomials and other algebraic functions are introduced by a purely algebraic elementary process based on factorization and double points. The basic idea goes back to René Descartes (1596—1650), who used double points to construct normals (and hence also tangents) for the ellipse and some other *special* algebraic curves. The implementation of this idea for *general* algebraic curves ran into major difficulties, and Descartes' method was eventually forgotten once the more analytic methods of Leibniz and Newton proved so enormously successful. It was discovered only recently that Descartes' ideas—properly reformulated—can be implemented in an elementary, transparent, and mathematically correct way. (See R. M. Range: *Where are Limits Needed in Calculus?* Amer. Math. Monthly **118** (2011), 404 - 417.) To summarize the method in the simplest case, the tangent to the graph of a polynomial P at the point $(a, P(a))$ is a line through $(a, P(a))$ that intersects the graph of P with "multiplicity greater than or equal to 2". It is then easily shown that there exists a unique line that satisfies this condition, and that its slope is given by $q(a)$, where q is the polynomial determined by the standard factorization $P(x) - P(a) = q(x)(x - a)$. Motivated by the simple polynomial case, the factorization result is easily extended to rational functions and their compositions, leading to the chain rule for derivatives. With just a bit more work this algebraic method is extended to inverse functions, products, and quotients, thereby obtaining all the familiar rules of differentiation. Most of this material could easily be presented in a high school algebra course, where it would provide a simple application of basic results about polynomials and their zeroes and multiplicities to the solution of a central and historically significant problem.

Most importantly, the simple factorization that is the heart of the method discussed in the Prelude is used to establish an estimate that exhibits in explicit form the continuity of an algebraic function f, that is, $|f(x) - f(a)| \leq K |x - a|$ for all x near a, where K is a suitable constant. Continuity is thus recognized as a fundamental property of all algebraic functions *before* there is any need to introduce the concept of a *limit*. As presented in the main part of this book, it is this estimate that motivates the concept of continuity and—most significantly for calculus—leads to the concept of derivative based on approximations and limits. In fact, when applied to the factor q in $P(x) - P(a) = q(x)(x - a)$, whose value $q(a)$ is the derivative $D(P)$ at the point a, the estimate $|q(x) - q(a)| \leq K |x - a|$ reveals the fundamental new idea that the value $q(a)$—that is, the derivative—can also be captured by $q(x)$ for $x \neq a$—that is, by the *average* rate of change—via an approximation process. This is the critical insight that opens the door to determining tangents and derivatives for non-algebraic functions.

The Prelude culminates with a preliminary investigation of the tangent problem for the simple exponential function $f(x) = 2^x$, where algebraic tools and evaluation no longer work. Motivated by the *approximation* procedure discovered for the algebraic case, one thus attempts to follow this alternate route. Numerical evidence quickly reveals that the slope of the tangent at the point $(0, 1)$ is approximated by numbers whose decimal expansions begins with 0.69314.... This unexpected and puzzling result makes it clear that new and more intricate phenomena occur as soon as one considers even the simplest non-algebraic functions. The stage is thus set for the main part of the book, that is, an introduction to the *analytic* version of calculus based on limits and continuity. Furthermore, since algebraic functions are already taken care of in the Prelude by simple algebraic methods, one can focus from the very beginning on the principal new ideas in their natural context, where they truly are indispensable.

From this perspective calculus, as part of *analysis* rather than algebra, begins with the simplest *non-algebraic* functions, such as exponential and trigonometric functions. These are the functions that occur most often in interesting applications involving growth or decay, or periodic phenomena, and their derivatives are given by simple differentiation formulas that however hide intriguing and deep phenomena, as evidenced by the surprising appearance of mysterious numbers such as 0.69314..., 3.14159..., 2.71828..., and so on. The importance of transcendental functions in calculus has of course been recognized for quite a while by many authors. Newer editions of classic textbooks are often offered in so-called "early transcendental" versions. Similarly, most texts designed for biology and/or business students also emphasize the elementary transcendental functions early on. Yet this emphasis typically just involves rearranging the order of sections, rather than a real shift in point of view. In contrast, in this book elementary transcendental functions are used systematically from the very beginning, i.e., after the Prelude, to develop the central new concepts of calculus.[1] Indeed, it is the author's view that the traditional introduction of derivatives of *algebraic* functions via limits, and the somewhat prominent role given the complicated product and quotient rules, lead to unnecessary detours and complications that delay and hinder the understanding of the main ideas.

Given the introductory nature of this book and its intended broad audience, the technical $\varepsilon - \delta$ definition of limits that features prominently in most analysis texts is not emphasized at all in this book. After all, calculus developed and flourished very well for over 200 years just based on a naïve

---

[1]Needless to say, a scientific calculator with graphing capabilities should be standard equipment for today's calculus students, just as decades ago slide rules and extensive tables were the standard tools used to find numerical values of the elementary transcendental functions.

understanding of limits and continuity. An intuitive understanding of continuity captured by the statement that $f(x) \to f(a)$ as $x \to a$, supported by the stronger explicit estimates available for all algebraic functions, is quite sufficient. For completeness' sake we introduce the standard precise definition of limit in one of the exercises and apply it in a couple of simple situations, but we do not dwell further on it. More advanced mathematics students, who will eventually have to learn this technical language, will have to consult any of the numerous texts in analysis. In that same spirit, proofs are often just discussed in a non-technical, though mathematically correct, outline. On the other hand, the *completeness* of the real numbers $\mathbb{R}$—a concept barely mentioned and/or relegated to an appendix in most introductory texts—is central for an understanding of limits and for analysis. Without it, there would be no assurance that the natural approximation processes that appear in calculus would have a limit that is part of our number system. Furthermore, in contrast to $\sqrt{2}$, for example, limits such as those denoted by $\ln 2$, e, π, and so on, are not even solutions of algebraic equations. For these reasons the author believes that completeness should not just be mentioned in passing and then ignored. Instead, it is at the core of our understanding of numbers as we use them in calculus once we go beyond algebraic functions. In particular, completeness is needed to identify limits with specific points on the number line, i.e., with some precise real numbers. Consequently, completeness is introduced early in Chapter I and formalized explicitly by the "Least Upper Bound Property". This geometric version appears as a natural property (i.e., an axiom) of the (number) line that is our standard model for $\mathbb{R}$. Given the importance of this property for analysis, the reader is often reminded of it along the way, especially when the correctness of certain intuitive arguments critically relies on the completeness of $\mathbb{R}$. Occasionally we also use completeness explicitly in the justification of statements when it might help to understand important principles.

Another significant difference to most calculus or analysis texts is the way "differentiability" is defined. Motivated by the central role of the factorization in the Prelude, one defines:

*A function f defined in a neighborhood of a point $a \in \mathbb{R}$ is **differentiable** at a if there exists a factorization*

$$f(x) - f(a) = q(x)(x - a), \text{ where } q \text{ is continuous at } x = a. \qquad (*)$$

The value $q(a)$ is called the derivative of f at a, and it is denoted by $D(f)(a)$ or $f'(a)$.

It follows directly from this definition that if f is differentiable at a, then its derivative $f'(a) = q(a)$ is well approximated by the values $q(x)$ for $x \neq a$ as $x \to a$, that is, by the average rates of change $[f(x) - f(a)]/(x-a)$ for $x \neq a$. In particular, one sees that this definition is equivalent to the standard one in terms of limits of difference quotients.

The above definition has been known and used occasionally for many years, but it still is not widely known, especially in the English language literature. To the author's knowledge it was first introduced by Constantin Carathéodory (1873—1950) in his classic text *Funktionentheorie* (Birkhäuser Verlag, Basel, 1950), and it has been used in a number of other German texts since the mid 1960s, both in real and in complex analysis. (See R. M. Range, op. cit.) Aside from the translation into English of Carathéodory's text (Chelsea Publishing Company, New York, 1956), the earliest English text known to the author that uses this formulation was published only in 1996. (A. Browder, *Mathematical Analysis*, Springer, New York, 1996.) A few years later it appeared also in the 3rd edition of the text by R. G. Bartle and D. R. Sherbert (*Introduction to Real Analysis*, 3rd. ed., John Wiley, New York 2000), and in the book by S. R. Ghorpade and B. V. Limaye (*A Course in Calculus and Real Analysis*, Springer, New York, 2006); these latter books make reference to Carathéodory. Still, differentiability continues to be *defined* via difference quotients, and it is then proved that this definition is equivalent to the formulation stated above. This latter version is then used in the proof of the chain rule and other results.

I believe that Carathéodory's definition has several advantages over the standard one, as follows.

- It is the natural generalization of the algebraic formulation.
- It avoids quotients with denominators that approach zero. We know that we cannot divide by 0, so—if at all possible—we should avoid anything that comes even close to it.
- It provides an easy and most natural proof of the chain rule by direct substitution, and of the inverse function rule (assuming the inverse of f is continuous at $f(a)$).
- It reduces technical details to simple standard properties of continuous functions.
- It naturally generalizes to the case of differentiable functions and mappings of several variables, thereby allowing a seamless transition from single to multivariable calculus.

- Last but not least it is a simple variation of the fundamental characterization of differentiability in terms of approximation by *linear* functions.

Regarding this last item, just rewrite the factorization (*) in the form

$$f(x) = [f(a) + q(a)(x - a)] + g(x)(x - a),$$

where $g(x) = q(x) - q(a)$. Clearly the continuity of q at $x = a$ is equivalent to $\lim_{x \to a} g(x) = 0$, that is, to the familiar property that characterizes the error term $g(x)(x - a)$ in the linear approximation for differentiable functions. This approximation property captures the critical geometric information that graphs of differentiable functions are, at the local level, essentially indistinguishable from their tangent *lines*. Rather than appearing—as in many standard calculus texts—as an after-thought that is mainly used to introduce "differentials" as a technique to approximate values such as $\sqrt{4 + 0.01}$ or $\sin(0.1)$, or to prove the chain rule, this approximation by linear functions is presented as the easy way to think of differentiability geometrically. Not only is this the property that is typically used as the defining one for functions of several variables, but it also very much enhances the understanding of some basic results. For example, it makes it clear that compositions and inverses of functions are the *natural* and more elementary operations to consider in calculus, rather than products and quotients. In fact, since the collection of *linear* functions is closed under composition and taking inverses, and since the verifications of the relevant differentiation rules are completely elementary for such functions, the extension of these properties to differentiable functions, which locally are essentially linear, hardly needs any further justification, at least at the conceptual level.[2] In contrast, the structures of product and quotient rules are necessarily quite a bit more complicated, since products and reciprocals of the approximating linear functions are no longer *linear*, so do not give potential linear approximations. Incidentally, the central role of the chain rule becomes even more evident if one observes that product and quotient rules can be viewed as simple special cases of the chain rule in several variables.

Let us mention a few other features of this book that are usually not found in standard calculus texts. For example, we follow the well-known practice to introduce the number $e = 2.7182...$ while searching for the base

---

[2] In fact, the definition of differentiability adopted here allows us to turn these intuitive arguments into precise proofs in a most elementary way.

for an exponential function $y = b^x$ with derivative 1 at $x = 0$. However, instead of using a common trial and error technique, we show that the value of the desired base is given by the exact formula $e = 2^{1/c}$, where c is the derivative of $y = 2^x$ at 0, whose existence is firmly motivated first by geometric and numerical evidence, and later verified exactly by explicitly using the completeness of the real numbers. Once e and the natural logarithm function are available, the number c is of course identified with $\ln 2$. Next, the derivatives of the sine and cosine functions are introduced via elementary geometric arguments based directly on the definition of these functions on the unit circle, rather than by the standard arguments that involve trigonometric addition formulas. Last but not least, the central fact that a function with derivative 0 at every point of an interval I is necessarily *constant* on I, is obtained by a direct intuitive (and mathematically correct) argument. (See R. M. Range, *On Antiderivatives of the Zero Function*, Math. Magazine **80** (2007), 387-390.) This avoids the unmotivated standard proof that involves a lengthy detour via existence of extrema, Rolle's Theorem, and the Mean Value Theorem, and it makes the critical role of completeness clearly visible. The basic intuitive argument is distilled into a formal Mean Value *Inequality*, which is all that is needed in order to discuss the relationship between properties of the derivative and geometric properties of the graph. Furthermore, this inequality readily implies the standard Mean Value Theorem for functions with continuous derivatives, a mild restriction that is insignificant for the purposes of this book.

As for applications of derivatives, we give priority to examples involving exponential growth models, simple initial value problems, and elementary periodic phenomena. Applications of calculus to graphing techniques are discussed thereafter, in a form that is much shorter than in traditional texts. Given today's highly sophisticated graphing calculators and computer algebra programs, it seems that these techniques are no longer as central as they used to be 30 years ago or so. A more significant difference involves the *early* introduction of higher order approximations by Taylor polynomials as a natural generalization of the linear approximation by the tangent line. This leads directly to Taylor series (i.e., power series), without the need for a separate detailed discussion of infinite series and all the convergence criteria that typically go with it. A formula for the remainder in the Taylor approximation and the related estimates are obtained later by a simple application of the Fundamental Theorem of Calculus and successive integrations by parts.

Another change from standard texts concerns the introduction and motivation of definite integrals. Rather then starting off with the new problem of calculating areas, which—taken by itself—is seemingly quite unrelated to derivatives, we consider the natural question of *reversing* the process of differentiation, i.e., how to recover the function if all one knows is its derivative.[3] This is first worked out in the context of motion, where the velocity is the known quantity, and hence the existence of an antiderivative, i.e., the distance function, is known *a priori*. The proof, of course, applies verbatim to the derivative $D(F)$ of any appropriate function F. Our presentation here has been inspired by the ideas of Qun Lin (*Calculus for High School*, People's Education Press (www.pep.com.cn), Beijing 2010). A precise version of the necessary uniformity condition is easily obtained for integrands $D(F)$ that have a *bounded* derivative. The process is then suitably modified to apply to functions f without any *a priori* knowledge of an antiderivative. By starting with an initial value and moving along short line segments with slopes given by the values of the function f at successive points, one readily obtains an explicit approximation procedure for constructing the values of a (potential) antiderivative of f. In essence, this is just the classical Euler method for solving differential equations, applied to the special case $y' = f(x)$. The approximating expressions are particular concrete realizations of *Riemann sums*, so that the limit that produces an antiderivative turns into a definite integral. In the case $f \geq 0$, these approximating expressions are readily interpreted geometrically as sums of areas of rectangles, thereby leading to the familiar approximation of the area under the graph of $y = f(x)$. This approach has the advantage that it directly ties antiderivatives to definite integrals, i.e., the heart of the matter is visible from the very beginning.

The proof of the integrability of *continuous* functions is quite subtle and technical, as it relies on the *uniform* continuity of a continuous function on a closed and bounded (i.e., compact) interval, a sophisticated concept that is difficult to formulate correctly without resorting to some version of the $\varepsilon - \delta$ machinery. Consequently many introductory calculus texts omit the proof or place it in an appendix or among the exercises. We bypass this difficulty by including a more elementary proof under the additional assumption that the integrand has a bounded derivative over the interval.[4] Note that all algebraic functions, as well as most combinations

---

[3]This approach is indeed much closer to the historical roots, as developed by Newton and Leibniz, than the emphasis on calculation of areas.

[4]Via the Mean Value Inequality this condition readily implies the Lipschitz continuity of the function, which is all that is needed. Rather than introduce a new definition, we

of elementary transcendental functions, satisfy this condition, so that for introductory purposes this restriction is not serious at all. Related simplifying techniques have also been investigated by H. Karcher (*Analysis mit gleichen Fehlerschranken*, Univ. Bonn, 2002) and M. Livshits (*You Could Simplify Calculus*, arXiv:0905.3611v1).

We conclude with some suggestions for instructors who want to use this book for a *non-traditional* one-semester (honors) calculus course, or for a first analysis course. The Prelude should be covered fairly quickly, say in at most three weeks, so as to leave ample time for the main topics. Besides the classical examples of tangents and their discussion for polynomials, one should definitely include early on the chain rule and inverse function rule (for rational functions) to highlight their elementary nature and emphasize their importance. On the other hand, the discussion of more general algebraic functions could be postponed until Section 6 in Chapter I. Similarly, product and quotient rules could also be deferred until they are taken up at the end of Chapter II in full generality, so as to avoid distracting complicated formulas in the early stages. However, the final section of the Prelude should be covered carefully, as it provides essential motivation for the main topics of the book. Among the basic concepts introduced and/or reviewed in Chapter I, completeness and the exponential functions are the most important ones. The latter functions are the primary examples used in the exploration of differentiability in Chapter II, which should be covered carefully. Section 2 in Chapter III includes fundamental results that are used in the remainder of the book, and the discussion on Taylor approximations in last section will be completed at the end of Chapter IV. The remaining sections in this chapter are pretty much independent of each other, and the instructor may choose to skip a few of them according to preference or if running short of time. In Chapter IV, Section 6 is somewhat theoretical and could be skipped. Section 7 includes, among others, the important example of a trigonometric substitution to find the antiderivative of the function given by $\sqrt{1-x^2}$ that arises in the calculation of the area of a disc. These results are not used thereafter, so this section could be left out. Integration by parts, however, is critical for completing the discussion of the convergence of the Taylor series in Section 9. The book concludes with a brief introduction to complex numbers and the application of Taylor series to the Euler Identity. This is a fitting grand finale for an introductory calculus course that should be included if at all possible.

---

prefer to formulate a sufficient condition in terms of known concepts.

Prelude to Calculus

1 Introduction

Differential calculus was developed in the 17th century in order to solve fundamental problems involving motion with variable velocity and the equivalent geometric problem of finding tangents to general curves. Tangents to simple special curves were already considered in antiquity, but their construction for general curves became possible only after the introduction of coordinates opened the door to using algebraic and analytic tools in the description of geometric properties. Similarly, motion subject to forces and acceleration could only be fully understood once the most creative minds of the 17th century were able to apply the new mathematical methods to real world phenomena, and expand their reach to new levels.

In this Prelude to Calculus we discuss tangents, their relationship to motion with variable speed, and all the standard rules of "differentiation" by means of elementary algebraic techniques familiar from high school algebra, without ever mentioning limits or other more advanced concepts. While the discussion is limited to the familiar *algebraic* functions (i.e., polynomial, rational, and root functions, and standard combinations of them), the simple proofs are presented in a form that will later readily generalize to exponential, trigonometric, and other more general functions that will be considered in the main part of this book. In the final section of this Prelude we attempt to use analogous methods for a simple exponential function, and we quickly recognize that some deeper new ideas are needed in order to deal with surprising new phenomena. This prepares the stage for the main topic of this book, that is, an introduction to the *analytic* version of calculus based on limits, and it will allow us to focus from the very beginning on the principal new ideas in their natural context, where

1

they truly are indispensable.

We shall freely use standard concepts and formulas familiar from typical high school geometry and algebra courses, including, for example, the slope of a line. The most important background material will be thoroughly reviewed in Chapter I, in a form that will include and highlight the critical concepts that are necessary to understand the new central ideas related to limits. The reader is encouraged to refer to appropriate sections in Chapter I as needed in order to follow the discussion in the Prelude.

2   Tangents to Circles

The construction of *tangent lines* to circles, parabolas, and similar classical curves has a long history, going back to Euclid (4th century B.C.), Apollonius (3rd century B.C.), and other Greek geometers over 2300 years ago. In antiquity a *tangent* was defined as "a line which touches a curve but does not cut it" [Victor J. Katz, *A History of Mathematics*, 3rd. ed., Addison-Wesley, New York 2009, p. 120]. The tangent appears to fit the curve near the point of contact in an optimal way. The situation is particularly simple for a circle C, where the tangent at a point P on C is that unique line that is perpendicular to the radial line connecting P to the center of the circle.

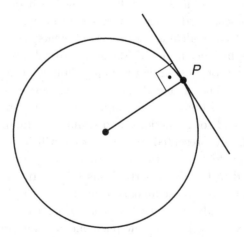

Fig. 1   Circle with tangent at point P.

In general, the line that is perpendicular to a tangent at a point P on a curve is called the *normal* to the curve at P. Circles are special, since all normals go through the center and consequently are easy to draw for any point P on the circle. However, when one considers more general curves, there is no obvious way to construct normals and/or tangents. Finding either one immediately determines the other.[1] The main problem then is to turn the intuitive but vague ancient idea of "tangent" recalled above into a precise definition that can be used to *identify* tangents and determine their slopes for arbitrary curves. Intuitively, we recognize that (in a small neighborhood) a tangent intersects the curve under consideration only at *one* point P—the *point of tangency*—while most small perturbations (i.e., changes) of the tangent will intersect the curve at *two* distinct points close to P. (See Figure 2.)

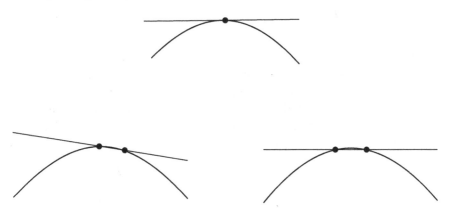

Fig. 2   Perturbation of a tangent reveals *two* points of intersection.

So the tangent intersects the curve at the single point P, which however covers *two* (or more) overlapping points that separate when the tangent is rotated just so slightly. The point P of tangency really accounts for *two* points of intersection that just happen to coincide in the special case of a tangent. We call such a point a "double point", or a point of "multiplicity two". Note that for any other line through P that "cuts" the curve—and hence does not fit our intuitive idea of a tangent— the point of intersection really gets counted only once. In Figure 3, the dashed lines are perturbations of such a line through P; they still intersect the curve only at one

---

[1]Typically, over the centuries, geometers have focused on *tangents*, although René Descartes (1596 - 1650), perhaps the best known mathematician and philosopher of the first half of the 17th century, preferred studying the *normal* to a curve. [V. Katz, op. cit., pp. 511-512]

point (at least in a neighborhood of P).

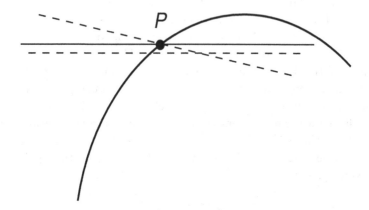

Fig. 3   Perturbations of a line that is not a tangent.

In certain situations the point of tangency may hide more than two points. In Figure 4 the horizontal line is tangent to the curve at P. Turning the tangent just so slightly counterclockwise will reveal two additional distinct points of intersection, for a total of *three* points. Such a point P is said to have "multiplicity three".

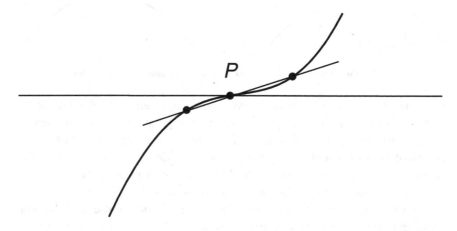

Fig. 4   A point of tangency of multiplicity three.

Based on these considerations we make the following geometric definition of a tangent.

Definition 2.1. *A tangent to a curve at the point P is a line that intersects the curve at that point with multiplicity two or higher, that is, a suitable arbitrarily small rotation of the line around P will separate P into two or more points of intersection.*

Now that we have a more precise definition of a tangent we can look for appropriate tools to identify such tangents, i.e., to find lines that intersect the curve with multiplicity two or higher. The introduction of *coordinates* by René Descartes in the 17th century was a major turning point, as it allowed mathematicians to translate geometric properties into algebraic properties involving numbers and equations, thereby making available algebraic methods for solving geometric problems. In particular, for a quadratic equation

$$x^2 + 2bx + c = 0$$

we are well familiar with the notion of a *double zero*, or zero of *multiplicity two*. This means that the two solutions $x_1 = -b + \sqrt{b^2 - c}$ and $x_2 = -b - \sqrt{b^2 - c}$ coincide; this occurs exactly when $b^2 - c = 0$. In this case, the equation takes the form

$$(x + b)^2 = 0,$$

which shows that the zero $x = -b$ has multiplicity 2, as the factor $(x + b)$ appears twice. Note that exactly in this case the x−axis is the tangent to the graph of $f(x) = (x + b)^2$. (See Figure 5.) Already at the dawn of

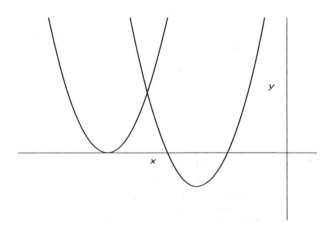

Fig. 5   Left: double zero; right: two distinct zeroes.

calculus Descartes used this insight to construct the normals to an ellipse.[2]

Let us apply this algebraic process to identify the tangents for a circle. To keep matters simple we place the center of the circle at the point $(0,0)$, i.e., at the center of the coordinate system, and we choose the radius to be 1, so that the equation of the circle is $x^2 + y^2 = 1$. Any (non-vertical) line through a fixed point (a, b) has an equation of the form $y - b = m(x - a)$, where m is the so-called *slope* of the line, which measures the inclination or direction of the line. This particular equation is known as the *point-slope form* of the line. (Note that b is not the "y-intercept" in this setting. Lines and their slopes are reviewed in detail in Section I.2.) Let us now choose (a, b) on the circle, so that $a^2 + b^2 = 1$, and consider lines through (a, b). (See Figure 6 below.)

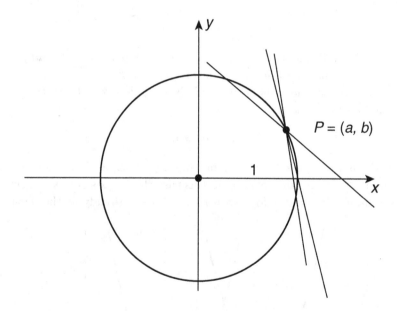

Fig. 6   Circle of radius 1 with lines through the point P.

We need to determine the slope m so that the line intersects the circle in a double point. This can be done by using simple familiar algebraic tools, as follows. The points of intersection (x, y) of the circle with such a line

---

[2]More precisely, Descartes constructed a *circle* which intersects the ellipse at a point P with multiplicity 2. The normal to that circle at P then coincides with the normal to the ellipse.

must satisfy the two equations

$$x^2 + y^2 = 1 \text{ and } y - b = m(x - a). \tag{P.1}$$

While the straightforward substitution of $y = b + m(x - a)$ into the first equation in (P.1) leads to a quadratic equation for x that can readily be solved by standard techniques, it is easier to take advantage of the fact that $x = a$ is one of the two solutions, i.e., the resulting equation must have a factor $(x - a)$. We use the equation $a^2 + b^2 = 1$ (the point (a, b) lies on the circle!) and subtract it from the left equation in (P.1). One obtains

$$x^2 - a^2 + y^2 - b^2 = 0,$$

which can be factored into

$$(x + a)(x - a) + (y + b)(y - b) = 0.$$

Now substitute $m(x - a)$ for $y - b$ and rearrange, so that the equation turns into the form

$$[(x + a) + (y + b)m] \, (x - a) = 0.$$

This clearly shows—as expected—that $x = a$ is one of the solutions, and that the other point of intersection (x, y) must satisfy

$$[(x + a) + (y + b)m] = 0.$$

Since we are looking for the slope m for which the point (a, b) is a *double* point of intersection, the second point (x, y) must be (a, b) as well, i.e., $x = a$ and $y = b$. Substituting these values into the last equation shows that m must satisfy

$$2a + 2bm = 0.$$

If $b \neq 0$, that is, if $(a, b) \neq (1, 0)$ or $(-1, 0)$, it follows that $m = -a/b$ is the slope of the unique line for which the point (a, b) of intersection with the circle is a *double* point. So $m = -a/b$ is the slope of the *tangent line* at the point (a, b). Note that this result confirms the classical construction: if we also assume that $a \neq 0$, the slope m_N of the *normal*, i.e., the radius line from the center $(0, 0)$ to (a, b), is b/a, and since $(-a/b)(b/a) = -1$, the tangent we determined algebraically is indeed perpendicular to the normal.[3]

---

[3] We use the fact that if m_1 and m_2 are the (non-zero) slopes of two lines, then the lines are perpendicular if and only if $m_1 m_2 = -1$. This result is discussed in Section I.2.3.

2.1 *Exercises*

1. Consider the curve given by $y = x^2$.

   a) Find the points of intersection of this curve with the line of slope m through $(0,0)$ given by $y = mx$.

   b) For which m is the point of intersection $(0,0)$ a *double* point?

2. This example illustrates a point of tangency of multiplicity 3. Consider the curve given by $y = x^3$. The line $y = 0$, i.e., the x-axis, is tangential to the curve and intersects the curve only at the point $(0,0)$.

   a) Set up the equation to determine the x-coordinates of all points of intersection of the curve with the line $y = mx$ with slope m.

   b) How many solutions are there in the case $m = 0$?

   c) How many solutions are there in the case $m < 0$?

   d) Find all points of intersection of the curve and the line in the case $m > 0$ (no matter how small). Are they all different? How many such points are there?

3. The (vertical) line given by $x = 1$ is the tangent to the circle $x^2 + y^2 = 1$ at the point $(1,0)$. Explain why any small perturbation of that line through $(1,0)$ is given by an equation $x = 1 + my$ for m close to 0. Find all points of intersection of the line $x = 1 + my$ with the circle C.

4. Modify the argument given in the text for the unit circle to find the equation of the tangent line to the ellipse given by the equation

$$\frac{x^2}{9} + \frac{y^2}{16} = 1,$$

   at the arbitrary point (a, b) on the ellipse.

5. Generalize problem 4 to the case of an arbitrary ellipse given by

$$\frac{x^2}{A^2} + \frac{y^2}{B^2} = 1,$$

   where $A, B > 0$ are the axis of the ellipse.

3 Tangents to Parabolas

Another classical curve studied extensively by Greek geometers is the parabola, which has the remarkable physical property that light rays that enter the parabola parallel to the axis of the parabola are reflected on the parabola so that they all go through one single point F on the axis, the so-called *focus*. (See Figure 7 below.) This property has important applications in optics; for example, the 3-dimensional version obtained by rotating the parabola around its axis provides the theoretical foundation for today's *parabolic telescopes.*

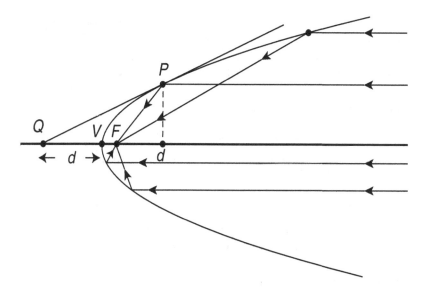

Fig. 7   Reflecting properties of the tangent of the parabola.

Parabolas arose in antiquity as special cases of so-called conic sections, that is, those curves that are obtained by intersecting a circular cone with a plane. Depending on the angle between the plane and the axis of the cone, these curves are either ellipses, parabolas, or hyperbolas. In particular, parabolas arise when the plane is parallel to the mantle of the cone. The great geometer Apollonius (3rd century B.C.) is credited with systematically recording the geometric definitions and known properties of the conic sections, and with discovering many additional properties. In particular, Apollonius discovered—in geometric language—a description of the parabola that is equivalent to the familiar algebraic formulation in Cartesian coordinates that we shall recall below. Most important for our discussion, based on this characterization, Apollonius deduced the following *geometric* construction of the tangent to a parabola at a point P. As shown in Figure 7, the (perpendicular) projection of P onto the axis of the parabola identifies a point at distance d from the vertex V. Consider the point Q on the extended axis that is at the same distance d from V on the opposite side. The tangent to the parabola at P is then that line through P that goes through the point Q.

Let us now translate geometry into algebra and apply the double point method—which was so successful for a circle—to determine the tangents to a parabola. We place the vertex V at the center $(0,0)$ of a Cartesian coordinate system and choose the axis of the parabola along the positive y-axis. The equation of the parabola is then $y = \lambda x^2$ for some $\lambda > 0$ that depends on the distance between the vertex and the focus. Let us fix a point (a, b) on the parabola. As before, any (non-vertical) line through (a, b) has an equation $y = b + m(x - a)$. Its points of intersection with the parabola are the solutions of

$$\lambda x^2 - b - m(x - a) = 0.$$

After replacing $b = \lambda a^2$ (the point (a, b) is on the parabola), this equation factors into

$$\lambda(x + a)(x - a) - m(x - a) = [\lambda(x + a) - m](x - a) = 0.$$

The two solutions are a and $m/\lambda - a$. Consequently (a, b) is a *double* point of intersection of the line with slope m precisely when $a = m/\lambda - a$, or $m = 2\lambda a$.

Example. At the point $(-1, 1)$ the slope of the tangent to the graph of $f(x) = x^2$ equals $2(-1) = -2$. Hence the equation of the tangent line at that point is $y = 1 + (-2)(x - (-1))$, or $y = 1 - 2(x + 1)$.

To complete the discussion, let us compare the algebraic result with the classical geometric construction of Apollonius. The projection of $(a, \lambda a^2)$ onto the axis of the parabola gives the point $(0, \lambda a^2)$ that is at distance $d = \lambda a^2$ from the vertex $V = (0, 0)$. According to Apollonius, the tangent at $(a, \lambda a^2)$ goes through the point $(0, -\lambda a^2)$, and consequently that tangent has slope $m = [\lambda a^2 - (-\lambda a^2)]/(a - 0) = 2\lambda a^2/a = 2\lambda a$. As expected, this agrees with result obtained by the double point method.

3.1   *Exercises*

1. Find the equation of the tangent line to the parabola given by $y = \frac{3}{2}x^2$ at the point $(2, 6)$.

2. Consider the point $(3, 9)$ on the parabola given by $y = x^2$.

   a) Determine the equation of the line through $(3, 9)$ with slope m.

   b) Substitute the equation in a) into the equation $y = x^2$ to obtain a quadratic equation in x of the form $x^2 + bx + c = 0$ for the x-coordinates of the points of intersection of the line with the parabola, where the coefficients b, c depend on m.

c) Determine m so that the discriminant $b^2 - 4c = 0$. Determine the solution(s) of the equation for this value m.

3. Consider the parabola $y = x^2/4$. Use the construction of Apollonius to find the points of tangency on the parabola for the tangents to the parabola through the point $(0, -6)$.

4. Find the equations of each of the tangents to $y = x^2$ that go through the point $(3, 0)$. (Hint: Make a sketch of the situation before starting any computations.)

4 Motion with Variable Speed

Before continuing with the study of tangents for other curves we first want to discuss the relationship of the tangent problem with a fundamental problem of motion. In fact, the search for a deeper understanding of motion and related phenomena in the physical world in the 17th century was arguably the major driving force that led to the development of calculus by Newton and Leibniz.

Experience shows that a stone that is dropped from the top of a building falls towards the ground at an *increasing* speed. The higher the building, the faster the stone will be falling just before impact. It was Galileo Galilei (1564 – 1642) who first analyzed the situation precisely in order to discover the underlying laws of motion. Rather than trying to explain the causes of phenomena by hidden actions of some mysterious higher being, Galileo thought to simply describe basic observations and use his analytical mind to distill the information into mathematical relationships. This shift from seeking to *understand the causes* of phenomena to the more modest goal to *describe them quantitatively*, turned out to be the breakthrough that— empowered by new mathematical tools—led to the amazing progress in mankind's understanding of the physical world since Galileo's days.

Based on numerous observations of falling stones[4] and balls rolling down inclined planes, and trusting that the observed motion is governed by simple principles, Galileo recognized in 1604 that the motion of a freely falling body is *uniformly accelerated*, i.e., the increase in speed over a time interval from t_1 to t_2 is a fixed multiple of the length $t_2 - t_1$ of that interval. In particular, if at time $t = 0$ the speed is zero, then at later times $t > 0$ the speed $v(t)$ equals $a \cdot t$ for a certain fixed number a, the so-called *acceleration*.

---

[4] It is often reported that Galileo carried out such experiments by dropping stones from the Leaning Tower of Pisa.

Another relevant quantity—more easily measurable than speed—is the distance that an object has moved in a given time interval.[5] Indeed, Galileo proved by geometric arguments that in the case of uniformly accelerated motion starting from rest, the ratio of the distances d_1 and d_2 traveled in corresponding times t_1 and t_2 equals the ratio of the *squares* of the times, i.e., $d_1 : d_2 = t_1^2 : t_2^2$. This translates into the formula $d(t) = ct^2$ for the distance $d(t)$ traveled in time t, where c is another constant. Galileo was able to confirm the validity of this latter formula in numerous experiments, thereby also obtaining a numerical value—which depends on the particular units chosen to measure distance and time—for the constant c. In the case of a freely falling object, and using today's standard units *meters* for distance and *seconds* for time, c is approximately 4.9 m/sec^2.[6]

Incidentally, Galileo's formula for a freely falling stone provides a practical technique to estimate the height of buildings or rock walls, as follows.

Example. Suppose a stone is dropped from the top of a building of unknown height H meters. Its height $h(t)$ above ground after t seconds is then given by the formula

$$h(t) = H - 4.9t^2 \text{ m},$$

where the minus sign accounts for the fact that the distance $d(t)$ traveled by the stone needs to be subtracted from the initial height H in order to get the height after t seconds. The rock hits the ground when $h(t) = 0$. Suppose this happens after t_0 seconds. Then $H - 4.9t_0^2 = 0$ implies that $H = 4.9t_0^2$ m. This formula is sometimes applied by rock climbers who need to estimate the height above a ledge in order to judge whether their rope is long enough to rappel down. Suppose a climber drops a stone, and by using a watch (a stop watch would be nice) she determines that the stone hits the ground after 3.5 seconds. By the preceding formula, the height above ground thus is approximately $5 \cdot (3.5)^2 \approx 61$ m, which is quite a bit more than her 50 m long rope. The climber thus decides not to rappel down at that location.

Returning to Galileo's result, the basic question that arises is how to derive a formula for the *velocity* $v(t)$ [7] of the falling stone at time t from

---

[5]For example, one could envision a long ruler placed vertically on the side of the building, with its initial point 0 placed at the top. A stop watch is started at the moment the stone is dropped, and one reads off the position of the falling stone against the ruler after $1, 2, ...$, seconds.

[6]The unit m/sec^2 for the constant c is a consequence of the relationship *distance* $= c \times (time)^2$. If *feet* is used instead of meters, the numerical value for c is approximately 16 ft/sec^2.

[7]*Velocity* is the term generally used in science for what common language calls *speed;*

the formula for the distance $d(t) = ct^2$. In particular, one needs to give precise meaning to the concept of *velocity at a single moment in time*. As commonly understood, velocity is a measure of the rate of change of position over time, that is

$$velocity = \frac{distance}{time}.$$

More precisely, for two distinct moments in time t_1 and t_2, the *average* velocity over the time interval $I = [t_1, t_2]$ (assume $t_1 < t_2$) is

$$v_I = \frac{d(t_2) - d(t_1)}{t_2 - t_1}, \tag{P.2}$$

where $d(t)$ is the distance traveled from the starting point $t = 0$, so that $d(0) = 0$. If the *average* velocity of a motion is independent of the time interval I, we say that the motion has *constant* velocity $v = v_I$. In the case of constant velocity, the velocity $v(t)$ at any moment t is always this same number v that equals the average velocity over any interval I, and it then follows easily that the distance $d(t)$ equals vt. However, in the case of the falling stone the velocity is *not* constant, so how do we define the velocity $v(t)$ at any particular moment? Intuitively, we agree that at any moment the falling stone is moving with a certain velocity, which increases with time until the stone hits the ground. Similarly in modern times, when traveling in a car, we do experience the (variable) velocity (or speed) at any moment, and the speedometer even gives us a number that measures this speed. If we apply the brakes, the speedometer indicates a decreasing speed. So what exactly is the speedometer measuring?

Notice that for a fixed moment t_0, while we agree that there is a velocity $v(t_0)$, surely we cannot compute the average velocity over the interval $[t_0, t_0]$ by formula (P.2), since this formula now gives the meaningless expression $\frac{0}{0}$. However, if we rewrite the equation that defines velocity as the product *distance = velocity × time*, then the problem becomes more manageable. In fact, let us consider the simple case considered by Galileo, i.e., $d(t) = ct^2$. If we fix a particular time t_0, then $d(t) - d(t_0) = ct^2 - ct_0^2$, which factors into

$$d(t) - d(t_0) = c(t + t_0)(t - t_0). \tag{P.3}$$

Note that if $t > t_0$ the factor $c(t + t_0)$ in this last formula obviously equals the average velocity over the time interval from t_0 to t. (Just divide both

---

velocity is allowed to be both positive and negative (or zero), with the sign accounting for the direction of motion along a line. More generally, when the motion is not constrained to a line, the velocity is represented by a so-called *vector*, a more complicated quantity that encodes, for example, the direction of the motion in space.

sides of (P.3) by $t - t_0 \neq 0$.) This also holds if $t < t_0$, where the time interval now goes from t to t_0. (See Problem 2 of Exercise 4.1.) Therefore, trusting in the consistency of the formula (P.3), we are led to define the velocity at t_0 by taking the value of this factor at $t = t_0$, i.e., we define

$$v(t_0) = c(t_0 + t_0) = 2ct_0.$$

Perhaps you have some doubts about the validity of this definition. After all, the basic formula *distance = velocity × time* reduces, in the case $t = t_0$, to the equation $0 = c(t_0 + t_0) \cdot 0$, which surely is correct, but then any other number k also satisfies the equation $0 = k \cdot 0$. So you may ask why do we single out the particular number $c(t_0 + t_0)$ among all the other possible numbers k that satisfy the equation?

One justification surely comes from the fact that $c(t_0 + t_0)$ is exactly that number that arises when t is replaced by t_0 in the algebraic formula $d(t) - d(t_0) = c(t + t_0)(t - t_0)$. Since this formula does represent a "universal truth", the value of $c(t + t_0)$ at $t = t_0$ should have an interpretation that is analogous to that for all other values t, that is, it should represent a velocity. And since only one moment in time t_0 is involved, it is reasonable to think of $c(t_0 + t_0)$ as the velocity at t_0.

Another justification is based on the geometric interpretation involving tangents to parabolas that we discussed earlier in Section 3. As we showed then (just replace $x = t$ and $y = d(t) = ct^2$), the line through the point (t_0, ct_0^2) with slope $2ct_0$ is the tangent to the graph of the function $d(t) = ct^2$, i.e., it is that line that fits the graph in an "optimal" way. Rephrasing this in the context of motion we thus can say that at the moment $t = t_0$, the *constant* speed motion $l(t) = ct_0^2 + 2ct_0(t - t_0)$ with velocity $2ct_0$ (i.e., the equation that defines the tangent) provides an optimal description of the motion given by $d(t) = ct^2$ at that moment. More precisely, this constant speed motion matches the given motion described by $d(t)$ at the moment t_0 "with multiplicity *two*", that is, at two points in time that just happen to coincide. Alternatively, think of a vehicle starting from rest at $t = 0$ under the same uniform acceleration as a falling stone, so that—according to Galileo—the distance traveled at time $t > 0$ equals $d(t) = ct^2$. At time t_0 the driver takes off his foot from the accelerator. Neglecting minor factors such as friction, air resistance, and so on, the car would continue rolling with *constant* velocity equal to $2ct_0$.

Finally we can also consider a *dynamic* point of view, which perhaps reflects most closely the crux of motion with variable speed, as follows. As we saw, for $t \neq t_0$ the value $q(t) = c(t + t_0)$ gives the *average* velocity during the time interval $[t_0, t]$ (or $[t, t_0]$ if $t < t_0$). Surely we expect that the velocity at t_0, no matter how defined, should be very close to the average velocity over very short time intervals, i.e., when t is very close to t_0, and furthermore, this approximation should improve as the time interval gets shorter, i.e., the closer t gets to t_0. The chosen value $v(t_0) = q(t_0)$ fulfills this expectation perfectly, since

$$|q(t) - q(t_0)| = |c(t + t_0) - 2ct_0| = |c|\,|t - t_0|. \tag{P.4}$$

Evidently formula (P.4) shows that when t is "very close" to t_0, then the average velocity $q(t)$ from t_0 to t is "very close" to $q(t_0)$ as well. For example, let us use meters and seconds, so that $c \approx 4.9$ m/sec^2. Suppose $t_0 = 5$ sec and $t = t_0 + 1/1000 = 5.001$ sec; then the average velocity $q(t)$ during the interval $[t_0, t]$ equals 4.9×10.001 m/sec, which differs from the velocity $q(5) = v(5) = 2 \times 4.9 \times 5$ m/sec by $4.9 \times 1/1000 = 0.0049$ m/sec. Stated differently, formula (P.4) gives a precise meaning to the intuitive statement that as t approximates t_0 (we write $t \to t_0$), then $q(t) \to q(t_0)$ as well. As we shall see later, the property we just discussed and that we encode in the statement

$$\text{if } t \to t_0, \text{ then } q(t) \to q(t_0),$$

is an elementary example of a fundamental abstract property that is known as *continuity*.

Our discussion shows that the concept of *instantaneous velocity*, i.e., velocity at a particular moment, is really just another version of the tangent problem. The techniques one develops in order to find the slope of tangents also allow us to define and calculate the velocity at a single moment in time. In particular, returning to Galileo's investigations of freely falling bodies, where $d(t) = ct^2$, we have determined that the velocity after t seconds is given by $v(t) = 2ct$. This confirms that the motion indeed is uniformly accelerated, with the acceleration a given by $2c \approx 9.8$ m/sec^2. Thus the distance formula under uniform acceleration takes the more informative form

$$d(t) = \frac{1}{2} \times acceleration \times t^2.$$

4.1 *Exercises*

1. Suppose a stone is pushed off a tower which is 60 m high. After how many seconds will the stone hit the ground?

2. Explain why the formula (P.2) gives the same value regardless of whether $t_1 < t_2$ or $t_2 < t_1$.

3. A coin dropped into a deep well hits water after 2.5 seconds. How deep is the well?

4. Let $f(x) = x^2 + 4x$.

    a) Establish an estimate $|f(x) - f(a)| \leq c\,|x - a|$ for $|x - a| \leq 1$ and some constant $c > 0$. (Hint: Factor $f(x) - f(a)$.)

    b) Explain why this implies that $f(x) \to f(a)$ as $x \to a$.

5   Tangents to Graphs of Polynomials

Before continuing with the tangent problem, let us review an important fundamental fact about zeroes of polynomials. Recall that a polynomial P is a function whose value at the real number x is given by a formula $P(x) = c_n x^n + c_{n-1} x^{n-1} + \ldots + c_1 x + c_0$, where the coefficients $c_0, \ldots, c_n$ are certain fixed numbers. If $c_n \neq 0$, the polynomial P is said to have degree n.

Proposition 5.1. *If the polynomial P of degree $n \geq 1$ has a zero at the point $x = a$, then $(x - a)$ is a factor of P, i.e., there exists a unique polynomial q of degree $n - 1$ such that*

$$P(x) = q(x)(x - a).$$

Proof 1. This is a well known simple consequence of the division algorithm for polynomials, as follows. By that algorithm, $P(x)/(x - a) = q(x) + R(x)/(x - a)$ for some polynomial q, where the remainder R is a polynomial of degree *less* than the degree of $x - a$, which is one. So R has degree 0 and hence must be a constant R_0. Thus $P(x) = q(x)(x - a) + R_0$, and evaluation at a shows that $0 = q(a) \cdot 0 + R_0$, so that $R_0 = 0$. This completes the proof of the proposition. ∎

Because this result is so important for our discussion, we shall also verify it by a different argument that does not rely on the division algorithm. The reader eager to proceed may surely skip this alternate verification.

Proof 2. We rewrite x as $(x - a) + a$ and note that

$$x^2 = [(x - a) + a]^2 = (x - a)^2 + 2(x - a)a + a^2\,.$$

More generally, for $k = 3, 4, 5, \ldots$ one similarly has

$$x^k = [(x-a)+a]^k = (x-a)^k + b_{k,k-1}(x-a)^{k-1}a^1 + \ldots + b_{k,1}(x-a)a^{k-1} + a^k$$

for some numbers $b_{k,1}, ..., b_{k,k-1}$. (The numbers $b_{k,j}$ can be described explicitly in terms of binomial coefficients: $b_{k,j} = \binom{k}{j}$.) All summands on the right side, except the last term a^k, contain the factor $(x-a)$. Consequently, $[(x-a)+a]^k = q_k(x)(x-a)+a^k$, where q_k is a polynomial of degree $k-1$. Therefore

$$\begin{aligned}
P(x) &= c_n[(x-a)+a]^n + c_{n-1}[(x-a)+a]^{n-1} + ... + c_1[(x-a)+a]^1 + c_0 \\
&= c_n[q_n(x)(x-a)+a^n] + c_{n-1}[q_{n-1}(x)(x-a)+a^{n-1}] + ... \\
&\quad ... + c_1[(x-a)+a] + c_0 \\
&= [c_n q_n(x) + c_{n-1}q_{n-1}(x) + ... + c_1](x-a)+ \\
&\quad ... + \left[c_n a^n + c_{n-1}a^{n-1} + ... + c_1 a + c_0\right] \\
&= q(x)(x-a) + P(a),
\end{aligned}$$

where $q(x) = [c_n q_n(x) + c_{n-1}q_{n-1}(x) + ... + c_1]$. Since $P(a) = 0$, the proposition is proved. ∎

Given the factorization $P(x) = q(x)(x-a)$, if q has a zero at some point $a^\#$, which means that P has zeroes both at a and $a^\#$, the proposition gives a factorization $q(x) = q^\#(x)(x-a^\#)$ for some other polynomial $q^\#$ of degree $n-2$, and consequently $P(x) = q^\#(x)(x-a^\#)(x-a)$. This result remains correct if the two points a and $a^\#$ happen to coincide, so that the zero at $a = a^\#$ is counted twice, that is, it has multiplicity 2 (or higher). One is thus led to the following definition.

Definition 5.2. *The polynomial P of degree n has a **zero of multiplicity** $\geq m$ at a for some m between 1 and n, if there exists a factorization*

$$P(x) = q_m(x)(x-a)^m$$

with some polynomial q_m of degree $n-m$.

We say that the zero at a has multiplicity *equal* to m if the multiplicity is $\geq m$ but not $\geq m+1$; clearly this occurs precisely when in the above factorization one has $q_m(a) \neq 0$.

We are now ready to apply the double point method to an arbitrary polynomial P, which we might as well assume to have degree ≥ 2. We fix a point $(a, P(a))$ on its graph. A non-vertical line through this point has equation $y = P(a) + m(x-a)$, and its points of intersection with the graph of P are the solutions of the equation

$$P(x) - [P(a) + m(x-a)] = 0. \tag{P.5}$$

We need to find the slope m so that this equation has a zero of multiplicity at least 2 at $x = a$. Since $P(x) - P(a)$ has a zero at a, Proposition 5.1 implies that $P(x) - P(a) = q(x)(x - a)$, and similarly it then follows that $q(x) - q(a) = k(x)(x - a)$, where q and k are polynomials of appropriate degrees. We want to emphasize that the polynomials q and k that are determined by these factorizations depend also on the point a that has been fixed. By combining the two factorizations one obtains

$$\begin{aligned}
P(x) - [P(a) + m(x - a)] &= q(x)(x - a) - m(x - a) = [q(x) - m](x - a) \\
&= [q(a) - m](x - a) + [q(x) - q(a)](x - a) \\
&= [q(a) - m](x - a) + k(x)(x - a)^2.
\end{aligned}$$

This representation shows that the equation (P.5) has a zero of multiplicity at least 2 at a if and only if $m = q(a)$.

We are thus justified in making the following definition that is just an algebraic version of the earlier geometric Definition 2.1.

Definition 5.3. *The **tangent line** to the graph of a polynomial P at the point $(a, P(a))$ is the (unique) line through $(a, P(a))$ that intersects the graph at that point with multiplicity at least 2. The slope of the tangent is called the **derivative** of P at a, and it is denoted by $D(P)(a)$, or also by $P'(a)$.*

This definition also applies if the graph of P is a line L, i.e., if P has degree ≤ 1. Note that in this case another line can intersect L with multiplicity greater than one only if the two lines coincide.

The preceding calculation proves the following elementary, but most important result.

Theorem 5.4. *The slope of the tangent line to the graph of P at the point $(a, P(a))$ is given by $q(a)$, where q is the polynomial factor in the representation $P(x) - P(a) = q(x)(x - a)$, that is, $D(P)(a) = q(a)$.*

We shall say that a function f is *algebraically differentiable* at the point a if the graph of f has a tangent line at $(a, f(a))$ according to the definition above, where the meaning of "multiplicity at least two" will have to be suitably modified according to the properties of f. Using this language, the result we just proved means that every polynomial is *algebraically differentiable* at every point. The process of finding the derivative (i.e., the slope) is also called *differentiation*.

Remarks on Notation. The symbol D refers to **D**erivative, and it is used to indicate that it is an operation applied to the polynomial P that results in a new function $D(P)$, the *derivative of* P. The symbol P' is often used for the sake of brevity. Historically, the derivative has also been denoted by the "differential quotient" dP/dx, a *formal* quotient of "differentials" dP and dx that were used to denote the vague concept of *infinitesimals*, or *infinitely small* quantities. This latter notation reminds us of the relationship of the derivative to the average rates of change $\Delta P/\Delta x$ that we encountered, for example, in the context of average velocity in the previous section. Since the approach chosen in this book emphasizes the factorization formula as a product, and since we avoid quotients that lead to $0/0$, we shall limit the use of the notation dP/dx mainly to applications, when we want to highlight the relevant variables under consideration and the interpretation of derivatives as rates of change.

As we just noted, the derivative $D(P) = P'$ of a polynomial P defines a new function given by $y = P'(x)$. By the rules established in Section 6, P' is again a polynomial. Consequently P' is also algebraically differentiable; its derivative $(P')'$ is written as P'' (or $D(D(P)) = D^2(P)$) and it is called the *second* derivative of P, or the *derivative of order two*. Similarly one can define derivatives of order three or higher, with analogous notations $P''' = P^{(3)}$, or $D^3(P)$, etc.

The factor q, whose value at a provides the critical piece of information to describe the tangent at the point $(a, P(a))$, can be computed by the division algorithm for polynomials, although that may not be the most practical approach for calculating the derivative. As we shall see in the next section, once we have developed some basic general rules, finding derivatives will turn out to be a quick and rather simple mechanical process. Let us discuss one such basic rule that is at the core of the differentiation rule for polynomials.

Example. For a positive integer n the derivative of $f(x) = x^n$ is obtained as follows. Fix a and factor $x^n - a^n = q(x)(x - a)$, where

$$q(x) = \sum_{j=0}^{n-1} x^{n-1-j} a^j. \qquad \text{(P.6)}$$

Then

$$D(f)(a) = q(a) = n\,a^{n-1}.$$

Consequently, the derivative of $f(x) = x^n$ at an arbitrary value x is given by $f'(x) = nx^{n-1}$, or $D(x^n) = nx^{n-1}$. Alternatively, we may simply write

$(x^n)' = nx^{n-1}$. This formula is called the *power rule* for differentiation. Note that in the case $n = 0$, i.e., if f is the constant function with value 1, one has $f(x) - f(a) = 0$, so that the factor q equals 0; this implies that f has derivative 0 at every point. Thus the power rule holds in this case as well.

Remark. The validity of formula (P.6) for q can readily be checked by multiplication. Another technique to find the value $q(a)$, i.e., the derivative, is based on replacing $x = a + h$, where $h = x - a$, and observing that

$$x^n - a^n = (a + h)^n - a^n$$
$$= a^n + nha^{n-1} + h^2 k(h) - a^n$$
$$= [na^{n-1} + h \cdot k(h)]h,$$

where k is a polynomial in h of degree $n - 2$ that also depends on the fixed number a. After substituting back $h = x - a$ in the last expression, one obtains

$$x^n - a^n = [na^{n-1} + (x - a)k(x - a)](x - a).$$

This shows that the factor q is given by $q(x) = na^{n-1} + (x - a)k(x - a)$. While this does not give the full explicit expression for q stated in (P.6), it does however imply the critical information that

$$D(x^n)(a) = q(a) = na^{n-1} + 0k(0) = na^{n-1}.$$

The algebraic differentiation process that we just discussed for polynomials extends immediately to *rational* functions $R = P/Q$ (i.e., quotients of polynomials) at any point a where R is defined, that is, where the denominator Q is non-zero. Given that $Q(a) \neq 0$, if $R(a) = 0$, then one must have $P(a) = 0$ as well, and hence $P(x) = q_P(x)(x - a)$ for some polynomial q_P. Consequently $R(x) = q(x)(x - a)$, where $q = q_P/Q$ is a rational function defined at a. If $R(a) \neq 0$, one obtains a corresponding factorization

$$R(x) - R(a) = q_R(x)(x - a) \tag{P.7}$$

with another *rational* function q_R defined at a. In analogy to the case of polynomials we say that a rational function R has a *zero at a of multiplicity* $\geq m$, where m is a positive integer, if $R(x) = k_m(x)(x - a)^m$ for some rational function $k_m(x)$ defined at a. By an argument analogous to the one used earlier for polynomials, it follows that a rational function R is *algebraically differentiable* at every point a where it is defined, i.e., its graph has a (unique) tangent line at the point $(a, R(a))$ defined by the property that it intersects the graph of R at $(a, R(a))$ with multiplicity at least two. The slope of the tangent (i.e., the derivative $D(R)(a)$ of R) equals the value $q_R(a)$, where q_R is the factor in equation (P.7), just as in the case of polynomials.

5.1 Exercises

1. Verify the following generalization of the power rule for differentiation. If c is a constant, then $D(cx^n) = cnx^{n-1}$ for $n = 1, 2, \dots$.
2. Show that if the rational function R is defined at the point $x = a$, then $D(cR)(a) = cD(R)(a)$ for any constant c.
3. Consider the rational function R defined by $R(x) = 1/x$ for all $x \neq 0$.

   a) If $a \neq 0$, show that $\frac{1}{x} - \frac{1}{a} = \frac{-1}{ax}(x - a)$.
   b) Use the result in a) to find the derivative of R at the point a.

4. Find the equation of the tangent line to the graph of $f(x) = 3x^4$ at the point $(1, 3)$. (Hint: Use Problem 1 above to find the slope of the tangent.)
5. Find the equation of the tangent line to the hyperbola described by $1 = xy$ at the arbitrary point $(a, 1/a)$ on the graph, where $a \neq 0$, by the following two methods.

   a) Write the equation of an arbitrary line through $(a, 1/a)$ with slope m. Substitute this into $1 = xy$ and determine m, so that the point of intersection $(a, 1/a)$ of the line with the hyperbola is a *double* point.
   b) Write the equation as $y = 1/x$ and use Problem 3 above to find the slope of the tangent directly.

6 Rules for Differentiation

In this section all functions will be assumed to be rational. As we shall see in the next section, the rules we develop in this section, as well as their proofs, will apply verbatim if rational functions are replaced by functions that are algebraically differentiable, which we defined earlier. Later, in the main part of this book, we will see that these rules also remain valid for arbitrary *differentiable* functions. In fact, except for a minor—though most critical—additional argument, the same proofs will work in that most general case.

6.1   *Elementary Rules*

We begin with the simplest rules, whose verification is straightforward.

Rule 0 (Power Rule). If $n \geq 0$ is an integer, then $(x^n)' = nx^{n-1}$. This is the rule we established already at the end of Section 5.

Rule I (Linearity).

(1) $D(cf) = c\, D(f)$ for any constant c.
(2) $D(f \pm g) = D(f) \pm D(g)$.

Rules 0 and I allow us to easily find the derivative of any polynomial P. **Examples.**

$$\text{i) } (4x^3)' = 4 \cdot (x^3)' = 4 \cdot 3x^2 = 12x^2.$$

$$\text{ii) } (3x^2 - 5x^4)' = (3x^2)' - (5x^4)' = 3x - 5 \cdot 4x^3.$$

$$\text{iii) } (5x^7 - 3x^6 + 2x^4 - 5x^2 + 7x - 4)' = 35x^6 - 18x^5 + 8x^3 - 10x + 7.$$

In general, if $P(x) = c_n x^n + c_{n-1} x^{n-1} + \ldots + c_1 x + c_0$, then

$$D(P)(x) = P'(x) = nc_n x^{n-1} + (n-1)c_{n-1}x^{n-2} + \ldots + c_1$$

is a polynomial of degree one less than the degree of P.

The verification of Rule I is straightforward. We prove Rule I.2, and leave Rule I.1 to the reader. Consider the factorizations $f(x) - f(a) = q_f(x)(x - a)$ and $g(x) - g(a) = q_g(x)(x - a)$. Then

$$(f + g)(x) - (f + g)(a) = f(x) - f(a) + g(x) - g(a)$$
$$= [q_f(x) + q_g(x)](x - a).$$

It follows that $(f + g)'(a) = [q_f + q_g](a) = q_f(a) + q_g(a) = f'(a) + g'(a)$. The proof with $-$ instead of $+$ works exactly the same way.

Rule II (Chain Rule). Recall that for two functions f and g, the composition $f \circ g$ of f and g is defined by evaluating first g and then inserting the output into f, i.e., $(f \circ g)(x) = f(g(x))$. Since we allow the functions to be rational, one must limit the input x to values a for which g is defined and so that f is defined at $b = g(a)$. (If both f and g are *polynomials*, there is no restriction on x.) The chain rule then states that

$$D(f \circ g)(a) = D(f)(b) \cdot D(g)(a), \text{ where } b = g(a), \text{ or}$$
$$(f \circ g)'(a) = f'(g(a)) \cdot g'(a).$$

By using functional notation, the chain rule can be written $D(f \circ g) = (D(f) \circ g) \cdot D(g)$. The crux of the matter is that the *derivative of a composition is the product of the derivatives.* The proof is very simple and natural. As before, we write $f(y) - f(b) = q_f(y)(y - b)$ and $g(x) - g(a) = q_g(x)(x - a)$, where q_f and q_g are the appropriate rational factors, and substitute $y = g(x)$ and $b = g(a)$ to obtain

$$(f \circ g)(x) - (f \circ g)(a) = f(g(x)) - f(g(a))$$
$$= q_f(g(x))(g(x) - g(a))$$
$$= q_f(g(x))q_g(x)(x - a).$$

Since $q_f(g(x))q_g(x)$ is a rational function defined at a, it follows that

$$(f \circ g)'(a) = q_f(g(a))q_g(a) = f'(g(a))g'(a),$$

as claimed.

Examples. i) Suppose $F(x) = (3x^3 - 5x^2 + 2)^{10}$. We could expand F into standard polynomial form by the binomial theorem and apply rules 0 and I to find the derivative. However, this involves a messy algebraic computation, and the simple structure of F and of its derivative would be lost. Instead, we note that F is the composition $F = f \circ g$ of the simpler functions $f(y) = y^{10}$ and $g(x) = 3x^3 - 5x^2 + 2$. By the chain rule it then follows that

$$F'(x) = [(3x^3 - 5x^2 + 2)^{10}]' = f'(g(x))g'(x)$$
$$= 10g(x)^9 g'(x)$$
$$= 10(3x^3 - 5x^2 + 2)^9(9x^2 - 10x).$$

ii) Let m and n be two positive integers. Then $(x^m)^n = x^{mn}$ by a standard rule for exponents. We calculate the derivative on the left by the chain rule, obtaining

$$[(x^m)^n]' = n(x^m)^{n-1}(mx^{m-1})$$
$$= nmx^{m(n-1)}x^{m-1}$$
$$= mnx^{mn-1},$$

where in the final step we have used another standard rule for exponents, i.e., $x^s x^t = x^{s+t}$. Note that the answer agrees with the direct application of the power rule to the right side x^{mn}.

6.2   *Inverse Function Rule*

Suppose the rational function R is one-to-one on the interval I, that is, if x_1 and x_2 are any two points in I with $R(x_1) = R(x_2)$, then $x_1 = x_2$. It then follows that R has an *inverse* function $x = S(y)$ defined on the set

$$J = R(I) = \{y : y = R(x) \text{ for some } x \in I\},$$

which satisfies $S(R(x)) = x$ for $x \in I$ and $R(S(y)) = y$ for $y \in J$. The following rule for the derivative of the inverse S is most natural, although its precise verification requires a little bit more work than the preceding rules.

Rule III. If $a \in I$ and $R'(a) \neq 0$, then S is algebraically differentiable at $b = R(a)$, and

$$D(S)(b) = \frac{1}{D(R)(a)}. \tag{P.8}$$

Remark. In explicit examples, such as the ones discussed below, one can typically check directly whether a function is one-to-one on a given interval (for example, one could apply the so-called "horizontal line test").[8] It is noteworthy that one can show that the condition $R'(a) \neq 0$ is already sufficient for R to be one-to-one on a suitably small interval that contains a. This will be discussed in greater generality in Chapter III.

Example. Consider the function $R(x) = x^2$ on the interval $I = \{x : x > 0\}$. R is one-to-one on I, and $R'(x) = 2x > 0$ for $x \in I$. So R has an inverse S given by $S(y) = \sqrt{y}$ that is defined on $J = \{y : y > 0\}$. Note that we can avoid any difficulties involving *irrational* numbers such as $\sqrt{2}$ if we limit x to just positive *rational* numbers, so that

$$J = \{y : y = x^2 \text{ for } x \text{ rational and } x > 0\}.$$

Rule III then implies that S is algebraically differentiable at any point $b = a^2$ with $a > 0$, and that, by (P.8),

$$D(\sqrt{y})(b) = \frac{1}{D(x^2)(a)} = \frac{1}{2a} = \frac{1}{2\sqrt{b}}.$$

Note that in exponential notation $\sqrt{y} = y^{1/2}$, so that the preceding result translates into

$$\left(y^{1/2}\right)' = \frac{1}{2}\frac{1}{y^{1/2}} = \frac{1}{2}y^{-1/2} = \frac{1}{2}y^{1/2-1}.$$

---

[8] The "test" states that a function $y = f(x)$ is one-to-one on the interval I if every horizontal line intersects the graph $\{(x, f(x)) : x \in I\}$ in at most *one* point.

This shows that the power rule **0** holds also for the exponent $1/2$. More generally, by applying the inverse function rule to $y = x^n$ with a positive integer n, one can check that the power rule remains valid for any exponent of the form $1/n$, that is,

$$D\left(y^{1/n}\right) = \frac{1}{n}y^{1/n-1} \text{ for all } y > 0. \tag{P.9}$$

(See Problem 5 in Exercise 6.5.)

The *proof* of the Inverse Function Rule **III** follows the familiar pattern.

Proof. Fix $a \in I$. By hypothesis,

$$R(x) - R(a) = q(x)(x - a),$$

where q is a rational function defined at $x = a$, and $q(a) = D(R)(a) \neq 0$. It follows that there exists an interval $I_a \subset I$ centered at a such that $q(x) \neq 0$ for $x \in I_a$. Therefore the rational function $1/q$ is defined on I_a as well, and it follows that

$$x - a = \frac{1}{q(x)}(R(x) - R(a)) = \frac{1}{q(x)}(y - b) \text{ for } x \in I_a. \tag{P.10}$$

By substituting $x = S(y)$ and $a = S(b)$, one obtains

$$S(y) - S(b) = \frac{1}{q(x)}(y - b) = \frac{1}{q(S(y))}(y - b). \tag{P.11}$$

If S were rational, the proof would be complete. However, since the inverse function S is not rational in general (i.e., it cannot be written as the quotient of two polynomials), some additional arguments are needed to show that S is algebraically differentiable. We note that for the rational function $1/q$ one has a factorization $(1/q)(x) - (1/q)(a) = k(x)(x-a)$, where k is rational as well. Hence

$$(1/q)(x) = (1/q)(a) + k(x)(x - a).$$

By substituting this into formula (P.11) and rearranging, one obtains

$$S(y) - S(b) - \frac{1}{q(a)}(y - b) = [k(x)(x - a)](y - b)$$

$$= \frac{k(x)}{q(x)}(y - b)^2 = [(\frac{k}{q}) \circ S](y)(y - b)^2,$$

where in the second equation we replaced $(x - a)$ by using formula (P.10). Since $[(k/q) \circ S]$—while not rational in general—is a well-defined composition of a rational function with S, this final formula shows that the line given by

$$L(y) = S(b) + \frac{1}{q(a)}(y - b)$$

does indeed intersect the graph of S at $y = b$ with "multiplicity at least 2". Indeed, this procedure naturally leads to the appropriate generalization of "multiplicity" from the known rational case to functions of a more general type. So S is algebraically differentiable at b with derivative $S'(b) = 1/q(a) = 1/R'(a)$, as claimed. ∎

In particular, we see that just as in the case of rational functions, the derivative $1/q(a) = 1/q(S(b))$ of S at b is precisely the value at b of the factor $1/q(S(y))$ in the relevant factorization (P.11).

Note that the inverse function rule becomes a special case of the chain rule once the latter has been extended to more general functions. In fact, if R and S are inverses of each other and algebraically differentiable at a and $b = R(a)$, respectively, the chain rule applied to the composition $S \circ R$—which satisfies $(S \circ R)(x) = x$—implies that

$$S'(R(a)) \cdot R'(a) = (S \circ R)'(a) = (x)'(a) = 1.$$

It follows that both $R'(a)$ and $S'(b) = S'(R(a))$ must be $\neq 0$, and the inverse function rule follows by dividing by $R'(a)$.

6.3 *Product Rule*

From the perspective of algebra, the product $f \cdot g$ of two functions defined by $(f \cdot g)(x) = f(x)g(x)$ might appear more natural and simpler than the composition $f \circ g$. However, for *derivatives*, the opposite is the case. Since by the chain rule the derivative of a composition is the product of the derivatives, we *cannot* expect the simple formula $D(f \cdot g) = D(f) \cdot D(g)$ for the product of two functions, because the right side is already "reserved". In fact, the rule for finding the derivative of a product is more complicated, as follows.

Rule IV (Product Rule).

$$D(f \cdot g) = D(f)\, g + f\, D(g).$$

Proof. Notice that rule **I.1** is a special case of the product rule: $(cf)' = c'f + cf' = cf'$, since $c' = 0$. For the proof of the product rule we suppose, as usual, that the two rational functions f and g are defined at the point $x = a$, and we rewrite the standard factorizations in the form

$$f(x) = f(a) + q_f(x)(x - a) \text{ and } g(x) = g(a) + q_g(x)(x - a).$$

Then

$$f(x)g(x) = f(a)g(a)+g(a)q_f(x)(x-a)+f(a)q_g(x)(x-a)+q_f(x)q_g(x)(x-a)^2.$$

It follows that the relevant factorization for $f \cdot g$ is given by

$$(fg)(x) - (fg)(a) = [g(a)q_f(x) + f(a)q_g(x) + q_f(x)q_g(x)(x-a)](x-a)$$
$$= q(x)(x-a),$$

where q denotes the rational function in the edged bracket [...]. Therefore $(fg)'(a) = q(a) = g(a)q_f(a) + f(a)q_g(a) = g(a)f'(a) + f(a)g'(a).$ ∎

Example. Let us take $f(x) = g(x) = x$. Then $(fg)(x) = x^2$, and hence $(fg)'(x) = 2x$. Since $f'(x) = g'(x) = 1$, clearly $f'(x)g'(x) = 1 \neq (fg)'(x)$. On the other hand, the product rule

$$(fg)'(x) = 1g(x) + f(x)1 = 1x + x1 = 2x$$

gives the correct derivative of fg. More generally, if $f(x) = x^n$ and $g(x) = x^m$ for two positive integers n and m, then, by the product and power rules,

$$(fg)'(x) = (x^n)'x^m + x^n(x^m)'$$
$$= nx^{n-1}x^m + x^n mx^{m-1}$$
$$= (n+m)x^{n+m-1}.$$

The answer agrees, as it should, with the direct application of the power rule to $x^{n+m} = x^n x^m$.

Example. Use the product rule to find the derivative of

$$f(x) = (x^3 - 4x + 1)(4x^5 + 2x^4 - x^3 + 20x).$$

Solution.

$$D(f)(x) = [D(x^3 - 4x + 1)](4x^5 + 2x^4 - x^3 + 20x)$$
$$+(x^3 - 4x + 1)D(4x^5 + 2x^4 - x^3 + 20x)$$
$$= (3x^2 - 4)(4x^5 + 2x^4 - x^3 + 20x)$$
$$+(x^3 - 4x + 1)(20x^4 + 8x^3 - 3x^2 + 20).$$

Do not simplify the answer any further.

6.4   Quotient Rule

The rule for differentiating the *quotient* of two functions is even more complicated then the product rule.   Let us first consider the simpler case of the *reciprocal* $1/f$ of a rational function f with $f(a) \neq 0$.   With q the appropriate rational factor that satisfies $f(x) - f(a) = q(x)(x - a)$, so that $q(a) = f'(a)$, it follows that

$$\frac{1}{f(x)} - \frac{1}{f(a)} = \frac{f(a) - f(x)}{f(x)f(a)} = \frac{-q(x)(x - a)}{f(x)f(a)}$$

$$= -\frac{q(x)}{f(x)f(a)}(x - a).$$

This factorization leads to the **reciprocal rule**

$$\left(\frac{1}{f}\right)'(a) = -\frac{q(a)}{(f(a))^2} = -\frac{f'(a)}{(f(a))^2},$$

or

$$D\left(\frac{1}{f}\right) = -\frac{D(f)}{f^2}.$$

Examples.   i) Let us apply the reciprocal rule to find the derivative of $y = 1/x$ at $x \neq 0$. It follows that

$$\left(\frac{1}{x}\right)' = -\frac{x'}{x^2} = -\frac{1}{x^2}.$$

ii) More generally, let m be any positive integer. Then

$$\left(\frac{1}{x^m}\right)' = -\frac{mx^{m-1}}{x^{2m}} = (-m)\frac{1}{x^{m+1}}.$$

By the definition of powers with negative exponents, this translates into

$$[x^{-m}]' = (-m)x^{-m-1}.$$

In this form the formula matches exactly the power rule **0** with exponent $n = -m$, i.e., for n a *negative* integer. By combining this last result with rule **0** one thus obtains the power rule

$$(x^n)' = nx^{n-1} \quad \text{for any integer } n \text{ and all } x \neq 0.$$

Of course the result holds also for $x = 0$ in the case $n \geq 0$. By combining this result with the formula (P.9) for $D(x^{1/n})$ for $x > 0$, and with the chain rule **II**, one verifies that the power rule

$$D(x^r) = rx^{r-1} \text{ for } x > 0$$

holds for any *rational* exponent r. (See Problems 7 and 8 of Exercise 6.5.)

iii) By the reciprocal rule, one obtains

$$D(\frac{1}{x^3+1}) = -\frac{3x^2}{(x^3+1)^2} \text{ for all } x \neq -1.$$

Remark. The reciprocal rule can also be obtained directly from the product rule. (See Problem 3 in Exercise 6.5.)

Finally, the general case of a quotient g/f of rational functions follows by combining the product rule **IV** with the reciprocal rule, as follows.

$$D\left(\frac{g}{f}\right)(a) = D[g \cdot \left(\frac{1}{f}\right)](a) = D(g)(a)\frac{1}{f(a)} + g(a)D\left(\frac{1}{f}\right)(a)$$

$$= D(g)(a)\frac{1}{f(a)} + g(a)\left(-\frac{D(f)(a)}{f(a)^2}\right).$$

Adding the two fractions gives the following formula.

Rule V (Quotient Rule).

$$D\left(\frac{g}{f}\right)(a) = \frac{D(g)(a)\ f(a) - g(a)\ D(f)(a)}{f(a)^2}$$

The expression in the numerator is very similar to the result of the product rule, except for the minus sign. It is thus very important to keep the order straight, i.e., to remember that differentiation begins with the numerator. Symbolically, if Num is the Numerator and Den is the Denominator, then

$$Quotient\ Rule: \quad \left[\frac{Num}{Den}\right]' = \frac{Num'\ Den - Num\ Den'}{Den^2}.$$

Example.

$$\left[\frac{x^3 - 4x^2 + 3x - 1}{x^2 - 9}\right]'$$

$$= \frac{(x^3 - 4x^2 + 3x - 1)'(x^2 - 9) - (x^3 - 4x^2 + 3x - 1)(x^2 - 9)'}{(x^2 - 9)^2}$$

$$= \frac{(3x^2 - 8x + 3)(x^2 - 9) - (x^3 - 4x^2 + 3x - 1)2x}{(x^2 - 9)^2} \text{ for all } x \neq \pm 3.$$

It is best to leave the answer in this last form which reflects the structure of the quotient rule, rather than to attempt any algebraic "simplification".

Remark. The quotient rule implies that the derivative $R'(x)$ of a rational function $R(x)$ is again a rational function that is defined wherever $R(x)$

is defined. Therefore R' is algebraically differentiable, and one can define its derivative R'', i.e., the second order derivative, as well as derivatives of higher order. All derivatives $R^{(n)} = D^n(R)$ are again rational with the same domain as R.

We conclude with an example that combines several rules of differentiation. It is best to proceed with one rule at a time, as appropriate, until all differentiations have been carried out. Moreover, do not attempt any simplifications neither during the calculations nor at the end.

$$\left(\frac{(x^3 + 2x)^6 \sqrt{x^2 + 1}}{4x + 5}\right)' =$$

$$\frac{D[(x^3 + 2x)^6 \sqrt{x^2 + 1}] \cdot (4x + 5) - [(x^3 + 2x)^6 \sqrt{x^2 + 1}] \cdot D(4x + 5)}{(4x + 5)^2} = (I)$$

by Rule V. Next,

$$(I) = \frac{[D((x^3 + 2x)^6)\sqrt{x^2 + 1} + (x^3 + 2x)^6 \, D(\sqrt{x^2 + 1})] \cdot (4x + 5)}{(4x + 5)^2}$$

$$+ \frac{-[(x^3 + 2x)^6 \sqrt{x^2 + 1}] \cdot D(4x + 5)}{(4x + 5)^2} = (II)$$

where we have used the Product Rule IV. Finally, by using Rules 0-III, one obtains

$$(II) = \frac{[6(x^3 + 2x)^5(3x^2 + 2)\sqrt{x^2 + 1} + (x^3 + 2x)^6(\frac{1}{2}\frac{1}{\sqrt{x^2+1}}2x)](4x + 5)}{(4x + 5)^2}$$

$$+ \frac{-[(x^3 + 2x)^6 \sqrt{x^2 + 1}]\, 4}{(4x + 5)^2}.$$

6.5   *Exercises*

1. Find the derivatives of the following functions:
   a) $P(x) = 4x^5 - 6x^4 - \frac{1}{5}x^3 + 3x^2 + 2$.
   b) $f(x) = 5x^3 + 7x^{1/2} - 3(x^2 + 1)^7$ for $x > 0$.
   c) $g(x) = 1/(3x^4 + 7x^2 + 2)^5$. (Hint: Use $1/(b^5) = b^{-5}$.)
   d) $k(x) = \frac{5x^3 - 2x}{4x^2 + x - 1}$.

e) $h(x) = 4/x^3 + 5x^{1/5} - 2\sqrt{x^4 + 2}$ for $x \neq 0$.

2. Find the derivative of $G(x) = (x^3 - 2x^2 + 4x)\sqrt{3x^2 + 1}$.

3. a) Derive the reciprocal rule for differentiation directly from the product rule by differentiating both sides of the equation $f \cdot (1/f) = 1$. Note that the reciprocal $1/f$ of a rational function f is rational as well, and therefore it is algebraically differentiable.

   b) Apply the analogous method to $f \cdot (g/f) = g$ to find the derivative of g/f.

4. Note that the reciprocal $1/f$ of a function f with $f(a) \neq 0$ can be written as the composition $1/f = g \circ f$, where $y(y) = 1/y$ for $y \neq 0$. Use the chain rule and power rule to prove the reciprocal rule for the derivative $D(1/f)(a)$.

5. Let n be a positive integer. The function $y = R(x) = x^n$ is one-to-one on the interval $I = \{x > 0\}$, with $R(I) = I$. Use the inverse function rule **III** to find the derivative of the inverse $x = S(y) = y^{1/n}$ on I. Verify that $D(y^{1/n}) = \frac{1}{n} y^{1/n\ -1}$.

6. This problem illustrates that root functions (i.e., the inverse function rule) can be presented just by using rational numbers, as follows. Restrict the domain of the function $R(x) = x^n$ in Problem 5 to the set $\mathbb{Q}^+ = \{r \in \mathbb{Q} : r > 0\}$ of positive rational numbers. Let S be the inverse of R restricted to the set $J = R(\mathbb{Q}^+)$. Find the derivative $D(S)(r^n)$ at the point $r^n = R(r)$. Note that the graph $\{(r, r^n) : r \in \mathbb{Q}^+\}$ is, for all practical purposes, indistinguishable from the familiar graph of R over the positive *real* numbers.

7. Prove that the power rule holds for arbitrary rational exponents $r = m/n$, $n > 0$. (Hint: Note that $f(x) = x^{m/n} = (x^{1/n})^m$ and apply the chain rule and the power rule for exponents $m \in \mathbb{Z}$ and $1/n$.)

8. Do Problem 7 by reversing the order in the composition, i.e., write $f(x) = (x^m)^{1/n}$.

7 More General Algebraic Functions

Notice that once one takes inverses of rational functions one ends up with functions that usually are no longer rational, but that are of a more general type. It is then natural to try to apply the differentiation rules we considered for rational functions to these new functions. More generally, let us consider the collection of functions $\mathcal{A}$ that are obtained from the rational functions by applying compositions and inverses, as well as the standard algebraic operations, a finite number of times, where the relevant functions are restricted to appropriate domains consisting of finite unions of open intervals, so that relevant quotients, compositions, and inverses are defined and algebraically differentiable on these intervals. For example, the sum of two functions f_1, $f_2 \in \mathcal{A}$ with domains Ω_1 and Ω_2 is defined on the domain $\Omega = \Omega_1 \cap \Omega_2$ provided Ω is not empty. Similar conventions need to be applied when one considers other algebraic operations involving functions in $\mathcal{A}$. Functions in $\mathcal{A}$ are also called *algebraic*.

. The most important fact is that the familiar factorization result for polynomials generalizes to functions $f \in \mathcal{A}$, as follows.

Lemma 7.1. *(Factorization Lemma.)* *If $f \in \mathcal{A}$ and a is in the domain of f, then there exists $q \in \mathcal{A}$ defined on the domain of f such that*

$$f(x) - f(a) = q(x)(x - a). \tag{P.12}$$

The proof of this statement basically involves checking through the proofs of the rules we discussed in the preceding section, where in each instance we were able to conclude that, given the factorization for the initial functions, one ends up with an appropriate factorization of the function that results by application of one or several of the admissible operations.

Based on the factorization result, it is clear how to generalize the notion of multiplicity of a zero a to the case of a function $f \in \mathcal{A}$.

In analogy to the case of polynomial and rational functions, successive application of the factorization lemma then implies the following result.

Corollary 7.2. *Given $f \in \mathcal{A}$ and the factorization (P.12), then*

$$f(x) - [f(a) + m(x - a)] = (q(a) - m)(x - a) + k(x)(x - a)^2$$

for some other $k \in \mathcal{A}$ that is defined on the domain of f.

Geometrically, this means that the line described by the linear function $y = f(a) + m(x - a)$ intersects the graph of $y = f(x)$ at $(a, f(a))$ with multiplicity at least two if and only if $m = q(a)$. Consequently, the line given by $y = f(a) + q(a)(x - a)$ is *the tangent* to the graph of f at $(a, f(a))$. This shows that the function $f \in \mathcal{A}$ is algebraically differentiable at a, with derivative $D(f)(a) = f'(a) = q(a)$, where q is defined by (P.12).

Furthermore, note that the differentiation rules I - V, including their verifications, only used the relevant factorizations and appropriate (algebraic) combinations of the functions and factors that arise. If the given functions are in $\mathcal{A}$, these combinations result in functions that remain within the class of functions $\mathcal{A}$. We thus obtain the following result.

All the rules for differentiation established in Section 6 remain valid for functions in the class $\mathcal{A}$ at all points in the domains of the respective functions.

It then follows that if $f \in \mathcal{A}$ is defined on the interval I, its derivative $D(f)$ defines a function on I that is again a member of the collection $\mathcal{A}$. Consequently one can define derivatives of higher order $D(D(f)) = f'', f''', ..., f^{(n)}, ...$ All derivatives $f^{(n)}$ are in the class $\mathcal{A}$ and have the same domain as the original function f.

Example. Let $f(y) = \sqrt{y}$ be the inverse of $y = x^2$ on $x > 0$. We already saw that f is (algebraically) differentiable on $I = (0, \infty)$, with

$$f'(y) = \frac{1}{2x} = \frac{1}{2\sqrt{y}}.$$

Let $g(x) = x^2 - 3x$. Since $g(x) > 0$ on $J = (-\infty, 0) \cup (3, \infty)$, the composition $(f \circ g)(x) = \sqrt{x^2 - 3x}$ is defined on J, is in $\mathcal{A}$, and is (algebraically) differentiable on its domain J, with

$$(f \circ g)'(x) = [(f' \circ g) \cdot g'](x)$$
$$= \frac{1}{2\sqrt{x^2 - 3x}}(2x - 3) \text{ for } x \in J.$$

One can then apply the rules from Section 6 to calculate $(f \circ g)'' = D[(f \circ g)']$ at points $x \in J$ as follows.

$$D[(f \circ g)'](x) = D\left[\frac{1}{2\sqrt{x^2 - 3x}}\right](2x - 3) + \frac{1}{2\sqrt{x^2 - 3x}}D(2x - 3)$$
$$= \left[-\frac{1}{4}(x^2 - 3x)^{-1/2 - 1}(2x - 3)\right](2x - 3) + \frac{1}{2\sqrt{x^2 - 3x}}2$$
$$= \frac{-(2x - 3)^2}{4(\sqrt{x^2 - 3x})^3} + \frac{1}{\sqrt{x^2 - 3x}}.$$

The structure of the formula for $D[(f \circ g)']$ is summarized by

$$D[(f \circ g)'] = D[(f' \circ g) \cdot g']$$
$$= [(f'' \circ g) \cdot g'] \cdot g' + (f' \circ g) \cdot g''.$$

As is well visible from this example, the calculation of successive derivatives of functions in $\mathcal{A}$, while based on repeated applications of the same basic differentiation rules, will very quickly result in more and more complicated functions in $\mathcal{A}$.

We conclude the discussion of algebraic functions with another important consequence of the factorization (P.12).

Theorem 7.3. *Given $f \in \mathcal{A}$ and a point a in the domain of f, there exist numbers $\delta > 0$ and K, such that one has the estimate*

$$|f(x) - f(a)| \leq K\,|x - a| \quad \text{for all } x \text{ with } |x - a| < \delta. \tag{P.13}$$

We had seen the significance of this kind of estimate already in Section 4, where it was used to recognize that the instantaneous velocity $v(t_0)$ is well approximated by average velocities over shorter and shorter time intervals around t_0. The crucial property expressed by the estimate (P.13) is that the values $f(x)$ approach $f(a)$ as $x \to a$, since clearly the left side of (P.13) becomes increasingly smaller as $|x - a| \to 0$. This is the essence of what is known as the *continuity* of the function f, a fundamental property that will be discussed more in detail in Chapter II. As we shall see in the next section, this approximation property is the critical ingredient that will allow us to study the tangent problem for more general functions that are not of algebraic type.

Proof. The proof of the theorem easily follows from the fact that functions $q \in \mathcal{A}$ are *locally bounded*, as follows: given a in the domain of q, there exist numbers $\delta > 0$ and K that depend on q and a, so that

$$|q(x)| \leq K \text{ for all } x \text{ with } |x - a| < \delta. \tag{P.14}$$

In order to prove the estimate (P.13), recall that by (P.12) one has $f(x) - f(a) = q(x)(x - a)$, where $q \in \mathcal{A}$ as well. Now use the above local bound (P.14) for the factor q to obtain

$$|f(x) - f(a)| = |q(x)|\,|x - a| \leq K\,|x - a|$$

for all x with $|x - a| < \delta$. ∎

To verify the existence of a local bound for functions in $\mathcal{A}$ is particularly simple for polynomials. Let $q(x) = c_n x^n + c_{n-1}x^{n-1} + ... + c_1 x^1 + c_0$ and choose *any* positive number δ. Standard estimations then imply that

$$|q(x)| \leq |c_n|\,\delta^n + |c_{n-1}|\,\delta^{n-1} + ... + |c_1|\,\delta^1 + |c_0| \text{ for all } x \text{ with } |x| \leq \delta,$$

that is, $|q(x)| \leq K$, with K equal to the constant on the right side of the preceding inequality. Things are a little bit more delicate in general. For example, note that the function q defined by $q(x) = 1/x$ is NOT bounded on the interval $(0, 1)$. However, if a is any point in the domain of q, then $a \neq 0$. Suppose $a > 0$, and take $\delta = a/2 > 0$. Then for all x that satisfy $|x - a| \leq \delta$ one has $x \geq a - \delta = a/2$, and therefore $|q(x)| = 1/x \leq 2/a$. The same sort of argument, choosing $\delta = |a|/2$, handles the case when $a < 0$. We shall discuss the proof of the estimate (P.14) in the general case in Chapter I.6.

7.1 *Exercises*

1. a) At which points is $f(x) = \sqrt{x^2 - 4}$ algebraically differentiable?

   b) Calculate $D(f)$ and $D^2(f)$. (Do not try any algebraic simplifications in the resulting formulas.)

2. Note that $g(x) = x^{1/3}$ is defined for all $x \in \mathbb{R}$. Show that g is algebraically differentiable at all $x \neq 0$ and find first and second derivative of g at such points.

3. Show that the function g in Problem 2 is NOT (algebraically) differentiable at $x = 0$. Reconcile this result with the (obvious) fact that the line $x = 0$ (the y-axis) is the tangent to the graph of g at $(0, 0)$.

4. Determine where

$$F(x) = \sqrt[4]{x^{1/2} \frac{4x - 1}{(x^3 + 3)^5}}$$

   is (algebraically) differentiable and find $D(F)(x)$ at those points.

8   Beyond Algebraic Functions

The discussion in the preceding sections has covered the differential calculus of algebraic functions. Only elementary algebraic tools were used, beginning with the basic factorization lemma for polynomials and the related concept of *multiplicity* of zeroes. These tools were then generalized in a natural and systematic way to all functions built up from polynomials by applying standard algebraic operations, including compositions and taking inverses, a finite number of times. No new results and concepts needed to be introduced beyond what is learned in typical high school algebra and geometry courses. In particular, we did not require any advanced concepts such as "limits" or "continuity", and no subtle properties of numbers were used beyond the basic arithmetic properties of the rational numbers, i.e., the quotients of integers. You may further have noticed that the formulas and other technical aspects really remained quite simple and natural until we got to the product and quotient rules. While the operations of taking products and quotients of functions are of course natural and useful, the complicated algebraic structure of the corresponding differentiation rules is quite surprising indeed, but it is important not to let these "unnatural" rules obscure the simplicity of the fundamental ideas.

In summary, the central ideas appear already at the very beginning, in the setting of the familiar polynomial functions. All subsequent work is just a variation of that theme, namely an enlargement by finite standard algebraic operations of the class of functions under consideration. The crux of the matter is the (algebraic) factorization

$$f(x) - f(a) = q(x)(x - a),$$

where the factor q is just another function of the same type as the original function f, which in principle can be computed explicitly, and that—most importantly—is well defined also at the point a by a unified algebraic formula. (See Lemma 7.1.) The value $q(a)$ is then the derivative $D(f)(a)$ of f at the point a. Depending on the setting, $q(a) = D(f)(a)$ gives the slope of the tangent line at the point $(a, f(a))$, the instantaneous velocity at time a, or, more generally, it can be viewed as an appropriate instantaneous rate of change at the input value a. From this point of view, the instantaneous velocity and other rates of change "at a single point" are captured by the *derivative* of the relevant functions. In particular, we do want to emphasize that many applications to classical topics in the physical sciences, such as velocity and acceleration, as well as to other areas, can be handled by the algebraic methods we have discussed so far, as long as the functions that are used to model the underlying phenomena are of algebraic type.

Unfortunately, the algebraic functions and the algebraic techniques we have discussed in this Prelude to Calculus are much too simple and limited in order to describe many of the fundamental phenomena of the real world. In response to this limitation the human mind, in its quest for deeper understanding, has created amazing new functions and abstract concepts that go well beyond the algebraic tools we have considered so far, and that— at its roots—require a sophisticated extension of the concept of number, resulting in the creation of the so-called *real* numbers that generalize the familiar fractions or *rational* numbers. As we shall see, the *real* story of differential calculus—in contrast to the elementary side discussed in this Prelude—begins when we reach beyond the algebraic functions and enter new uncharted territory.

Among the familiar phenomena that transcend algebraic methods are *periodic events*, such as the revolution of planets around the sun, waves in various media (e.g. sound waves or electromagnetic waves), or the fine structure of electrons circling the nucleus of an atom, and problems related to *growth and decay*, as they arise, for example, in the areas of biology (growth of populations), finance (compound interest), or physics (radioactive decay). The relevant simplest mathematical functions that need to be considered—such as trigonometric, exponential, and logarithm functions—have long been known, but they cannot be captured by finite algebraic formulas, concepts, and techniques. To highlight this fact, these functions and their close "relatives" are usually referred to as the elementary *transcendental* functions.

The more complex nature of these *transcendental* functions shows up clearly as soon as one investigates the tangent problem for these functions. To be specific, let us consider the simple exponential function $f(x) = 2^x$ that is used to describe a process in which the output doubles whenever the input is increased by one unit. In fact, by one of the basic rules of exponents, f satisfies

$$f(x + 1) = 2^{x+1} = 2^x 2^1 = 2f(x) \text{ for any } x.$$

It follows that if n is a positive integer, then $f(x + n) = 2^n f(x)$. Let us recall the definition of 2^x in the case where the exponent x is a rational number. (This was already used in Section 6 in the discussion of the power rule for differentiation in the case of *rational* exponents.) If m and n are integers, with $n > 0$, then $f(m/n) = 2^{m/n} = \sqrt[n]{2^m}$, i.e., $\gamma = 2^{m/n}$ is that (unique) positive number γ that satisfies $\gamma^n = 2^m$. It follows that γ can also be written as $\gamma = (2^{1/n})^m$. We must emphasize that—even though the same operation of "exponentiation" is used—the *exponential* function $f(x) = 2^x$ and the *power* function $y = x^2$, or more generally, $y = x^{m/n}$ are very different. The latter $y = x^{m/n}$ is of *algebraic* type, and its derivative was handled by finite algebraic methods in Section 6, while the exponential function $f(x) = 2^x$, as we shall see, forces us to come to grips with amazing new phenomena.

The graph of $y = 2^x$ for $x \in \mathbb{Q}$, which is easily produced with a graphing calculator (see Figure 8), looks just like an unbroken line that has been gently bent in the same direction across its total length according to some hidden rule.

Compared to the graphs of polynomials or rational functions, things could not get any simpler, short of just considering lines. And yet, this

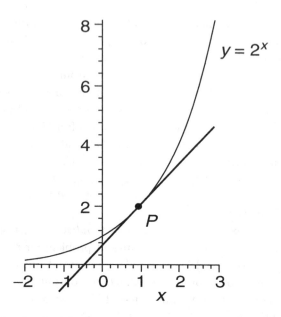

Fig. 8   Graph of the exponential function $y = 2^x$ with a tangent.

simplicity hides remarkable new phenomena that come to light as soon as one attempts to determine the tangent line at some point P on the graph. Figure 8 certainly suggests that there indeed is a line that fits our intuitive concept of *tangent* line—a line that *touches* the graph but does not *cut* it. Again, as we had seen in Section 2, this geometric feature is made precise by observing that small rotations of the tangent reveal that the point of tangency P is indeed a *double* point. In order to investigate the slope of the tangent more in detail, we simplify by choosing $P = (0, 1)$. Proceeding along the familiar path that was so successful in the case of polynomials and other algebraic functions, we look for a factorization

$$f(x) - f(0) = q(x)(x - 0), \text{ i.e., } 2^x - 1 = q(x)x.$$

Unfortunately, there is no obvious explicitly known factor q defined at $x = 0$ that fits this factorization. In particular, there is no *algebraic* function $q(x)$ that does the job. Furthermore, searching for some explicit expression for q built up from 2^x that would provide an unambiguous natural definition for $q(0)$ turns out to be futile. Of course, as long as $x \neq 0$, the value $q(x)$ is completely determined by the formula

$$q(x) = \frac{2^x - 1}{x} \text{ for } x \neq 0,$$

but this is useless for $x = 0$, since the formula would result in the meaningless expression $0/0$. Hence there is no way to evaluate $q(0)$, which—by analogy to the case of algebraic functions—would produce the value of the slope of the tangent, i.e., the derivative of $f(x) = 2^x$ at $x = 0$. However, the discussion of instantaneous velocity in Section 4 provides an important clue about how we might proceed. Recall the insight—based on an intuitive understanding of instantaneous velocity consistent with our experience—that the velocity $v(t_0)$ at a single moment t_0 should be approximated as closely as desired by the *average* velocity over smaller and smaller time intervals $[t_0, t]$. As we had seen, this important approximation property was made precise by a suitable simple estimate. In fact, at the end of the last section we generalized this estimate to all algebraic functions in the class $\mathcal{A}$. (See Theorem 7.3.)

If q were algebraic, the estimate $|q(x) - q(0)| \leq K |x - 0|$ (see equation (P.13)) would imply that $q(x) \to q(0)$ as $x \to 0$. In the case at hand q is of course not algebraic, and furthermore, we do not even have any clue for the value $q(0)$. The geometric version of this idea in the present setting suggests that the missing value $q(0)$ for the slope of the tangent should be approximated by the slope of lines through $(0, 1)$ and a second *distinct* nearby point $(x, 2^x)$ on the graph as $x \neq 0$ approaches 0. (See Figure 9.) In fact, for $x \neq 0$, the slope of such a line is given precisely by the quotient $q(x)$.

It certainly looks very plausible that the unknown slope m of the *tangent* can be approximated by $q(x)$ as the *non-zero* value of x gets closer and closer to 0. In Figure 9, as $x > 0$ moves closer and closer to 0, the point $(x, 2^x)$ glides towards $(0, 1)$ along the curve that is the graph of $f(x) = 2^x$, so that the line through $(0, 1)$ and $(x, 2^x)$ slowly turns in clockwise direction. In contrast to the situation in Section 4, where the value $v(t_0)$ of the instantaneous velocity was known to us by algebra, the present situation is more complicated, as we do not know a value m for the slope of the tangent, nor do we even have any obvious guess for it. We are literally shooting in the dark. Lacking a value for m, there is no way to estimate $|q(x) - m|$ as in the case of the velocity in Section 4. The best we can do is to analyze the behavior of the average rate of change $q(x)$ as the non-zero value x approaches 0.

Modern technology has created powerful tools that make this analysis easy and quick. A good programmable calculator would serve the purpose; a computer that runs one of the powerful computer algebra programs such

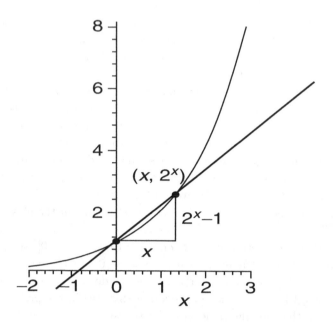

Fig. 9   Secant to $y = 2^x$ of slope $q(x) = (2^x - 1)/x$ for $x > 0$.

as *Maple* or *Mathematica* would be even better. Equipped with such tools, we can readily evaluate q for very small non-zero points x, and thereby obtain numerical approximations for the elusive slope m. Table P.1 shows the values $q(x_k)$ for $x_k = 10^{-k}$, $k = 1, 2, ..., 10$, evaluated to ten decimal places.

x_k	$q(x_k) = (2^{x_k} - 1)/x_k$
10^{-1}	0.7177346253
10^{-2}	0.6955550056
10^{-3}	0.6933874625
10^{-4}	0.6931712037
10^{-5}	0.6931495828
10^{-6}	0.6931474207
10^{-7}	0.6931472045
10^{-8}	0.6931471829
10^{-9}	0.6931471808
10^{-10}	0.6931471805

Table P.1. Approximation of slope of tangent to 10 digits.

It appears that the values $q(x_k)$ approach a number whose decimal expansion begins with $0.69314...$. Who could have guessed this by just looking at Figure 9? Let us increase the precision by evaluating $q(x_k)$ to 30 digits for $k = 11, ..., 20$. The result is shown in Table P.2.

x_k	$q(x_k) = (2^{x_k} - 1)/x_k$
10^{-11}	0.69314718056234757448682867899
10^{-12}	0.69314718056018553592419127767
10^{-13}	0.69314718055996933206792803208
10^{-14}	0.69314718055994771168230171247
10^{-15}	0.69314718055994554964373908055
10^{-16}	0.69314718055994533343988281736
10^{-17}	0.69314718055994531181949719104
10^{-18}	0.69314718055994530965745862841
10^{-19}	0.69314718055994530944125477215
10^{-20}	0.69314718055994530941963438652

Table P.2. Approximation of slope of tangent to 30 digits.

Consistent with the geometric interpretation, the numerical data does provide evidence that the values $q(x)$ approximate some "number" m_2 as $x \to 0$ that lies between 0.6931471 and 0.6931472, or—more precisely— between 0.69314718055994530 and 0.69314718055994531. However, even though we could narrow the interval that contains m_2 as far as we wish, limited only by the available computing technology, no precise familiar value seems to emerge from this process. For example, no periodicity appears in the decimal expansions displayed above, so it is not clear at all whether m_2 is a rational number.[9] And if m_2 is not rational, what type of "number" is it? Is it some "irrational" number that is the root of a polynomial equation with integer coefficients, analogous to the positive number λ that satisfies $\lambda^2 - 2 = 0$ and which is denoted by $\sqrt{2}$? Or does m_2 even transcend such "algebraic" numbers? We really cannot answer these questions at this time.

What is clear, however, is that the tangent problem for the simple natural function $f(x) = 2^x$ leads us into new, unknown territory. At the most fundamental level we are not even sure whether our basic concept

---

[9] Recall that a number is rational if and only if its decimal expansion is finite or periodic.

of number—which includes "irrationals" such as $\sqrt{2}$ beyond the familiar rational numbers—is sufficient to describe the truly complex phenomena that have come to light, and ultimately capture the "correct" value of the slope.

In order to answer some of these questions that are central for an understanding of basic growth phenomena, we need to take a few steps back and first build an appropriate foundation. This foundation should include, in particular, an understanding of the critical properties of the *number system* that we are using, of the basic concept of *function*, and of the *approximation process* that has emerged, first in an elementary and post-facto version in the study of tangents and of instantaneous velocity in the algebraic setting, and now in the far more intriguing form that arises in the study of tangents to the graph of a simple exponential function. We will therefore begin the main part of this book with an exploration of these foundations. We will try to focus on the principal ideas without getting entangled in technicalities. However, the reader needs to be willing to think carefully and not be deterred by some mathematical abstractions, as we try to describe one of the amazing creations of the human mind that has developed into an indispensable fundamental tool for understanding the world around us.

8.1 *Exercises*

1. Let $m = m_2$ denote the elusive number that measures the slope of the tangent to $f(x) = 2^x$ at $(0,1)$. Show that if the analogous approximation process is worked out at the arbitrary point $(a, 2^a)$ on the graph of f, it leads to the apparent result that the slope of the tangent at this arbitrary point is given by $m_2 2^a$. (Hint: Consider a second point $(a + h, 2^{a+h})$ with $h > 0$ and use $2^{a+h} = 2^a 2^h$ by a basic property of exponentials.

2. Use a scientific calculator or appropriate computing software to investigate, as in the preceding discussion, numerical approximations to the slope of the tangent to the graph of $g(x) = 10^x$ at the point $(0,1)$. Try to estimate the first 4 digits of that slope.

Chapter I

The Cast: Functions of a Real Variable

In this chapter we introduce and discuss in some detail the basic objects of study in calculus. Some of this material may be familiar to the reader, but other parts will be new, and particular attention should be given to the latter. As we realized at the end of the Prelude, the required foundations will include some material that is not part of typical high school courses. Consistent with the goals of this book, rather than aiming for *technical* completeness, we shall focus on the *key concepts and ideas,* and we will emphasize those aspects that are most important for an understanding of calculus.

I.1   Real Numbers

I.1.1   *Rational Numbers*

As we stated in the Preface, *Calculus* provides the mathematical ideas, tools and techniques to analyze rates of change in very general settings. These concepts have proved extremely useful for modeling phenomena in the natural sciences, including physics, chemistry, biology, as well as many areas of the social sciences. The quantities involved, such as time, distance, velocity, population size, blood pressure, profits, rate of inflation, invento- ries, etc., are usually described and measured by *numbers.* Relationships between different quantities are then expressed by *functions* of one or sev- eral variables, where each of the input variables, as well as the output of the function takes on numerical values.

Thus *numbers* are an important ingredient, and we need to have a solid understanding of their basic properties. While most people typically only have to deal with *rational* numbers, that is, with fractions, we recognized at the end of the Prelude that investigations of tangents for simple exponential functions require us to consider a number concept that is sufficiently broad to include the "limits" of certain natural approximation processes. This leads us to consider the intriguing and elusive property known as "completeness" that is usually not part of high school algebra. *Completeness*, while not always mentioned explicitly, is the critical ingredient without which the fundamental concept of limit—so central for the ideas of calculus—would more often than not lead nowhere and hence be meaningless.

Before getting to that new deep idea, let us first quickly review the basics of the *rational* number system. We are all familiar with the collection of *counting* numbers $1, 2, 3, ...$, also known as the set of *natural* numbers $\mathbb{N}$, and with the basic arithmetic operations of addition and multiplication. When reversing addition and multiplication, one sees that it is important to extend the natural numbers first by adding 0 and all their "negatives", thereby obtaining what is called the set $\mathbb{Z}$ of *integer* numbers, and next, by considering the reciprocals of non-zero integers and the related fractions. One thus obtains the set of *rational* numbers, such as $-2, 4/7, 0, -1/3$, and so on. More precisely, the set $\mathbb{Q}$ of rational numbers is given by

$$\mathbb{Q} = \{\frac{m}{n}, \text{ where } m, n \text{ are integers, with } n \neq 0\}.$$

There is a slight complication, due to the fact that certain fractions are "equivalent", i.e., they may look different but in fact represent the same number. For example, $2/3 = 6/9 = (-4)/(-6)$, or $5 = 5/1 = 10/2$. The general rule is that

$$\frac{m}{n} = \frac{p}{q} \text{ precisely when } mq = pn.$$

In particular, a rational number is not changed if the numerator and denominator are multiplied by the same *non-zero* number. This property is used extensively in calculations involving rational numbers.

Every non-zero number $q \in \mathbb{Q}$ has a unique *multiplicative inverse*, denoted by q^{-1} or $1/q$, defined by the property that $q \cdot q^{-1} = 1$. If $q = m/n \neq 0$, then $m \neq 0$ and $q^{-1} = n/m$. Division by a number n is just multiplication by its reciprocal (or multiplicative inverse) $1/n = n^{-1}$. You should be familiar with these matters and with the arithmetic operations on rational numbers. One important, though often forgotten or overlooked fact is that the denominator of a rational number is never allowed to be zero. In other words, 0 does not have a multiplicative inverse or reciprocal, that is, we **never divide by zero!**

Perhaps it is instructive to review the simple reasoning that forces us to ban division by zero forever. As you know, a basic arithmetic rule states that $0 \cdot a = 0$ for any number a. So, if m and p are any two integers, one has $m \cdot 0 = p \cdot 0$ and hence, if one could divide by 0 (i.e., multiply by a reciprocal 0^{-1} of 0) on both sides of this equation, one would get $m = p$, i.e., all integers would be equal. This is clearly absurd. Stated differently, the rules of arithmetic are incompatible with division by zero.

Most of elementary arithmetic never needs anything more complicated than rational numbers, which do include all numbers with finite decimal representation. That is why most people typically never learned, or do not recall much, if anything, about any other numbers, such as $\sqrt{2}$ or π. Fractions—in contrast to integers—are already hard enough, yet it is not easy to avoid them in ordinary life. For example, any measurement typically involves fractional components, whether expressed with a decimal point, or whether involving outright fractions such as 9/16 inches. Or consider a cooking recipe formulated to serve 4 people, and you want to use it to serve 6 people.

Any ruler serves as a concrete model for the fundamental process of identification of numbers with points on a line. The number line, our favorite model for the set of numbers, is just a ruler without any end on either side, that is, an infinitely long ruler. Once the numbers 0 and 1 are marked on it, all other rational numbers correspond to a unique spot on the line. (See Figure I.1.) Metric rulers highlight fractions with denominators 10,

Fig. I.1   A small section of the number line.

100, 1000, ..., while anglo-saxon rulers favor fractions with denominators 2, 4, 8, 16, and so on. In any case, every rational number corresponds to a specific point on the number line. Conversely, every point on the number line or on a ruler "seems" to be identified with a rational number. However, as we shall see shortly, this latter statement is not correct.

In the following, we think of "numbers" to be just points on the ruler, and all operations on numbers follow the same rules that we are familiar with from the system of rational numbers.

I.1.2   *Order Properties*

The next important property of numbers concerns their *order*. Moving from 0 towards 1 identifies the preferred direction of the line (usually displayed from left to right, or from bottom to top). This direction reflects the natural order in the numbers: numbers a to the right of zero, i.e., on the side where 1 is, are said to be *greater than zero* ($a > 0$), and are also called *positive*, those on the left are said to be *less than zero* $(a < 0)$ and are called *negative*. Also, $a < 0$ if and only if its additive inverse $-a$ is positive. If b is further to the right than a, then $b - a > 0$ and one writes $b > a$ (*b is greater than a*) or, equivalently, $a < b$ (*a is smaller (or less) than b*). For any number a exactly one of the following three distinct possibilities must hold: either $a = 0$, or $a > 0$, or $a < 0$. More generally, given any two numbers a, b one of the following must hold: either $a = b$, or $a < b$, or $a > b$. It is convenient to introduce the notation $a \leq b$ (or, equivalently, $b \geq a$) to mean that either $a = b$ or $a < b$.

The natural ordering interacts with arithmetic operations according to precise rules. First of all, given two positive numbers $a, b > 0$, then their sum $a+b$ and their product ab are positive, as is readily seen on the number line. Other rules then follow. For example, assume that $a < b$ and let c be any other number. Then $a+c < b+c$. The analogous multiplicative version $ac < bc$ however holds only if c is positive! If $c < 0$, the inequality reverses, i.e., one has $ac > bc$, or, equivalently, $bc < ac$. For example, multiplication of $2 < 3$ by (-1) on both sides gives $-3 < -2$, since on the number line the number -2 is to the right of -3. (See Figure I.2.)   These rules are not at

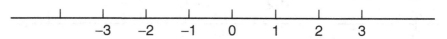

Fig. I.2   Positions of -3 and -2 compared to 2 and 3.

all arbitrary, but they are a consequence of the need to work with numbers in a logically consistent manner. For example, the rule that $(-1) \cdot a = -a$ (i.e., that unique number that satisfies $a + (-a) = 0$) is verified as follows. Notice that

$$(-1) \cdot a + a = (-1) \cdot a + 1 \cdot a = ((-1) + 1) \cdot a = 0 \cdot a = 0,$$

where we have used the distributive property and the fact that $(-1)+1 = 0$ since -1 is the additive inverse of 1. The equation $(-1) \cdot a + a = 0$ then implies that $(-1) \cdot a$ is the additive inverse of a, i.e., the number denoted by $-a$. This rule is at the heart of the familiar rules for multiplication

negative × *positive* = *negative*, and *negative* × *negative* = *positive*.

For example, if $a, b > 0$, then $ab > 0$, and $-a < 0$. Hence $(-a)b = [(-1)a]b = (-1)(ab)$, where we have used the associative property of multiplication. It then follows that $(-1)(ab) = -(ab) < 0$, since $ab > 0$.

We also recall that the *absolute value* $|a|$ of the number a is defined by

$$|a| = \begin{cases} a & \text{if } a = 0 \text{ or } a > 0 \\ -a & \text{if } a < 0. \end{cases}$$

So $|a| \geq 0$ for any number a. For example, $|-4| = 4$, $|4| = 4$. Geometrically, $|a|$ measures the *distance* between a and 0 on the number line. More generally, if a, b are two numbers, $|a - b| = |b - a|$ measures the distance between the two corresponding points on the number line.

The reader should be well familiar with all the standard rules of arithmetic and inequalities involving (rational) numbers. In particular, we recall the so-called "triangle inequality"[1] and its variations, which states

$$|a + b| \leq |a| + |b|.$$

Replacing b with $-b$ results in the equivalent estimate

$$|a - b| \leq |a| + |b| \text{ for all } a \text{ and } b.$$

By applying the former estimate to $|a| = |(a - b) + b|$, one obtains $|a| \leq |a - b| + |b|$, i.e., $|a| - |b| \leq |a - b|$. This same estimate holds if a and b are interchanged, so that $|b| - |a| \leq |b - a| = |a - b|$. It follows that

$$||a| - |b|| \leq |a - b|.$$

These estimates will be used extensively throughout this book.

I.1.3 *Irrational Numbers*

We now want to focus on a much more subtle property of the "points" on the number line that is not readily visible. It turns out that as one looks more closely at the number line (say with a super magnifying glass), one would see an unimaginable quantity of "tiny holes" scattered among the rational numbers. This "empty" space, i.e., points on the number line that are not rational, is indeed "real", that is, it cannot just be ignored. This

---

[1] This name originates by considering the distance of points in the plane; in that setting the corresponding inequality states that in a triangle the length of one side is less than or equal to the sum of the lengths of the other two sides.

fact, first recognized by Greek philosophers in the 4th century B.C., ranks among the great discoveries of the human mind. It shattered the belief that all observed quantities could be measured by integers and ratios between them (i.e., fractions).

Fact: *The diagonal in a square of side one cannot be measured exactly by a ruler that just includes rational numbers.*

We analyze the simple argument. Consider the unit square placed so that one side covers the number line from 0 to 1, as shown in Figure I.3. By Pythagoras' Theorem, the length d of its diagonal satisfies $d^2 = 1^2 + 1^2 = 2$. The diagonal is rotated onto the number line, thereby identifying a point at distance d from 0. The startling fact, as we shall explain in a moment,

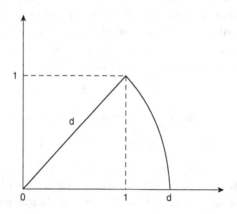

Fig. I.3   Diagonal d in a square of sides 1.

is that there is NO fraction $d = m/n$ that satisfies $d^2 = 2$. We are thus forced to conclude that the point $d = \sqrt{2}$ on the number line cannot be represented by a rational number, i.e., $\sqrt{2}$ is an "irrational" number. *The number line contains more than just rational numbers!*

So how do we see that no rational number m/n satisfies $(m/n)^2 = 2$? Suppose we had integers m, n that satisfy this equation. Clearly we must have $n \neq 0$, and we can also assume that $m, n > 0$. By basic number facts, we can cancel all common factors in numerator and denominator. Suppose that has been done, so that m, n cannot both contain a factor 2, i.e., we can assume that m and n are not *both* even. From $(m/n)^2 = 2$ one obtains $m^2 = 2n^2$, so m^2 is even. Since the square of an odd number is odd (check it!), this implies that m itself must be even, so $m = 2p$ for some integer p.

Therefore $(2p)^2 = 4p^2 = 2n^2$. After dividing by the common factor 2 on both sides one gets $2p^2 = n^2$, so n^2 is even. Again, this implies that n itself is even. So the assumption $(m/n)^2 = 2$ leads to the conclusion that both m and n are even, but that had been ruled out at the very beginning by canceling all common factors! This contradiction shows that it is impossible to find integers m, n with $(m/n)^2 = 2$, i.e., the number $\sqrt{2}$, which surely exists on the number line (the diagonal of the unit square!), is not *rational*.

Note that on a *practical* level the matter that $\sqrt{2}$ is not a rational number is not that important. We can always approximate the length of the diagonal by rational numbers to any desired degree of accuracy, say, by 1.414, or 1.41421, and so on. Again, this explains why in everyday life the matter is largely ignored. Any calculator displays $\sqrt{2}$ as a rational number, typically showing 8 or 12 digits, and no one worries that this is not "exact". Yet for theoretical considerations, and in particular for understanding and formulating basic concepts of calculus, it is important to be confident that the rational approximations really do approximate a concrete and precise point $d = \sqrt{2}$ on the number line, rather than some "hole" in the line.

$\sqrt{2}$ is just one particular example of a number that is not rational. It is a solution of the polynomial equation $x^2 - 2 = 0$. More generally, an *algebraic number* is a number x that satisfies a polynomial equation

$$a_n x^n + a_{n-1} x^{n-1} + a_{n-2} x^{n-2} + \dots + a_1 x + a_0 = 0,$$

where the coefficients $a_0, \dots, a_n$ are integers, with $a_n \neq 0$. Every rational number p/q satisfies the equation $q\,x - p = 0$, where p, q are integers, and hence is algebraic. On the other hand, as we just saw, algebraic numbers like $\sqrt{2}$ or $\sqrt{3}$ are *not* rational. More generally, for a rational number $r = p/q \geq 0$, its *nth* root $\sqrt[n]{r}$, n a positive integer, is algebraic (a solution of $qx^n - p = 0$) but will not be rational in most cases. Still other algebraic numbers are not even found on the number line: for example, there is no point on the number line that solves the equation $x^2 + 1 = 0$.[2] This latter phenomenon leads us to consider "complex" numbers, but we shall not pursue this right now.

Unfortunately, algebraic number points on the line still do not capture the vastness of space on the line. In fact, filling in the potholes in the rational line by algebraic numbers hardly makes a dent. This amazing fact was discovered in 1874 by the German mathematician Georg Cantor (1845

---

[2] Note that any point a on the number line satisfies $a^2 \geq 0$, so $a^2 + 1 \geq 0 + 1 = 1 > 0$, and hence $a^2 + 1 \neq 0$.

- 1918).[3] This discovery is perhaps even more surprising than the discovery of irrational numbers, and yet it is even more removed from everyday life. Furthermore, while it was quite elementary—at least with hindsight—to recognize that $\sqrt{2}$ is not rational, it requires quite a bit more ingenuity to identify specific points on the number line that are not even algebraic. Such points are called *transcendental* numbers. Certainly the reader has encountered at least one such number in high school mathematics, the famous number *Pi*, or π, the ratio between the circumference and the diameter of a circle. Because of its concrete geometric visualization—specifically, π equals the length of the circumference of a circle of diameter 1—this number "exists" just as well as any other number on the line. (Just think of your ruler as a flexible thin wire that you can wrap around a circle with diameter 1.) Approximate values for π, such as 22/7, were already known in antiquity, and it has been known since the middle of the 18th century that π is not a rational number. However, it was only verified comparatively recently in 1882 by Ferdinand Lindemann (1852 - 1939) that π is not even algebraic.

I.1.4 *Completeness of the Real Numbers*

As it became visible in Section 8 of the Prelude, a thorough discussion of tangents and derivatives reveals new phenomena that require a more extensive number system than just the rationals. In contrast to high school mathematics and algebra, one must include *all* points on the line, and most definitely the non-algebraic, i.e., the transcendental ones. The most important functions in calculus and its many applications, such as exponential, logarithm, and trigonometric functions, literally thrive on transcendental numbers. But most importantly, the fundamental idea of limit, which distinguishes calculus from algebra, requires that the system of numbers used is sufficiently "complete", so that one is guaranteed that in a wide variety of situations the limiting processes that need to be considered converge to a definite point on the line, which more often than not will turn out not to be a rational or not even an algebraic number.

So how do we recognize if and when we have filled in all the potholes in the number line? Well, we take a leap of faith, and *postulate* that the num-

---

[3] Cantor found a way to distinguish different orders of infinity. He showed that the set of algebraic numbers is "countable" (the simplest type of infinity that is modeled by the set ℕ of counting numbers), while the set of points on the "complete" number line is "uncountable", that is, it corresponds to a much higher order of infinity. We shall discuss this more in detail in the next section.

ber line is *complete* in a precise technical sense. Just as Euclid formulated his famous *axioms* to describe the properties of *lines* and of other basic geometric objects, we now add the requirement that a line—and consequently the "numbers" corresponding to it—satisfies the so-called *completeness axiom*, sometimes also referred to as the *continuity axiom*. A line, as drawn on paper, conveys the intuitive idea of something that can be drawn with a *continuing* stroke of a pen. We idealize by trusting that indeed there are no holes at all in the line. (Note: This really is a major idealization: the concrete physical line and the underlying paper have vast gaps of empty space between the molecules and atoms that make up its matter.) This idealization entails the statement that all holes are completely filled in by points, without any gaps whatsoever. By introducing a ruler, points on the line are related to numbers. The totality of numbers so obtained is the set of all *real numbers*, denoted by $\mathbb{R}$. Every point on the "complete" line corresponds to exactly one *real* number and vice-versa. The rational numbers $\mathbb{Q}$ are just a small proper subset of the set $\mathbb{R}$ of real numbers, though a very important one indeed. On the one hand the rational numbers have a simple representation as fractions, on the other hand the rational numbers are densely and evenly distributed along the line. In particular, every real number can be approximated to any desired accuracy by a rational number. More precisely, if $a \in \mathbb{R}$, for each positive integer n one can find a *rational* number s_n, such that the distance between a and s_n is smaller than 10^{-n}, i.e., so that $|a - s_n| < 10^{-n}$.

So what exactly is the set of real numbers? This question has no simple answer. A geometric model for $\mathbb{R}$ is given by the "complete" or "continuous" number line. Another concrete representation of real numbers is through their (usually infinite) decimal expansion. As you may recall, rational numbers are precisely those numbers whose decimal expansion is either finite or periodic. One could say that $\mathbb{R}$ is the set of all "possibly infinite decimal expansions", although it takes much detailed work to give this a precise meaning.

What ultimately matters are the *properties* of real numbers. As far as arithmetic and order, the relevant properties are exactly those one is familiar with from the rational numbers. The critical new feature is the *completeness* of the real numbers, i.e., that property that ensures that there are no hidden holes in the number line. This property, that is, the *completeness axiom*, needs to be formulated precisely so that it can be used effectively in arguments involving real numbers. Over the years, mathematicians have introduced several different equivalent formulations for this axiom. We shall focus on one of them, the so-called *Least Upper Bound Property*, which has a simple geometric visualization, and which can readily be applied in a variety of situations.

Suppose $S \subset \mathbb{R}$ is a set of numbers. We say that a number M is an *upper bound for S* if $s \leq M$ for every $s \in S$.

The Least Upper Bound (=LUB) Property of $\mathbb{R}$ (A version of the Completeness Axiom). *For every non-empty set $S \subset \mathbb{R}$ that has an upper bound, there exists a number $L_S \in \mathbb{R}$ that has the following properties:*

i) L_S is an upper bound for S.

ii) Any number $c < L_S$ is not an upper bound for S, i.e., given $c < L_S$, there exists $s \in S$, such that $c < s$.

Clearly the properties i) and ii) characterize the number L_S as the smallest (or *least*) upper bound for the set S. The situation is visualized on the number line by starting with an upper bound, i.e., a point M to the right of all points in S. We then move the point M to the left as long as it is possible to keep it to the right of S, i.e., we want that the new points $M^{\#} \leq M$ are still upper bounds for S. Clearly S provides a barrier for this process on the left, and therefore it intuitively looks very plausible that this process must stop at a "smallest" upper bound L_S. As expressed by the *LUB property*, the completeness axiom simply ensures that the real number line indeed matches what our intuition clearly expects. In other words, the completeness property ensures that the process of decreasing the upper bounds as much as possible really ends at a point $L_S \in \mathbb{R}$, rather than at some "hole" in the number line.

We note that the properties i) and ii) imply that there can be **only one** least upper bound L_S for the set S. In fact, if $r \in \mathbb{R}$ is not equal to L_S, then either $r < L_S$, so that r is not an upper bound of S by ii), or $r > L_S$, so that r could not satisfy ii), since L_S would be an upper bound for S that is smaller than r.

In order to better understand the significance of the LUB property, we shall analyze why the set $\mathbb{Q}$ of rational numbers does **not** have the LUB property. Let us consider the set $A \subset \mathbb{Q}$ defined by

$$A = \{r \in \mathbb{Q} : r > 0 \text{ and } r^2 < 2\}.$$

This set is clearly bounded. We claim that there is NO *rational* number b that satisfies the properties of a "least upper bound" in $\mathbb{Q}$, i.e., the rational numbers $\mathbb{Q}$ are not complete in the precise sense described by the completeness axiom. In fact, any positive *rational* number $b \in \mathbb{Q}$ must satisfy

$b^2 < 2$ or $b^2 > 2$. (Remember: We cannot have $b^2 = 2$ for $b \in \mathbb{Q}$!) One then shows by elementary, though somewhat tedious arguments involving inequalities that the following is true. If $b^2 < 2$, then $(b + r)^2 < 2$ as well for any sufficiently small rational $r > 0$, that is, $b + r$ is a rational number in A that is *larger* than b, so that b is *not an upper bound* for A. And if $b^2 > 2$, then b *is* an upper bound for A, but one also has $(b - r)^2 > 2$ for any sufficiently small rational $r > 0$. Therefore any such number $b - r$ is *smaller* than b and still an upper bound for A, so that b *is not the* **smallest** *upper bound*. We thus have verified that any *rational* b is definitely not a *least upper bound* for A. Hence there is no smallest upper bound for A within the rational numbers. On the other hand, if one considers A as a subset of $\mathbb{R}$, then completeness of $\mathbb{R}$ implies that A has a least upper bound $L_A \in \mathbb{R}$. By what we just saw, neither $(L_A)^2 < 2$ nor $(L_A)^2 > 2$ can hold for the least upper bound of A. Therefore $(L_A)^2 = 2$. In other words, we just verified how the *completeness* of $\mathbb{R}$ implies the "existence" of $\sqrt{2}$ inside the real numbers.

Note that the least upper bound L_S of a set S may or may not be an element of S. For example, the least upper bound L_A of the set A we just considered is not contained in A, which only consists of rational numbers. On the other hand the set $B = \{x \in \mathbb{Q} : x \leq 0\}$ contains its least upper bound 0, which in this case happens to be a rational number!

The LUB property implies that $\mathbb{R}$ also satisfies the analogous *Greatest Lower Bound Property*: If the non-empty set $S \subset \mathbb{R}$ has a *lower* bound l, i.e., if there exists a real number l, so that $l \leq s$ for all $s \in S$, then there exists a (unique) *greatest* lower bound $G_S \in \mathbb{R}$ for S. This means that G_S is a lower bound for S with the property that any number $c > G_S$ is **not** a lower bound for S, that is, there exists $s \in S$ so that $s < c$. (See Problem 13 of Exercise I.1.6.)

I.1.5 *Intervals and Other Properties of* $\mathbb{R}$

From now on we shall assume that the set of real numbers $\mathbb{R}$ is **complete**. In particular, every non-empty *bounded* subset S of $\mathbb{R}$ (that is, S has both an upper and a lower bound) has a *least* upper bound, denoted by $\sup S$ (supremum of S) and also a *greatest* lower bound, denoted by $\inf S$ (infimum of S).

Just as we had seen that completeness implies the existence of a positive real number labeled $b = \sqrt{2}$ that satisfies the equation $b^2 = 2$, one can show the following more general result.

Lemma 1.1. *For every positive integer n and for any real number $a > 0$, there exists exactly one real number $b > 0$ that satisfies $b^n = a$.*

This number is given by $b = \sup\{r \in \mathbb{R} : r^n < a\}$. It is called the *nth root* of a, and it is denoted by $b = \sqrt[n]{a}$, or $a^{1/n}$. For $n = 2$ one simply writes $\sqrt[2]{a} = \sqrt{a}$. Note that the symbol $\sqrt{a}$ denotes the *positive* solution of $x^2 = a$; the other (negative) solution of this equation is then the number $-\sqrt{a}$.

The following result will turn out to be quite useful in many applications.

Lemma 1.2. *If $c > 0$ is any real positive number, then there exists a natural number n such that $0 < 1/n < c$.*

Proof. While this may appear obvious, the *proof* for arbitrary *real* $c > 0$ does in fact involve the completeness axiom. We shall first verify the following equivalent property, also known as the *Archimedean Property* of the real numbers.

The set $\mathbb{N}$ of natural numbers is NOT bounded in $\mathbb{R}$, i.e.,

$\mathbb{N}$ *does not have any upper bound.*

Stated differently, the symbol ∞ (= *infinity*), which is commonly used to label "something" that is larger than any natural number, cannot be identified with a real number.

We prove this latter result by contradiction. Assume $\mathbb{N}$ had an upper bound in $\mathbb{R}$. By the LUB property there then exists a real number L that is the *least* upper bound for $\mathbb{N}$. Hence the number $L - 1 < L$ is not an upper bound for $\mathbb{N}$, i.e., there exists $m \in \mathbb{N}$ such that $L - 1 < m$. It then follows by the properties of order that $L < m + 1$. We have thus found a natural number $m + 1 \in \mathbb{N}$ that is *larger* than L, so L could not be an upper bound for $\mathbb{N}$. We end up with a hopeless contradiction. This means that our initial assumption cannot be correct, and therefore $\mathbb{N}$ does not have an upper bound.

Returning to an arbitrary real number $c > 0$, the number $1/c$ is also real and positive. As we just saw, there exists a natural number $n > 1/c$. By the order properties, this implies that $0 < 1/n < c$, and we have verified the Lemma. ∎

We review some standard useful notations. Given two numbers $a < b$, the *open interval* with *boundary points* a, b equals the set

$$(a, b) = \{\lambda \in \mathbb{R} : a < \lambda < b\}.$$

If one adds the boundary points to (a, b) one obtains the *closed* interval

$$[a, b] = \{\lambda \in \mathbb{R} : a \leq \lambda \leq b\}.$$

Notice that a is the greatest lower bound of both (a, b) and $[a, b]$, and similarly b is the least upper bound of each set. Intervals with *boundary points* a, b are examples of *bounded* sets. More generally, every bounded set of numbers is contained in some bounded interval. Sometimes one considers unbounded intervals such as $\{\lambda \in \mathbb{R} : a < \lambda\}$, which—in analogy to the notation for bounded intervals—we also denote by (a, ∞). As previously noted, ∞ is just a *symbol* and not an element of $\mathbb{R}$; consequently we do *not* call ∞ a *boundary* point of (a, ∞). Correspondingly, the interval $[a, \infty) = \{\lambda \in \mathbb{R} : a \leq \lambda\}$ is a closed interval, as it contains its (only) boundary point a. Similarly, $\mathbb{R}$ itself can be identified with the interval $(-\infty, \infty)$. Note that the interval $(-\infty, \infty)$ is *open* (it does not contain any boundary points), and since there are NO boundary points to include, it is also said to be *closed*. If this sounds strange, think of a door that stands alone, without any frame and wall around it.

Given a point $a \in \mathbb{R}$, one often needs to identify intervals centered at a that satisfy specific properties. It is convenient to introduce the following notation: given $\delta > 0$, the symbol $I_\delta(a)$ denotes the set $\{x \in \mathbb{R} : |x - a| < \delta\}$, which can also be written in interval notation as $I_\delta(a) = (a - \delta, a + \delta)$. Informally, we shall also say that $U \subset \mathbb{R}$ is a *neighborhood* of a if there exists a positive δ, so that $I_\delta(a) \subset U$. Note that $I_\delta(a)$ is then a neighborhood of any of its points $x \in I_\delta(a)$.

Finally, we discuss another special property of the complete real numbers $\mathbb{R}$ that is often used to prove the existence of specific numbers that are required to satisfy certain properties. For example, suppose we want to find explicit rational approximations for $\sqrt{2}$. We begin by choosing $r_0 = 1$ and $s_0 = 2$, so that $r_0 < \sqrt{2} < s_0$. We then take the midpoint $3/2$ of r_0 and s_0. Since $(3/2)^2 = 9/4 > 2$, we have $\sqrt{2} < 3/2$. Set $r_1 = 1$ and $s_1 = 3/2$, so that $r_1 < \sqrt{2} < s_1$; note that $s_1 - r_1 = 1/2$. By continuing this process we obtain rational numbers r_n and s_n for each $n = 2, 3, ...$, so that $r_n < \sqrt{2} < s_n$ (we cannot have equality since $\sqrt{2}$ is not rational) and $s_n - r_n = 1/2^n$. More in detail, suppose we have found r_{n-1} and s_{n-1} with the desired properties. We then choose the midpoint m_n between r_{n-1} and s_{n-1}. If $m_n < \sqrt{2}$, we set $r_n = m_n$ and $s_n = s_{n-1}$; if $m_n > \sqrt{2}$, we set $r_n = r_{n-1}$ and $s_n = m_n$. In either case we will have $r_n < \sqrt{2} < s_n$, and $s_n - r_n = 1/2(s_{n-1} - r_{n-1}) = 1/2^n$. Each interval $[r_n, s_n]$ is contained in the preceding one. We claim that $\sqrt{2}$ is the only number that is contained in each interval $[r_n, s_n]$. In fact, if $\lambda \neq \sqrt{2}$, we will show that λ is not contained in $[r_n, s_n]$ if n is sufficiently large. Since $c = \left| \sqrt{2} - \lambda \right| > 0$, by the Lemma there exists a natural number $n^* > 1$, so that $1/n^* < c$, and therefore one also has $1/2^{n^*} < c$. Since $\sqrt{2} \in [r_{n^*}, s_{n^*}]$, every other number $x \in [r_{n^*}, s_{n^*}]$ satisfies $\left| \sqrt{2} - x \right| \leq s_{n^*} - r_{n^*} = 1/2^{n^*} < c = \left| \sqrt{2} - \lambda \right|$; it follows that $\lambda \notin [r_{n^*}, s_{n^*}]$, as required.

We generalize this approximation process as follows. Suppose for each $n = 1, 2, 3, \ldots$ we are given a *closed bounded* interval $[a_n, b_n]$ so that

$$[a_1, b_1] \supseteq [a_2, b_2] \supseteq \ldots \supseteq [a_n, b_n] \supseteq [a_{n+1}, b_{n+1}] \supseteq \ldots .$$

We call such a sequence a *nested sequence* of closed bounded intervals. The following result states an intuitively obvious property in a precise form. Recall that the symbol $\varnothing$ denotes the "empty set", that is, a set that does not contain any elements at all.

Theorem 1.3. *If $I_n = [a_n, b_n], n = 1, 2, \ldots$, is a nested sequence of closed, bounded intervals in $\mathbb{R}$, then*

$$F = \bigcap_{n=1}^{\infty} [a_n, b_n] \neq \varnothing,$$

i.e., there exists at least one real number $c \in \mathbb{R}$ that is contained in each interval $[a_n, b_n]$.

While this result may appear obvious to you, the situation is not quite so simple. For example, let us take the *open* intervals $I_n = (0, 1/n)$ for $n = 1, 2, 3, \ldots$, which satisfy $I_n \supset I_{n+1}$ for each n. Since each interval I_n contains only positive numbers, clearly neither 0 nor any *negative* number is contained in $\cap \, I_n$. On the other hand, if c is any *positive* real number, then we know that there is $n^* \in \mathbb{N}$ with $1/n^* < c$, which means that $c \notin (0, 1/n^*)$, and therefore $c \notin \cap_{n=1}^{\infty} I_n$. We conclude that

$$\bigcap_{n=1}^{\infty} (0, 1/n) = \varnothing .$$

Similarly, $J_n = [n, \infty)$ for $n = 1, 2, \ldots$ defines a nested sequence of closed intervals that are NOT bounded. By the Archimedean property it easily follows that $\cap_{n=1}^{\infty} [n, \infty) = \varnothing$. So the particular hypothesis for the intervals in the theorem are indeed essential. More significantly, the theorem is false if we just consider *rational* numbers. For example, recall the rational numbers r_n and s_n introduced in the example before the theorem. Let $I_{n,\mathbb{Q}} = \{x \in \mathbb{Q} : r_n \leq x \leq s_n\}$. Clearly $I_{1,\mathbb{Q}} \supseteq I_{2,\mathbb{Q}} \supseteq I_{3,\mathbb{Q}} \supseteq \ldots$ is a nested sequence of "non-empty closed bounded intervals of rational numbers". Since $I_{n,\mathbb{Q}} \subset \mathbb{Q}$ by construction, it follows that $\cap_{n=1}^{\infty} I_{n,\mathbb{Q}} \subset \mathbb{Q}$ as well. And since the only number $\lambda \in \mathbb{R}$ that satisfies $r_n \leq \lambda \leq s_n$ for all $n \in \mathbb{N}$ is the number $\sqrt{2}$, and $\sqrt{2}$ is not rational, i.e., $\sqrt{2} \notin \mathbb{Q}$, it follows that

$$\bigcap_{n=1}^{\infty} I_{n,\mathbb{Q}} = \varnothing.$$

We see that the validity of the theorem must rest on the completeness of the real number.

To prove the theorem, observe that for any nested sequence of intervals $[a_n, b_n]$, $n = 1, 2, 3, ...$, one must have

$$a_1 \leq a_2 \leq ... \leq a_n \leq a_{n+1} \leq ... \leq b_{n+1} \leq b_n \leq ... \leq b_2 \leq b_1.$$

Therefore the set $A = \{a_1, a_2, ...\}$ of left boundary points is not empty and it is bounded above by any of the right boundary points. It then follows that the real number $\sup A$—here we use *completeness*—is contained in the interval $[a_n, b_n]$ for each n. A more detailed outline of this argument is given in Problem 15 of Exercise I.1.6.

To summarize, the *completeness axiom* provides the firm foundation that supports our sometimes faulty and vague geometric intuition, and it ensures that many problems, algebraic or more general, do indeed have solutions within the set of real numbers. Along the way we will see many applications of completeness, often via the theorem we just discussed. One of the most surprising consequences is a proof of Georg Cantor's amazing discovery that the set of real numbers is of a much higher order of infinity than the set of natural numbers $\mathbb{N}$. More precisely, a set S is said to be *countable* if its elements can be arranged in a suitable order, so that they can be "counted", that is, if one can write $S = \{s_1, s_2, s_3, ...\}$. Clearly every finite set is countable, and the set of natural numbers $\mathbb{N}$ is the prototype of countable sets that are not finite. By writing $\mathbb{Z} = \{0, 1, -1, 2, -2, 3, -3, ...\}$ one sees that the set of integers is countable as well. Similarly, one can show that the set of rational numbers is countable (see Problem 16 of Exercise I.1.6), and Cantor proved that even the set of all *algebraic* numbers is countable.

Theorem 1.4. *(G. Cantor). The set $\mathbb{R}$ of real numbers is NOT countable.*

Proof. We prove this by contradiction. Suppose that $\mathbb{R}$ is countable. That means that we can arrange $\mathbb{R}$ in a counting sequence $\mathbb{R} = \{c_1, c_2, c_3, ...\}$. Every number $\lambda \in \mathbb{R}$ must eventually appear in this sequence, that is, there must be some $k \in \mathbb{N}$ so that $\lambda = c_k$. We will show that this assumption is incompatible with the completeness of $\mathbb{R}$, so that the statement that $\mathbb{R}$ is countable must be false. We construct a nested sequence of non-empty closed bounded intervals I_n as follows. Choose any interval $I_1 = [a_1, b_1]$ with $a_1 < b_1$, so that $c_1 \notin I_1$. Next, choose $I_2 = [a_2, b_2] \subset I_1$ with $a_2 < b_2$, so that $c_2 \notin I_2$. This is easily done as follows. If $c_2 \notin I_1$, just choose $I_2 = I_1$; if $c_2 = a_1$, let a_2 be any point with $a_1 < a_2 < b_1$ and take $b_2 = b_1$; and if $a_1 < c_2 \leq b_1$, choose $a_2 = a_1$ and b_2 so that $a_1 < b_2 < c_2$. Next, choose $I_3 = [a_3, b_3] \subset I_2$ with $a_3 < b_3$, so that $c_3 \notin I_3$, and so on. This process can be continued, provided the interval $I_n = [a_n, b_n]$ is always chosen with $a_n < b_n$. By Theorem 1.3 there exists a real number $\lambda \in \cap_{n=1}^{\infty} [a_n, b_n]$. As we observed earlier, according to our hypothesis, this number λ must occur at some place in the ordering of $\mathbb{R}$, i.e., $\lambda = c_k$ for some $k \in \mathbb{N}$. But then $\lambda = c_k \notin I_k$ by the construction of I_k, and this contradicts $\lambda \in \cap_{n=1}^{\infty} [a_n, b_n]$. We thus conclude that $\mathbb{R}$ is not countable.[4]   ∎

I.1.6   *Exercises*

Solve the inequalities in Problems 1 through 5. Write each solution set as an interval.

1. $1 - 6x > 2$
2. $3 + 4x < 1$
3. $-6 < 5 - 2x < 2$
4. $|x + 5| \leq 2$
5. $|5x - 4| < 4$

Simplify the following expressions by eliminating the absolute value sign.

6. $|(-3)(5 - 9)|$
7. $|(-2)^3|$
8. $-|2 - 5|$
9. $|(-1)^{2n}|$, where n is a positive integer.
10. Use the fact that $(-1)a = -a$ to verify that the product of two negative numbers is positive.
11. Explain by using Problem 10 why there is no *real* solution of the equation $x^2 = a$ for $a < 0$.
12. Modify the argument used to show that $\sqrt{2}$ is not rational to show that $\sqrt{3}$ is not rational either. More generally, show that if an integer $p > 0$ is not a perfect square (i.e., if p is not equal to m^2 for some integer m), then $\sqrt{p}$ is not rational.

---

[4]The author learned of this proof from the text of Bartle and Sherbert (op. cit.). It differs from Cantor's orginal proof, which was based on what has become known as "Cantor's Diagonal Sequence" argument.

13. In analogy to the LUB property, one can define the Greatest Lower Bound Property of $\mathbb{R}$ as follows: A set S of numbers is bounded from below if there is a number $l \in \mathbb{R}$ so that $l \leq s$ for all $s \in S$. Such l is called a lower bound for S. A number G_S is called a **Greatest Lower Bound** for S if G_S is a lower bound for S, and any number $c > G_S$ is *not* a lower bound. Verify that the real numbers $\mathbb{R}$ satisfy the **Greatest Lower Bound property**, that is, *each non-empty set in $\mathbb{R}$ that is bounded from below has a unique Greatest Lower Bound.* (Hint: If l is a lower bound for S, then $(-l)$ is an upper bound for the set $S^* = \{s : -s \in S\}$.)

14. Find the least upper bound and greatest lower bound for the set $S = \{1 - \frac{1}{n} : n = 1, 2, 3, ...\}$.

15. Suppose $I_n \subset \mathbb{R}$ is a closed bounded interval $[a_n, b_n]$ for $n = 1, 2, ...$ so that $I_1 \supseteq I_2 \supseteq ...I_n \supseteq I_{n+1} \supseteq ....$ Show that $\cap_n I_n \neq \emptyset$ by completing the following steps.

    a) Let $A = \{a_n : n = 1, 2, ...\}$. Show that each right endpoint b_n is an upper bound for A.

    b) Explain why $\sup A \leq b_n$ for each $n \in \mathbb{N}$.

    c) Show that b) implies that $\sup A \in [a_n, b_n]$ for each $n \in \mathbb{N}$, and consequently $\sup A \in \cap_{n=1}^{\infty} I_n$.

    d) More generally, prove that the closed interval $[\sup A, \inf B] \subset \cap_{n=1}^{\infty} I_n$, where $B = \{b_n : n = 1, 2, ...\}$ is the set of right endpoints.

16. Consider the set $\mathbb{Q}^+ = \{\frac{m}{n} : m, n \text{ positive integers}\}$. Arrange $\mathbb{Q}^+$ in the following pattern.

    line 1: $\dfrac{1}{1}, \dfrac{2}{1}, \dfrac{3}{1}, ..., \dfrac{m}{1}, ...$

    line 2: $\dfrac{1}{2}, \dfrac{2}{2}, \dfrac{3}{2}, ..., \dfrac{m}{2}, ...$

    line 3: $\dfrac{1}{3}, \dfrac{2}{3}, \dfrac{3}{3}, ..., \dfrac{m}{3}, ...$

    $\vdots$

    line n: $\dfrac{1}{n}, \dfrac{2}{n}, \dfrac{3}{n}, ..., \dfrac{m}{n}, ...$

    $\vdots$

    a) Use this pattern to show that $\mathbb{Q}^+$ is countable. (Hint: Start "counting" in the upper left corner, then take 1/2 and next 2/1, and continue by moving along the diagonals parallel to the first one, skipping any number that has already been covered, and so on.)

b) Show that this implies that $\mathbb{Q}$ is also countable. (Hint: Look at how we saw that $\mathbb{Z}$ is countable.)

I.2   Functions

Most readers will probably be familiar with the concepts in this section. Still, for completeness' sake we include a brief review to help the readers refresh their memory.

I.2.1   *Functions of Real Variables*

A **function** is a rule or machine that assigns a specific output to a given input. Stated in this form, this is a very general concept. We shall primarily consider functions whose inputs and outputs are subsets of real numbers.

Such functions are often described by some mathematical formula. The simplest examples are the constant functions. Fix a number $c \in \mathbb{R}$; the constant function f_c assigns to each input $x \in \mathbb{R}$ the output c, in other words, $f_c(x) = c$ for all inputs x.

A less trivial example describes the conversion of degrees Fahrenheit to degrees Celsius that is given by the formula

$$C = C(F) = \frac{5}{9}(F - 32).$$

Here the input is a temperature measured in degrees Fahrenheit, and the output is the corresponding temperature in degrees Celsius.

The area A of a disc is a function of its radius r; it is given explicitly by the formula $A(r) = \pi r^2$.

A function with only finitely many inputs (not too many) is often described by a table that lists the input values and the corresponding output values. (See Figure I.4.)

Mo, 4/3	Tu, 4/4	We, 4/5	Th, 4/6	Fr, 4/7	Sa, 4/8	Su, 4/9
52°	50°	47°	49°	45°	46°	47°

Fig. I.4   Average temperature as a function of the days of one week.

Sometimes the rule is not given by a single formula. For example, the federal income tax $T(x)$ due on a taxable income of \$ x is described in tax tables; the mathematical formula for $T(x)$ varies, depending on the level of income x.

The "function" keys on a scientific calculator provide other concrete examples of functions. For example, the key $\boxed{x^2}$ identifies the squaring function. After entering a number such as 2.1 (the input) into the calculator, pressing the $\boxed{x^2}$ key results in the calculator displaying the output 4.41 ($= 2.1^2$). Note that input and outputs of calculators are numbers with finite decimal expansion, i.e., rational numbers. So the square root function identified by the key $\boxed{\sqrt{}}$ on a calculator (sometimes invoked by $\boxed{inv}$ followed by $\boxed{x^2}$) only provides an *approximation* of the abstract function.

Let us fix some basic terminology and notation. The collection of numbers that can be taken as input is called the *domain* of the function f and it is denoted by $dom(f)$. If the function machine is denoted by the letter f, one writes symbolically $f : \Omega \to \mathbb{R}$ to indicate that f is a function with $dom(f) = \Omega$ that takes its values, i.e., outputs, in the real numbers. The value of the function f at the input $x \in \Omega$ is denoted by $f(x)$. The set of all output values $f(\Omega) = \{f(x) : x \in \Omega\}$ is called the *image* of f, or, more precisely, the image of Ω under f. It is common practice to denote a function f also by the symbol $y = f(x)$, or simply by $f(x)$, thereby blurring the distinction between the *function machine* f and the *output* of f at x. Since x is just a symbol for the *unspecified* variable input, this abuse of notation is not fatal. On the other hand, if a specific value is replaced for x, say $x = 2$, the symbol $f(2)$ definitely should not be used to denote the function f. The symbol $f(2)$ uniquely identifies just a single number, namely the output of f corresponding to the input 2.

The most important fact to remember about functions is that for every input from an appropriate domain there is **exactly one output**. One is free to choose the notation for the input and output variables. In general discussions mathematicians like to use x and y, although in applications other letters may be chosen to help identify the meaning of the variable. For example, the letter t is often used for a variable that corresponds to time.

I.2.2 *Graphs*

Real valued functions of a real variable can be visualized by their graphs. We first need to recall the concept of a **rectangular (or Cartesian) coordinate system** in the plane. One fixes a pair of perpendicular number lines, each one with the number 0 at the point of intersection. (See Figure I.5.)

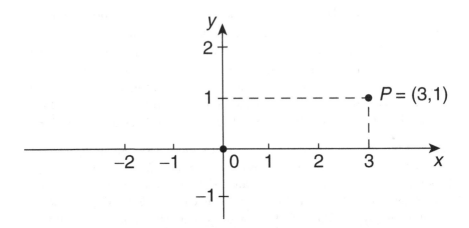

Fig. I.5   A Cartesian coordinate system.

On the horizontal axis (labeled x), numbers are increasing to the right, on the vertical axis (labeled y), numbers are increasing towards the top. According to Figure I.5, each point P in the plane determines an *ordered* pair of numbers (a, b), its *coordinates* with respect to the given coordinate system. (The notation (a, b) does not refer to an open interval here; the context should make clear what is meant in a particular instance.) Conversely, every *ordered* pair (a, b) of numbers determines a unique point P in the plane; one writes $P = (a, b)$. By convention, the first number in an ordered pair refers to the horizontal axis. The point of intersection of the two axes has coordinates $(0, 0)$. It is often called the *origin* of the coordinate system and is denoted by O.

By means of a coordinate system, geometric properties can be translated into algebraic or analytic statements involving the coordinates. Conversely, algebraic formulas involving the coordinates (x, y) can be interpreted geometrically.

Example. The distance between two points $P_1 = (x_1, y_1)$ and $P_2 = (x_2, y_2)$ is given by

$$dist(P_1, P_2) = \sqrt{(x_2 - x_1)^2 + (y_2 - y_1)^2}.$$

This formula is a consequence of Pythagoras' Theorem (See Figure I.6.)

Example. The equation $x^2 + y^2 = r^2$ describes the set

$$\{P : [dist(P, O)]^2 = r^2\}.$$

This set is a circle of radius r centered at the origin. (See Figure I.7.)

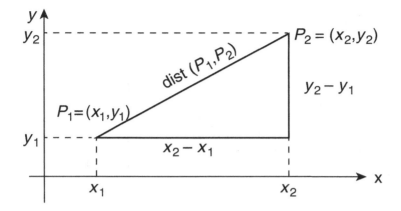

Fig. I.6 Length of hypothenuse measures the distance $d(P_1, P_2)$.

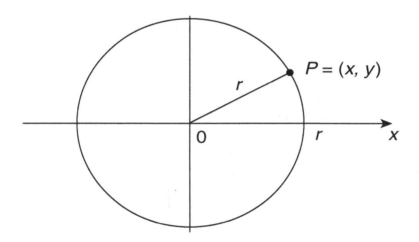

Fig. I.7 Circle of radius r centered at O.

Given a function $f : I \to \mathbb{R}$ defined on the interval I, its *graph* is the set

$$Graph\,(f) = \{(x, y) : y = f(x), x \in I\}.$$

Figure I.8 illustrates the concept. Notice that the graph of f is a "curve" with the distinctive property that the vertical line through any point $x \in I$ meets the graph in exactly one point $(x, f(x))$, corresponding to the fact that a function has exactly one output for each input. This is the so-called *vertical line test* for graphs of functions. Every curve that satisfies the

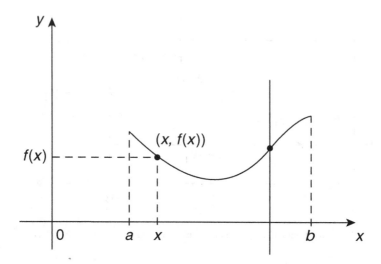

Fig. I.8   Graph of the function $y = f(x)$.

vertical line test can be viewed as the graph of a function. Notice that a (complete) circle fails the vertical line test, and hence is not the graph of a function. (See Figure I.9.)

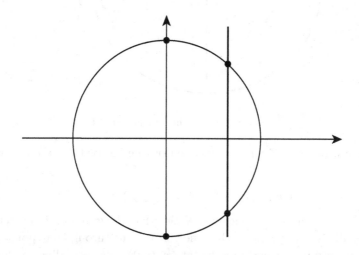

Fig. I.9   A circle fails the *vertical* line test.

I.2.3 Some Simple Examples

The temperature conversion function $C = C(F)$ mentioned earlier is an example of a *linear function*

$$y = f(x) = mx + b,$$

where m and b are constants. Algebraically, the formula for a linear function is given by a polynomial of degree 1. The name reflects the fact that the graph of a linear function is a line. We shall discuss this relationship more in detail in the next section.

Slightly more complicated algebraically are **quadratic functions**, i.e., functions of the form

$$f(x) = ax^2 + bx + c,$$

where a, b, c are constants with $a \neq 0$. The graphs of such functions are parabolas, the classical curves we had discussed in Section 3 of the Prelude.

An important application of quadratic functions arises in the description of a stone dropped from a tower. (See Prelude, Section 4.) If the height of the tower is H meters, the stone falls towards the ground according to the following law: its height $s(t)$ in meters after t seconds is given by the formula, i.e., by the function

$$s(t) = H - 4.9t^2 .$$

This relationship can be tested experimentally. The rock hits the ground when $s(t) = 0$. See Figure I.10 for the graph of s corresponding to $H = 100$.

This formula reflects the fundamental physical fact that near the surface of the earth the gravitational force is approximately constant. By Newton's Law of Motion, the acceleration of an object is a constant multiple of the force acting on it; hence the motion of a freely falling object close to the ground is uniformly accelerated. This was the key discovery of Galileo mentioned in Section 4 in the Prelude. We shall discuss motion with constant acceleration more in detail in Section 4 of Chapter III.

More generally, one can consider $f(x) = ax^n$, where n is a positive integer, and $a \neq 0$ is a fixed real number. Such functions are called **power functions.** The characteristic feature is that the input variable is in the base of the power, while the exponent is fixed, in contrast to *exponential functions* discussed later on, where the input variable is in the exponent. By adding up power functions with non-negative integer exponents one obtains the familiar polynomial functions P, which are defined by a formula

$$P(x) = a_n x^n + a_{n-1} x^{n-1} + ... + a_1 x + a_0 .$$

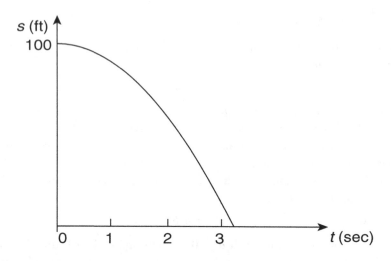

Fig. I.10   Height of a falling stone as a function of time t.

One says that P has degree n if the coefficient a_n of the highest power is not zero. Accordingly, quadratic functions are polynomials of degree 2.

As we saw in Section 7 of the Prelude, most algebraic expressions involving one variable can be used to define a function. More details will be discussed in Section 6 of this chapter later on.

Often the formula under consideration is meaningful only for inputs that are restricted by certain conditions, so that the domain will be a *proper* subset of the real numbers.

Example. The function $y = \sqrt{x}$ has as its domain the set of nonnegative numbers $\{x \geq 0\}$. Its graph is shown in Figure I.11.   As we saw in Section 6 of the Prelude, the function $f(x) = \sqrt{x}$ is algebraically differentiable at all points $x > 0$, but not at 0. Consequently we shall exclude 0 from the domain when considering derivatives for this function..

Remark on notation. As already mentioned earlier, the square root symbol $\sqrt{a}$ for $a > 0$ represents a unique number, namely that *positive* number $c > 0$ that solves the equation $x^2 - a = 0$. The other solution of this equation is the *negative* number denoted by $-\sqrt{a}$.

I.2.4 *Linear Functions, Lines, and Slopes*

We shall now review *linear* functions f, which are given by a formula $f(x) = mx + b$, i.e., polynomials of degree ≤ 1, more in detail. Clearly the constant

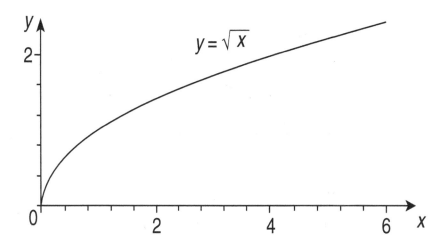

Fig. I.11   Graph of the function $y = \sqrt{x}$ for $x \geq 0$.

b is the value of f for $x = 0$, i.e., it identifies the point $(0, b)$ where the graph of f intersects the y−axis. To understand the geometric meaning of m, fix a point (x_1, y_1) on the graph and let $(x, y) = (x, mx + b)$ be any other point on the graph with $x \neq x_1$. Then

$$\frac{\Delta y}{\Delta x} = \frac{y - y_1}{x - x_1} = \frac{mx + b - (mx_1 + b)}{x - x_1} = \frac{m(x - x_1)}{x - x_1} = m.$$

This shows that in the right triangle in Figure I.12 the angle α is independent of the particular point (x, y), so that all such points lie on the unique line through (x_1, y_1) that forms an angle α with the horizontal x-axis.   The ratio $m = (y - y_1)/(x - x_1)$—sometimes also referred to as the "rise over the run"—is called the *slope* of the line. It measures the inclination of the line, and—for those familiar with basic trigonometry—it relates to the angle α by the formula $\tan \alpha = m$.

Conversely, if we begin with the line through (x_1, y_1) shown above, the ratio $(y - y_1)/(x - x_1)$ is independent of any other point (x, y) on the line since the right triangles obtained by different points (x, y) all have the same angle α at the point (x_1, y_1), so that they are *similar* to each other. If we denote this constant ratio by m and solve the equation $(y - y_1)/(x - x_1) = m$ for y, one obtains $y - y_1 = m(x - x_1)$, which implies $y = mx + (y_1 - mx_1)$. We thus see that the given line is the graph of a linear function.

As we saw in the Prelude, the so-called *point-slope form* $y - y_1 = m(x - x_1)$ of a line, which can also be written as the function f defined by $f(x) = y_1 + m(x - x_1)$, is most useful in describing tangents to curves.

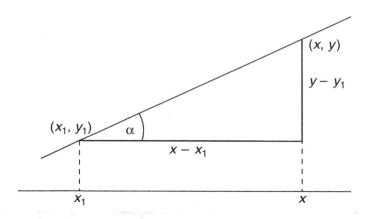

Fig. I.12   Constant $\Delta y/\Delta x$ implies constant angle α.

Note that *horizontal* lines (i.e., those lines parallel to the x-axis), which are the graphs of constant functions, are precisely those lines that have slope 0. In contrast, observe that the *vertical* line through the point (x_1, y_1) definitely is not the graph of a function f (the vertical line test fails); also, the formula for the slope becomes meaningless in this case, since $x = x_1$ for all points (x, y) on this line. Therefore the slope of vertical lines is NOT defined.

Example. Let us find the equation of the line of slope -2 that passes through the point $(4, 3)$. If (x, y) is any point on the line different from $(4, 3)$, then

$$\frac{y - 3}{x - 4} = -2, \text{ or } y - 3 = (-2)(x - 4).$$

The advantage of the second version is that it makes sense also for $x = 4$. Solving for y gives

$$y = (-2)(x - 4) + 3, \text{ i.e.,}$$
$$y = -2x + 11.$$

While the last equation identifies the y intercept 11, the coordinates of the original point $(4, 3)$ have been lost in the process.

Example. Find the equation of the line that goes through the points $(-1, 1)$ and $(2, 3)$. (See Figure I.13.)

Solution. We first use the given points to calculate the slope $m = \frac{3-1}{2-(-1)} = \frac{2}{3}$. Now one can readily write down the point-slope form of the

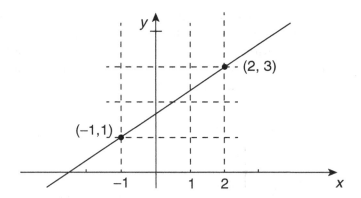

Fig. I.13   Line through $(-1, 1)$ and $(2, 3)$.

equation by using either one of the two points. By choosing the first point, one obtains

$$y - 1 = \frac{2}{3}(x - (-1)).$$

If the other point is chosen, the resulting equation is $y - 3 = 2/3(x - 2)$, which is easily transformed into the previous equation by adding 2 on both sides.

Example. We want to see how one can find the conversion formula from degrees Fahrenheit F to degrees Celsius C that was mentioned earlier. The relevant property is that the conversion is uniform, regardless of the temperature level. More precisely, for any change in temperature the corresponding change ΔC in Celsius measurement is a constant multiple $\Delta C = m\Delta F$ of the change ΔF measured in Fahrenheit degrees. This means that the relationship between the two temperature scales is a linear one, that is, C is a linear function

$$C = mF + b$$

of F. The constants m and b can be determined from the known temperature values at any two distinct points. For example, the freezing point of water is 0^0C , or 32^0F, while the boiling point (at sea level) is 100^0C, or 212^0F. Thus the two points $(32, 0)$ and $(212, 100)$ lie on the graph of this function. The slope of the line through these points is $(100 - 0)/(212 - 32) = 100/180 = 5/9$. Hence the equation of this line in point-slope form, i.e., the desired conversion formula, is

$$C = C - 0 = \frac{5}{9}(F - 32).$$

Finally, we examine the relationship between the slopes m_1 and m_2 of two perpendicular lines, neither of which is vertical. The triangle corresponding to the line with slope m_2 shown in Figure I.14 is obtained by rotating the other triangle by 90^0 in the mathematically positive direction. This has the effect of interchanging the "rise" and the "run", whereby the

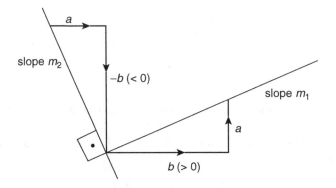

Fig. I.14   Slopes a/b and $-b/a$ of two perpendicular lines.

original (positive) run b results in the negative rise $-b$ after rotation. Thus $m_1 = a/b$ and $m_2 = (-b)/a$, and it follows that

$$m_1 m_2 = -1 \text{ for any two perpendicular lines.}$$

Example. Let us determine the slope of the tangent to the circle of radius 1 at the point $P = (1/2, \sqrt{3}/2)$. We know that the tangent is perpendicular to the radius through P. That radius has slope $m_1 = (\sqrt{3}/2)(1/2) = \sqrt{3}$. The slope m of the tangent must satisfy $m_1 m = -1$, and therefore $m = -1/\sqrt{3}$. The point-slope form of the tangent line thus is $y - \sqrt{3}/2 = (-1/\sqrt{3})(x - 1/2)$.

I.2.5   *Exercises*

1. Identify the points $P = (-3, 4), Q = (0, 3)$ and $R = (4, -2)$ in a Cartesian coordinate system.
2. Find the distance between P and R in Problem 1.
3. Sketch the graph of the function $y = \frac{1}{2}x^2 - 2$.
4. Use the conversion formula given in the text to find the temperatures in Celsius degrees corresponding to 95^0F and 212^0F.

5. If the temperature is 25^0C, what is the temperature in degrees Fahrenheit? More generally, write out explicitly a conversion formula from degrees Celsius to Fahrenheit.

6. Consider the following example of a tax function T. For incomes $x \leq$ $\$10,000$, the tax $T(x) = 0$. If $\$10,000 \leq x \leq \$30,000$, the tax is 10% of the income above $\$10,000$, for $\$30,000 \leq x \leq \$50,000$, the tax is $\$2,000+20\%$ of the amount over $\$30,000$, and for incomes $x \geq \$50,000$, the tax is $\$6,000 + 30\%$ of the amount above $\$50,000$.

   a) Write out formulas for $T(x)$. There will be different expressions for the various income levels.

   b) Sketch the graph of the function T.

7. Sketch the set of points $\{(x,y) : x = y^2\}$ in a Cartesian coordinate system. Is the curve so obtained the graph of a function? Explain.

8. Do Problem 7 with the set $\{(x,y) : y \leq 0 \text{ and } x = y^2\}$. Write a formula for the function whose graph is given by this set.

9. a) Find the slope of the line that goes through the points $(3,1)$ and $(6,-2)$.

   b) Find the equation of the line in a).

10. Consider the line given by $y = 3x + 2$. Find the coordinates of the points on the line at distance 4 from the point $(0,2)$. (Hint: Make a sketch!)

11. Find the equation of the line that is perpendicular to the graph of $y = \frac{1}{3}x + 4$ and that goes through the point $(1,2)$.

I.3 Simple Periodic Functions

Certain phenomena keep repeating a particular pattern over time. Typical examples include the motion of a pendulum, the bouncing motion of a spring, the rotation of the earth around its axis, the (regular) heart beat of a person, waves in the ocean, sound waves in the air, and electromagnetic waves as they appear in the propagation of light or radio signals. The functions used to model such phenomena must be "periodic", that is, they must exhibit the repetition of basic patterns. More precisely, one says that a function f is periodic with period ω if

$$f(x + \omega) = f(x) \text{ for all } x \text{ in the domain of } f.$$

In particular, this implies that if $x \in dom(f)$, then $x+\omega$ must be in $dom(f)$ as well.

I.3.1   *The Basic Trigonometric Functions*

Except for constant functions, none of the familiar algebraic functions such as polynomials, root functions, and so on (see Prelude) are periodic. Other mathematical concepts are required to produce concrete precise examples of periodic functions. We shall now examine two of the most useful periodic functions that have found wide applications in the natural sciences and in mathematics, and that are fundamental for studying general periodic functions. These are the (trigonometric) functions *sine* and *cosine*. They have been used for thousands of years in many practical applications, mainly involving relationships between angles of a triangle and ratios of appropriate sides, in order to solve geometric problems or carry out large scale measurements. Many readers will be familiar with them from high school trigonometry courses, where geometric applications and numerous formulas and identities are emphasized. However, we shall not require any such prior knowledge; instead, we shall start from the beginning and focus on the essential features that are most useful for the applications in calculus.

In order to exhibit the periodic behavior of the *sine* and *cosine* functions it is best to describe them in the context of the unit circle in the plane rather than through triangles and ratios of their sides. We consider the unit circle $x^2 + y^2 = 1$ in a Cartesian coordinate system. Beginning at the point $(1, 0)$, given a number s we measure the distance $|s|$ along the circle, moving counterclockwise—this is known as *mathematically positive*—around the circle if $s > 0$, and moving clockwise if $s < 0$, thereby reaching a well defined point $P(s)$ on the circle. (See Figure I.15.)   The precise concept of distance along a curve is actually quite complicated, but we can intuitively visualize the process by using a measuring tape, i.e., a flexible number line, placing its 0 at the point $(1, 0)$ on the circle, wrapping it around the circle, and then reading off the distance on the tape. Naturally, an infinitely long measuring tape can be wrapped around the circle numerous times. One full tour around the circle takes us back to $P(0) = (1, 0)$. This occurs after having moved a distance s corresponding to the circumference of the circle, i.e., when $s = 2\pi$. So $P(2\pi) = P(0)$. Similarly, if one moves a distance 2π around the circle clockwise, one obtains $P(-2\pi) = P(0)$. Starting from an arbitrary point $P(s)$ and moving farther along the circle a distance 2π also takes us once around the circle back to $P(s)$, so that

$$P(s + 2\pi) = P(s) \text{ for every } s.$$

We have thus constructed a *periodic* function with period 2π (≈ 6.28) whose values, however, are not real numbers but points in the plane. Writing

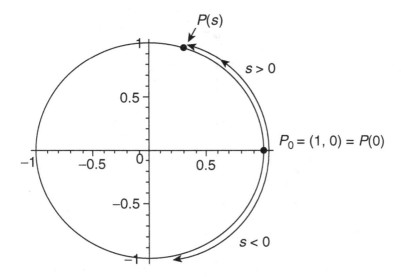

Fig. I.15   Arc of length s on the circle of radius 1.

$P(s) = (x(s), y(s))$, one sees that the coordinate functions $x(s)$ and $y(s)$ satisfy the same periodicity relation

$$x(s + 2\pi) = x(s), \quad y(s + 2\pi) = y(s).$$

Because of their importance, these periodic functions are given the special names *cosine* and *sine*, abbreviated as

$$\cos s = x(s) \text{ and } \sin s = y(s) .$$

The point $P(s)$ is thus described by

$$P(s) = (\cos s, \sin s) .$$

Since the reflection of the point $P(s)$ across the x-axis gives the point $P(-s)$, it follows immediately that

$$\cos(-s) = \cos s \text{ and } \sin(-s) = - \sin s.$$

Figure I.16 illustrates the connection of these functions with the right triangle with vertices

$$(0,0), (\cos s, 0), \text{ and } P(s) = (\cos s, \sin s)$$

and hypotenuse of length 1, and with the angle $\theta(s)$ at the vertex $(0,0)$. As long as the angle $\theta(s)$ is between 0^0 and 90^0 one sees that

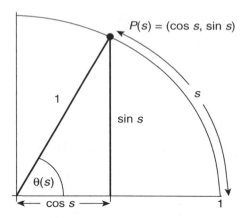

Fig. I.16   Right triangle with legs cos s and sin s.

$$\sin s = \frac{\sin s}{1} = \frac{opposite\ leg}{hypotenuse}, \quad \cos s = \frac{\cos s}{1} = \frac{adjacent\ leg}{hypotenuse}.$$

These are the classical formulas that have long been used to define the basic trigonometric functions of an angle of a right triangle. Note that the values of the ratios are independent of the radius of the circle as long as the angle $\theta(s)$ is kept fixed, since the resulting triangles are similar. In order to recognize the periodic nature of the trigonometric functions it is important to consider arbitrary real arguments s as inputs, as introduced here, rather than being restricted to angles in a triangle.

I.3.2   *Radian Measure*

When the input of the trigonometric functions *sine* and *cosine* is interpreted as an angle, it becomes important to specify how angles are to be measured. The number s (i.e., the distance along the unit circle from the point $(1,0)$ to the point $P(s)$) is known as the *radian measure* of the angle $\theta(s)$ formed by the ray from $(0,0)$ to $P(s)$ and the positive x-axis. (See Figure I.16.) Note that $(0,1) = P(\pi/2)$, so that the right angle between the (positive) coordinate axis has radian measure $\pi/2$. The familiar *degree measure* for angles is based on dividing a right angle into 90 equal pieces, each piece identifying an angle of 1^0 ($= 1$ degree), so that radian measure $\pi/2$ corresponds to 90^0. In general

$$s \text{ radians correspond to } (\frac{180}{\pi}s) \text{ degrees, and}$$

α degrees correspond to $s = \dfrac{\pi}{180}\alpha$ radians.

While we will occasionally use degree measure for angles in some applications, we shall only use the radian measure of angles as the input of trigonometric functions.

Except for very special choices of inputs there is no direct computational procedure to determine exact numerical values of the sine and cosine functions. Of course, approximate values can be obtained from careful graphs of the points $P(s)$ on the unit circle, or from measuring the length of the sides in appropriate right triangles. As we will see later, other analytical approximations can be obtained with the tools of calculus. At a practical level, years ago people had to use tables and slide rules to look up appropriate values of trigonometric functions. Technology has now improved, and the common method to determine (approximate) values for these functions uses scientific calculators. (**Warning:** It is important to set the calculator to the appropriate mode (degree or radian), so that the given input is understood correctly. In most scientific calculators the default mode is degree mode.)

We emphasize that for the purposes of calculus it is more convenient to use the radian measure of an angle as the input in trigonometric functions. Radian measure is based on intrinsic geometric concepts, while degree measure is based on the (arbitrary) partition of a full circle into 360 equal parts.

I.3.3 *Simple Trigonometric Identities*

Since $\cos s$ and $\sin s$ are the coordinates of the point $P(s)$ on the unit circle, one has the obvious fundamental trigonometric identity

$$(\sin s)^2 + (\cos s)^2 = 1 \text{ for all } s \in \mathbb{R}.$$

Other basic relations follow from the geometric observation that $P(s+\frac{\pi}{2}) = (-\sin s, \cos s)$ for all $s \in \mathbb{R}$. (See Figure I.17.)

In terms of coordinates, this means that

$$\cos(s + \frac{\pi}{2}) = -\sin s, \text{ and}$$

$$\sin(s + \frac{\pi}{2}) = \cos s.$$

Replacing s by $-s$ in these formulas one obtains

$$\cos(\frac{\pi}{2} - s) = -\sin(-s) = \sin s,$$

$$\sin(\frac{\pi}{2} - s) = \cos(-s) = \cos s.$$

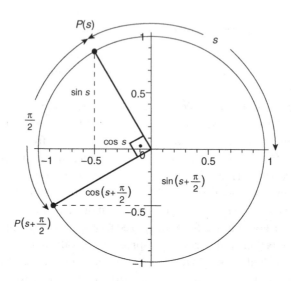

Fig. I.17   Location of $P(s)$ and $P(s + \pi/2)$ on the circle.

These latter formulas express the trigonometric functions of the complementary angle $\pi/2 - s$ in a right triangle in terms of the opposite trigonometric functions of the original angle.

These formulas are useful in order to translate known statements about one of the trigonometric functions into statements about the other function.

Rather than memorizing all these formulas—it is easy to get mixed up with the minus sign—one should clearly understand the geometric construction that defines the point $P(s) = (\cos s, \sin s)$ on the unit circle.

There are many other identities for trigonometric functions. To keep matters simple, we just recall the *addition formula*

$$\sin(s + t) = \sin s \cos t + \cos s \sin t$$

for the sine function that is discussed in high school trigonometry courses. From this identity the corresponding addition formula for the cosine function is readily obtained by using the simple formulas we mentioned earlier, as follows.

$$\cos(s + t) = \sin(\frac{\pi}{2} - (s + t)) = \sin((\frac{\pi}{2} - s) + (-t))$$
$$= \sin(\frac{\pi}{2} - s)\cos(-t) + \cos(\frac{\pi}{2} - s)\sin(-t)$$
$$= \cos s \cos t - \sin s \sin t.$$

Other formulas will be reviewed as needed in appropriate places later on.

In trigonometry courses one usually introduces other functions that are simple algebraic combinations of the two basic functions sine and cosine. For example, the *tangent* function is defined by

$$\tan s = \frac{\sin s}{\cos s} \text{ for all } s \text{ with } \cos s \neq 0.$$

Note that $\cos s = 0$ precisely when $P(s)$ lies on the y-axis, i.e., when $s = \frac{\pi}{2} + k\pi$, k any integer. One easily checks that $\tan(s + \pi) = \tan s$ for all $s \in dom \tan$, i.e., the tangent function is periodic with period π. In terms of the sides of a right triangle with hypotenuse 1 (see Figure I.18), the tangent of an angle α of s radians is

$$\tan s = \frac{opposite \ side}{adjacent \ side} = \frac{b}{a}.$$

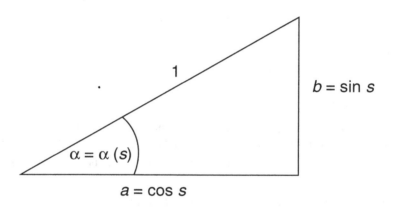

Fig. I.18   Right triangle with angle α of s radians.

In this book we shall mainly use the sine and cosine functions.

I.3.4   *Graphs*

The graphs of the sine and cosine functions are shown in Figures I.19 and I.20. Rough approximations of these graphs can be obtained by inspection of the coordinates of $P(s)$ as the point moves around the circle. Note that because of the periodicity, only points for $0 \leq s \leq 2\pi$ need to be sketched. This part of the graph is then repeated on adjacent period intervals, and so on. Of course, graphing calculators or computers are the most efficient tool for graphing these functions.

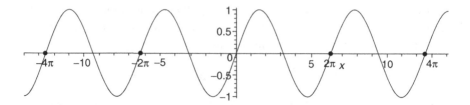

Fig. I.19   Graph of $y = \sin x$.

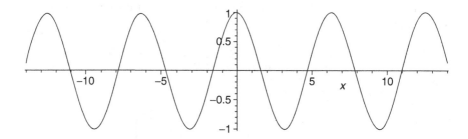

Fig. I.20   Graph of $y = \cos x$.

I.3.5   *Exercises*

1. Show that if ω is a period for the periodic function f, then 2ω and $-\omega$ are also periods, i.e., $f(x + 2\omega) = f(x)$ and $f(x - \omega) = f(x)$ for all x. More generally, verify that any integer multiple $k\omega$ is also a period.
2. Find

   a) $\sin(k\pi)$ and $\cos(k\pi)$ for any integer k, and
   b) $\sin(\frac{\pi}{2} + k\pi)$ and $\cos(\frac{\pi}{2} + k\pi)$ for any integer k.

3. Identify the point $P(\pi/4) = (x, y)$ on the unit circle and calculate its coordinates x, y, making use of the fact that $x = y$. Use this to evaluate $\sin \frac{\pi}{4}$ and $\cos \frac{\pi}{4}$.
4. Use the addition formulas for sine and cosine to find formulas for $\sin(2s)$ and $\cos(2s)$ in terms of $\sin s$ and $\cos s$.
5. Find the radian measure of $30^0, 45^0$, and 60^0.
6. What is the degree measure of the angle of $\pi/12$ radians?
7. a) Determine the mode of your scientific calculator as follows. Enter $\sin(1.57)$. Explain why the result is either close to 1 or close to 0,

depending on the mode. Use this fact to determine the mode your calculator is set for.

b) Find sine and cosine of 15^0. (Make sure that the calculator is set to degree mode.)

c) Compare the results in b) to the values of $\sin 15$ and $\cos 15$. (Use radian mode.)

8. Use a graphing calculator to display the graphs of the sine function and of $f(x) = \sin 3x$ and $g(x) = \sin(\frac{x}{3})$ in one window.

I.4   Exponential Functions

We had already mentioned power functions of the form $f(x) = ax^n$, where n is a positive integer. Here the input variable x is in the *base*. This makes the evaluation of such functions quite easy, since only basic arithmetic operations are involved. Furthermore, as seen in the Prelude, finding the derivatives of such functions just involves elementary algebra. The situation is quite different if the base is kept fixed, and the input variable occurs in the *exponent*. Such functions are called *exponential* functions. In the last section of the Prelude we recognized that finding tangents to the graphs of such functions involves considerably more complicated phenomena and new mathematical concepts. In this section we carefully discuss exponential functions, with particular attention to their definition for *real* numbers, in order to lay the foundations for the study of their tangents in Chapter II.

I.4.1   *Compound Interest*

It turns out that exponential functions are most important for many applications. A typical example involves the calculation of compound interest in the area of finance. Suppose a bank pays interest on a savings account at the rate of 6% per year, compounded annually. This means that at the end of a year the interest earned during the past year is added to the principal. More precisely, if $A(k)$ is the balance (in \$s) on the account at the end of year k, then $A(k+1) = A(k) + 0.06 \cdot A(k) = A(k)(1+0.06)$ (assuming there have been no other deposits or withdrawals). It follows that after t years $(t = 1, 2, ...)$ the value of the account is given by

$$A(t) = Q\,(1+0.06)^t = Q\,1.06^t,$$

where $Q = A(0)$ is the amount deposited at the beginning, i.e., when $t = 0$. Since *annual* compounding is assumed, only *integer* values for t would seem to matter. Still, it is natural to ask how much the initial deposit would have grown after $1/2$ year, or after one month, that is, when $t = 1/12$, and so on. You may recall that powers with fractional exponents involve roots. For example, one has $2^{1/2} = \sqrt{2}$. So numerical evaluation, even for simple rational inputs, gets quite difficult. Values such as $2^{\sqrt{2}}$ or 2^{π} are even more complicated. Explicit numerical calculations with exponential functions usually require the use of a scientific calculator.

I.4.2   *The Functional Equation*

Because a thorough familiarity with exponential functions is critical for the study of calculus and for many of its most important applications, we shall carefully review the basic steps involved in the definition. In particular, we shall emphasize the importance of the *functional equation* that characterizes them, and show how the definitions for different classes of numbers follow from simple general principles.

Starting at the beginning, the meaning of the power b^n for positive integers $n = 1, 2, 3, \ldots$ is just a shorthand notation for repeated multiplication

$$b^n = b \cdot b \cdot \ldots \cdot b, \text{ the factor } b \text{ appearing } n \text{ times.}$$

Examples. $3^4 = 3 \cdot 3 \cdot 3 \cdot 3 = 81, \qquad (1/2)^2 = (1/2) \cdot (1/2) = 1/4,$

$$\pi^3 = \pi \cdot \pi \cdot \pi \approx (3.14)^3 \approx 30.959.$$

Note that $1^n = 1$ and $0^n = 0$ for all $n \in \mathbb{N}$. From now on we shall only consider the case when the base b is different from 0 and 1, as otherwise there would be nothing interesting to say.

If m, n are two positive integers, the basic definition and a simple counting argument show that

$$b^{m+n} = b^m \cdot b^n \tag{I.1}$$

Since $mn = n + n + \ldots + n$ (m summands n), it follows that

$$b^{mn} = b^{n+n+\ldots+n} = b^n \cdot b^n \cdot \ldots \cdot b^n \ (m \text{ factors})$$

$$= (b^n)^m.$$

Another useful formula that can easily be checked states that $(bc)^n = b^n c^n$. However, since two different bases are involved, this formula will not be so relevant for the discussion that follows.

The basic principle that controls the generalization of b^n to exponents u other than just positive integers is the desire to keep matters simple, that is, to stick to the same familiar rules as much as possible. More concretely, if we consider the function $E_b(u) = b^u$, then the rule (I.1) states that

$$E_b(u + v) = E_b(u)E_b(v) \tag{I.2}$$

whenever u and v are positive integers. This is an example of a *functional equation*. It states an internal law of the function under consideration. The basic principle requires that this internal law remains valid for all numbers u and v.

I.4.3 *Definition of Exponential Functions for Rational Numbers*

We shall now step by step extend the domain of $E_b(n) = b^n$ from positive integers to other numbers, always staying "within the law", i.e., by observing the functional equation (I.2).

First we want to define $E_b(0)$. Since $b = b^1 = E_b(1) = E_b(0 + 1)$, we apply the law to get $E_b(0 + 1) = E_b(0)E_b(1) = E_b(0)b$. Since $b \neq 0$, the equation $b = E_b(0)b$ implies that we must define $E_b(0) = 1$.

Next we take a positive integer n, and we try to define $E_b(-n)$. Again, according to the law (I.2),

$$E_b(-n + n) = E_b(-n)E_b(n),$$

and we also know that $E_b(-n + n) = E_b(0) = 1$. The equation $1 = E_b(-n)E_b(n)$ then implies that $E_b(-n)$ must be the multiplicative inverse of $E_b(n) \neq 0$, i.e.,

$$E_b(-n) = \frac{1}{E_b(n)}, \text{ also written } [E_b(n)]^{-1}.$$

In fact, it is this conclusion that justifies the notation b^{-1} for the reciprocal, that is, for the multiplicative inverse $\frac{1}{b}$. So we see that the law requires that $b^{-n} = \frac{1}{b^n}$ for a positive integer n. We have thus extended the definition of $E_b(m)$ to arbitrary integers $m \in \mathbb{Z}$ in the only way that is consistent with (I.2). It is easy to now check that the functional equation (I.2) remains valid for all $u, v \in \mathbb{Z}$.

The next extension involves the definition of $E_b(u)$ for a rational number u. We first consider $u = 1/n$, where n is a positive integer. The law requires that

$$E_b(1) = E_b(n \cdot \frac{1}{n}) = E_b(\frac{1}{n} + \frac{1}{n} + \dots + \frac{1}{n}) \quad (n \text{ summands})$$

$$= E_b(\frac{1}{n}) \cdot E_b(\frac{1}{n}) \cdot \dots \cdot E_b(\frac{1}{n}) = [E_b(\frac{1}{n})]^n .$$

By taking $n = 2$, one sees that the last quantity is positive, so $b = E_b(1) > 0$. Extension of the law to rational numbers thus requires that the base b must be positive. Furthermore, it also follows that $E_b(1/n)$ must be a real number that solves the equation $x^n = b$. As we noted earlier, it is a consequence of the completeness of $\mathbb{R}$ that this equation has a unique *positive* solution denoted by $\sqrt[n]{b}$. One therefore defines

$$E_b(1/n) = b^{1/n} = \sqrt[n]{b}.$$

Note that when n is even the equation $x^n = b$ has two real solutions ($\sqrt[n]{b}$ and $-\sqrt[n]{b}$); the definition chosen for $E_b(1/n)$ selects the positive solution $\sqrt[n]{b}$.

The case of an arbitrary rational number $u = m/n$ now follows easily, since the law requires that $E_b(m/n) = E_b(m\frac{1}{n}) = [E_b(\frac{1}{n})]^m$. A slight modification of the last argument shows that one also has $E_b(m/n) = \sqrt[n]{E_b(m)}$. To summarize, it follows that one must define

$$E_b(\frac{m}{n}) = b^{\frac{m}{n}} = [\sqrt[n]{b}]^m = \sqrt[n]{b^m} \text{ for any } n \in \mathbb{N} \text{ and } m \in \mathbb{Z}.$$

So, following the law (i.e., the equation (I.2), we have now extended the domain of E_b to all rational numbers. It would appear that the resulting function still obeys the law. While one could try to make a legal argument for this based on some higher principles, mathematicians prefer to check the validity of the functional equation for rational numbers by a more precise argument. In essence, this involves some routine verifications that can safely be skipped, as no surprises appear. We shall therefore assume from now on that the exponential function $E_b(u) = b^u$ is defined for any rational number u, and that the functional equation

$$E_b(u + v) = E_b(u)E_b(v)$$

and the related equation

$$E_b(u \cdot v) = E_{E_b(u)}(v) = E_b(u)^v = (b^u)^v$$

hold for all $u, v \in \mathbb{Q}$.

I.4.4 *Properties of Exponential Functions*

We now list a few important properties of exponential functions that easily follow from the functional equation and the definitions discussed in the preceding section. While we only consider rational inputs at this time, all results will remain valid for arbitrary real numbers as inputs, as will be considered in the next section.

We fix the base $b > 0$. First of all, $E_b(u) \neq 0$ for all u. This follows from $E_b(-u) \cdot E_b(u) = E_b(-u + u) = E_b(0) = 1$. Furthermore, the definition of $E_b(m/n)$ implies that $E_b(u) > 0$.

Assume $b > 1$. Then

$$E_b(u) = b^u > 1 \text{ for all } u > 0, \text{ and } E_b(u) = b^u < 1 \text{ for } u < 0.$$

In fact, if one had $\lambda = E_b(1/n) \leq 1$ for some $n \in \mathbb{N}$, it would follow that $b = \lambda^n \leq 1$, which contradicts the assumption $b > 1$. Consequently $E_b(u) > 1$ for all $u > 0$, and hence $E_b(-u) = 1/E_b(u) < 1$. If $0 < b < 1$, corresponding properties follow by using $E_b(-u) = 1/E_b(u) = E_{1/b}(u)$, where now $1/b > 1$.

The graphs of exponential functions are most easily obtained by means of a graphing calculator. Note that only rational numbers, or rational approximations to irrational numbers can be processed by a calculator or computer. Figure I.21 shows the graph of E_2.

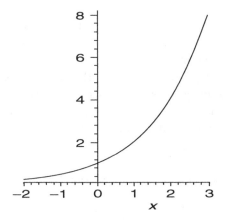

Fig. I.21 Graph of $y = E_2(x) = 2^x$.

The graphs of E_b for $b > 1$ look very similar to the graph of E_2. Figure I.22 shows three more such graphs.

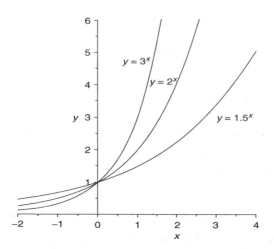

Fig. I.22   Exponential functions with base $b > 1$.

The graphs suggest the following fact.

If $b > 1$, the exponential function is *strictly increasing*:

$$\text{If } u < v, \text{ then } E_b(u) < E_b(v).$$

In fact, write $v = u + \gamma$, where $\gamma > 0$. Then $1 < E_b(\gamma)$, and multiplication on both sides of this inequality with the positive number $E_b(u)$ gives $E_b(u) < E_b(u)E_b(\gamma) = E_b(u + \gamma) = E_b(v)$.

If $0 < b < 1$, the situation is reversed: it follows that E_b is strictly *decreasing*, i.e., if $u < v$, then $E_b(u) > E_b(v)$. Since $E_b(-u) = b^{-u} = (1/b)^u = E_{1/b}(u)$, the graph of E_b ($b < 1$) is obtained by reflecting the graph of $E_{1/b}$ ($1/b > 1$) on the y-axis, that is, by replacing u with $-u$. Figure I.23 shows some graphs of exponential functions with base $b < 1$.

Finally we note the following property that is intuitively obvious from the graphs of the exponential functions.

$$E_b(u) \to 1 \text{ as } u \to 0.$$

Anticipating the discussion of limits in Section II.4.1, we write this property in the form

$$\lim_{u \to 0} E_b(u) = 1. \tag{I.3}$$

Let us verify a precise version of this statement. We want to show that the distance between $E_b(u)$ and 1 can be made arbitrarily small by choosing u sufficiently close to 0. Assume $b > 1$. Then the number 1 is a lower bound

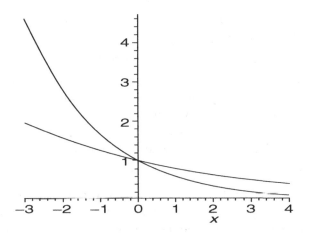

Fig. I.23   Exponential functions with base $b < 1$.

for the set $\{E_b(1/n) : n \in \mathbb{N}\}$, and therefore the *greatest* lower bound λ for this set, i.e., the infimum $\lambda = \inf\{E_b(1/n) : n \in \mathbb{N}\}$ must be ≥ 1. Since $\lambda \leq b^{1/n}$ implies $\lambda^n \leq b$ for all n, one clearly must have $\lambda = 1$. Hence, given any small number $\varepsilon > 0$, the number $1 + \varepsilon$ is *not* a lower bound for $\{E_b(1/n)\}$, so that there exists a natural number n_0 such that $E_b(1/n_0) < 1 + \varepsilon$; since for $u > 0$ the values $E_b(u)$ decrease as $u \to 0$, we have $1 < E_b(u) \leq E_b(1/n_0) < 1 + \varepsilon$ for all $0 < u \leq 1/n_0$ as well. By taking reciprocals, it then follows that $1 > E_b(-u) > 1/(1 + \varepsilon)$ for all such u. Since $1/(1 + \varepsilon) = 1 - \varepsilon/(1 + \varepsilon) > 1 - \varepsilon$, one obtains altogether that

$$1 - \varepsilon < E_b(u) < 1 + \varepsilon \text{ for all } u \text{ with } |u| < 1/n_0, \qquad (\text{I.4})$$

that is,

$$|E_b(u) - 1| < \varepsilon \text{ for all } u \text{ with } |u| < 1/n_0.$$

This shows that no matter how small ε is chosen, the distance between $E_b(u)$ and 1 will be smaller than ε provided u is sufficiently close to 0. These estimates therefore give precise meaning to the statement (I.3). If $b < 1$, note that $b^u = 1/E_{1/b}(u)$, where now $1/b > 1$; the result then follows by applying the estimate (I.4) to $E_{1/b}(u)$ and taking reciprocals.

We also identify some properties of exponential functions for large input values. If $b > 1$, the graph suggests that b^x grows larger than any fixed large number M if x is chosen sufficiently large. In fact, if $\delta = b - 1 > 0$, then $b = 1 + \delta$. It readily follows from the binomial theorem that for $n \in \mathbb{N}$ one has $b^n = (1 + \delta)^n > 1 + n\delta$. Given M, it follows that if $n > M/\delta$, then

$b^n > M$. Since E_b is strictly increasing $(b > 1)$, one has $b^u > M$ for all $u \geq n > M/\delta$. One writes

$$\lim_{x \to \infty} b^x = \infty \text{ when } b > 1.$$

We emphasize once again that ∞ is *not* a number. The preceding statement is just a convenient shorthand notation to refer to a quantity (b^x in the case at hand) that grows without any bound as x gets larger. Since $b^{-x} = 1/b^x$, it then follows that $b^{-x} \to 0$ as $x \to \infty$. We write this symbolically as $\lim_{x \to \infty} b^{-x} = 0$, a statement that is also written as

$$\lim_{x \to -\infty} b^x = 0 \text{ when } b > 1.$$

By considering numerical examples, one recognizes that exponential functions b^x with base $b > 1$ in fact grow much faster than any particular power function x^k, no matter how large the exponent k is chosen. More precisely, if $b > 1$ and k is any fixed positive integer, then

$$\frac{b^x}{x^k} \to \infty \text{ as } x \to \infty, \text{ or } \lim_{x \to \infty} \frac{b^x}{x^k} = \infty.$$

This latter result can also be established for $x \in \mathbb{N}$ by using the binomial theorem. (See Problem 6 of Exercise I.4.6.)

I.4.5   *Exponential Functions for Real Numbers*

Recall that the collection of all real numbers is visualized by the (continuous) number line, which has no gaps whatsoever. The graph of an exponential function E_b (for rational inputs) is given by a curve that is obtained by gently bending the "line" consisting of the rational numbers according to a particular rule. Extending the definition of E_b to all real numbers should result in a graph that to the eye looks exactly like the graph corresponding to rational inputs, where now the curve is assumed to be completely filled in without any gaps whatsoever, i.e., the curve should satisfy the same "continuity property" as the real line $\mathbb{R}$.

Let us consider the case with base $b > 1$, so that E_b is strictly increasing. If $x \in \mathbb{R}$, the point $(x, E_b(x))$ on the graph should arise as the "least upper bound" of the set of points $\{(r, E_b(r)) : r \in \mathbb{Q} \text{ and } r < x\}$, where the order relation on the curve is the natural one coming from the order relation on the number line. More precisely, given $x \in \mathbb{R}$, choose a natural number n with $x < n$. For any $r \in \mathbb{Q}$ with $r < x$ one then has $r < n$ as well, and hence $E_b(r) < E_b(n)$. Therefore the set $S(x) = \{E_b(r) : r \in \mathbb{Q} \text{ with } r < x\}$ has the upper bound $E_b(n)$, and consequently, by the completeness axiom, it has a *least* upper bound $\sup S(x)$ in $\mathbb{R}$.

Definition 4.1. *Assume $b > 1$. For $x \in \mathbb{R}$ one defines*

$$E_b(x) = \sup\{E_b(r) : r \in \mathbb{Q} \text{ with } r < x\}.$$

An analogous definition involving the greatest lower bound is made in the case $0 < b < 1$.

Remark. In the case $q \in \mathbb{Q}$ to begin with, b^q is already defined, and since E_b is increasing, b^q clearly is an upper bound for $S(q) = \{E_b(r) : r \in \mathbb{Q}$ with $r < q\}$. Indeed, b^q is the *least* upper bound of this set; this follows from (I.3), since $E_b(q - 1/n) = E_b(q)E_b(-1/n) \to E_b(q)$ as $n \to \infty$. In other words, if q is rational, one has $\sup\{E_b(r) : r \in \mathbb{Q}$ with $r < q\} = b^q$, so that the above definition is consistent with the original definition of the exponential function for rational inputs.

A somewhat more concrete interpretation of the definition of $E_b(x) = b^x$ for arbitrary *real* x is as follows. Choose any (increasing) sequence $\{r_n : n = 1, 2, 3, ...\}$ of rational numbers that approximates x, i.e., with $r_n \to x$; then b^{r_n} is defined, and it follows from the preceding discussion that the sequence $\{b^{r_n}\}$ approaches a certain number, namely the supremum of $\{b^{r_n}\}$. This latter value is taken as the definition of b^x. So b^x can be approximated by the values b^{r_n} as r_n approximates x.

Given this definition of E_b for arbitrary input $x \in \mathbb{R}$, one can then show that the functional equation (I.2) remains valid for arbitrary *real* numbers u, v, and that all the other properties that were listed in the preceding section for rational inputs still hold in the more general case. We shall not go into the details of these technical verifications. Most important for our purposes are the functional equation and the understanding that the graph of $y = E_b(x)$ for $x \in \mathbb{R}$ looks just like a bent copy of the *complete* number line, as shown in Figures I.22 and I.23.

At the intuitive level, the function $E_b(u) = b^u$ has been defined in such a way that its graph is a curve that is obtained by simply bending the number line so that it fits through all the points $\{(r, b^r) : r \in \mathbb{Q}\}$. The continuity *axiom* (i.e., completeness) of $\mathbb{R}$ and the definition of E_b for arbitrary real numbers ensure that there are no "gaps" in the graph, just as there are no gaps in the number line.

Let us summarize the basic properties of exponential functions.

Properties of Exponential Functions of Real Numbers

(1) For any $b > 0$ the exponential function $E_b(x) = b^x$ is defined for all real numbers and $E_b(0) = 1$.

(2) $E_b(x) > 0$ for all $x \in \mathbb{R}$, so that, in particular, $E_b(x)$ is never zero.

(3) E_b satisfies the functional equation

$$E_b(u + v) = E_b(u)\, E_b(v) \text{ for any } u, v \in \mathbb{R}.$$

(4) If $u, v \in \mathbb{R}$, then $E_b(u \cdot v) = [E_b(u)]^v$.

(5) If $1 < b$, then E_b is *strictly increasing* (i.e., $b^c < b^{c'}$ whenever $c < c'$), and if $0 < b < 1$, then E_b is *strictly decreasing* (i.e., $b^c > b^{c'}$ whenever $c < c'$).

(6) If $b \neq 1$, and if $c \neq c'$, then $b^c \neq b^{c'}$.

Remark. Functions that have the property stated in (6). are said to be *one-to-one*. (See Definition 5.2 below.)

I.4.6 *Exercises*

1. Evaluate the expressions 5^{-3}, $27^{1/3}$, $4^{-2} \cdot 2^4$, $32^{3/5}$, $6^6 \cdot 6^{-4}$ by applying appropriate functional equations. (Do NOT use a calculator!)

2. Use a calculator to find approximate values for $10^{0.6}$, $3^{\sqrt{3}}$, pi^π, $[\sin(1)]^{3.2}$.

3. Simplify as much as possible:

$$\text{i)} \quad \frac{b^{-3} b^5 b^{1/2}}{b^{3/2} b^{-2}} \qquad \text{ii)} \quad \frac{\sqrt[5]{c}\,\sqrt[2]{c^4}}{c^{-2/5}\sqrt[3]{c^2}} \qquad\qquad (b, c > 0).$$

4. Use a graphing calculator to plot the functions $f(x) = x^{10}$ and $g(x) = 2^x$ in one window for $-1 \le x \le 5$.

    a) Which function grows faster?

    b) What are the solutions of the equation $f(x) = g(x)$? (Use graphing techniques.)

    c) Are you sure to have found *all* solutions in b)? How does the graph relate to the statement that exponential functions grow faster than power functions?

    d) Investigate the behavior of f and g by changing the viewing window.

5. Use a scientific calculator to estimate the value of $x^{20} \cdot 2^{-x}$ as $x \to \infty$, i.e., as x gets larger and larger.

6. a) Suppose $b > 1$, i.e., $b = 1 + \delta$, where $\delta > 0$. Fix a positive integer k.

    Show that if $n \ge k + 1$, then $b^n > 1 + \binom{n}{k+1} \delta^{k+1}$, where $\binom{n}{k+1} =$

$\frac{n!}{(k+1)!(n-k-1)!}$ is the binomial coefficient. (Hint: Expand $(1+\delta)^n$ by the binomial theorem.)

b) Show that there exist a constant $c_k > 0$, such that $\frac{n!}{(k+1)!(n-k-1)!} \geq n^{k+1} \cdot c_k$ for all $n \geq 2k$.

c) Use a) and b) to show that $\frac{b^n}{n^k} \to \infty$ as $n \to \infty$.

7. A roast is taken out of an oven at 350^0 F at time $t = 0$ and set on the counter to cool off. Its temperature (in degrees F) after t hours is given by $T(t) = 350 - Q \cdot (1 - 2^{-t})$ for some constant Q.

a) Make a rough sketch of the graph of T . (Do NOT use a graphing calculator!)

b) Give an interpretation of the number Q. (Hint: See what happens after a long time.)

I.5   Natural Operations on Functions

Given two or more functions, there are various ways in which they can be combined to build up new functions. Operations such as addition or multiplication are possible whenever the functions take on values in the real numbers, or in other sets for which certain algebraic operations are well defined. We shall consider such algebraic operations in the next section. On the other hand, there are some even simpler operations that relate directly to the function concept, without requiring any additional structures, as follows: apply one function *after* another function, or, if possible, "reverse" a function. As we recognized in the Prelude, the corresponding differentiation rules are particularly simple and natural. In this section we briefly review these operations, mainly to recall the basic notions, and to introduce logarithm functions.

I.5.1   *Compositions*

Given two functions f and g, it seems reasonable to first apply one function and then the other one, thereby obtaining a new function. For this to make sense, the output values of the first function, say g, must be among the possible inputs for f. In formulas, if $x \in dom(g)$, one needs $g(x) \in dom(f)$, so that one can consider $f(g(x))$. The assignment $x \to f(g(x))$ defines a new function that is denoted by $f \circ g$, and that is called the *composition of g with f*. Note that the language suggests that one begins with the function

g, although the chosen notation $f \circ g$ displays f as the first function as we read from left to right. The apparent confusion is resolved by observing that in standard functional notation the input variable is placed after (i.e., to the right of) the function symbol, as shown in the following formal definition.

Definition 5.1. *Suppose Ω is the domain of g and that the image $g(\Omega)$ of Ω is contained in $\mathrm{dom}(f)$. Then the **composition** $f \circ g$ is the function with domain Ω defined by*

$$(f \circ g)(x) = f(g(x)) \text{ for } x \in \Omega.$$

Examples.

(1) If f is defined by $f(u) = u^3$, and $g(x) = \cos x$, then $f \circ g$ is defined by $(f \circ g)(x) = (\cos x)^3$ for all $x \in \mathbb{R}$. Note that one can also apply first f and then g, resulting in the function $g \circ f$, with $(g \circ f)(x) = g(f(x)) = \cos(x^3)$. Clearly $f \circ g \neq g \circ f$.

(2) $h(u) = \sqrt{u}$ has domain $\{u : u \geq 0\}$, and $g(x) = 4 - x^2$ has domain $\mathbb{R}$. In order for $h \circ g$ to be defined at $x \in \mathbb{R}$, one needs that the output $u = g(x) = 4 - x^2 \geq 0$; this holds precisely if $x^2 \leq 4$, i.e., if $-2 \leq x \leq 2$. So $h \circ g$ has domain $I = [-2, 2]$, and $(h \circ g)(x) = \sqrt{4 - x^2}$.

(3) It is often useful to recognize how more complicated function are built up by composition of simpler functions, and to identify the simpler pieces. For example, $F(x) = 3^{x^2+1}$ is the composition $f \circ g$ of $g(x) = x^2 + 1$ with $f(u) = 3^u$.

(4) Composition of more than two functions is defined in an analogous manner.

  a) If $f(u) = \sin u$, $g(x) = x^2 + 4$, and $h(t) = 2^t$, then $f \circ g \circ h$ is defined by

  $$(f \circ g \circ h)(t) = f(g(h(t))) = \sin((2^t)^2 + 4).$$

  b) The function $F(t) = [\cos(t^3 + t)]^2$ is the composition $f \circ g \circ h$ of $f(u) = u^2$, $g(x) = \cos x$, and $h(t) = t^3 + t$.

Notice that it follows readily from the definition that it does not matter in which way the functions are grouped, i.e., $(f \circ g) \circ h = f \circ (g \circ h)$. We see that composition satisfies the *associative* law familiar from addition and multiplication of numbers. On the other hand, as we already noted, composition is **not** commutative.

I.5.2 *Inverse Functions*

Another general principle to obtain new functions involves putting a function in "reverse". Given

$$f : x \to f(x) = u,$$

consider

$$g : u \to x, \text{ where } x \text{ satisfies } f(x) = u.$$

In order for this assignment to define a function, one needs that there is exactly one output $g(u)$ for the input u, i.e., given u there is **only one** value x that satisfies $f(x) = u$. Functions f with this property are called one-to-one, or 1-1 in short version. Let us state the formal definition.

Definition 5.2. *The function $f : \Omega \to \mathbb{R}$ is one-to-one on its domain Ω if the equation $f(x_1) = f(x_2)$ for $x_1, x_2 \in \Omega$ implies that $x_1 = x_2$. Alternatively, f is one-to-one if $x_1 \neq x_2$ implies that $f(x_1) \neq f(x_2)$.*

Geometrically, if $\Omega \subset \mathbb{R}$, and f is real-valued, then f is one-to-one if its graph satisfies the "horizontal line test": any horizontal line meets the graph of f at most in one point. (See Figures I.24 and I.25.)

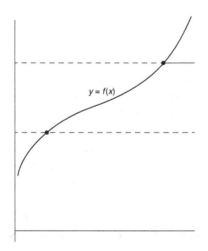

Fig. I.24 f is one-to-one.

Example. We had seen in Section I.4 that all exponential functions $E_b(x) = b^x$ with $b \neq 1$ are one-to-one on $\mathbb{R}$.

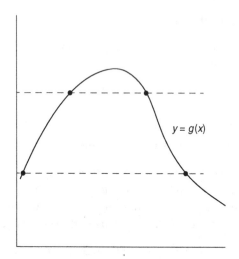

Fig. I.25　*g* is NOT one-to-one.

Definition 5.3. *If* $f : \Omega \to \mathbb{R}$ *is one-to-one on its domain* Ω*, then the function* $g : f(\Omega) \to \Omega$ *defined by*

$$g(u) = x, \text{where } x \in \Omega \text{ is the unique value that satisfies } f(x) = u,$$

is called the inverse function of f*.*

We note that if g is the inverse of f, then g is also one-to-one on its domain, with f the inverse of g. Furthermore, $g \circ f$ is the identity function $\mathrm{id}_\Omega(x) = x$ on the domain Ω, and $f \circ g$ is the identity function on $dom(g) = f(\Omega)$, i.e., $f \circ g(u) = u$ for $u \in f(\Omega)$.

The inverse function g of f is often also denoted by the symbol f^{-1}, since g acts like a "multiplicative inverse" of f if composition is viewed as "multiplication" of functions. Great care must be used not to confuse this with the reciprocal $1/f$ of f, defined by $(1/f)(x) = 1/f(x)$ provided $f(x) \neq 0$, since $1/f(x)$ is also denoted by $f(x)^{-1}$.

Example. The function $f(x) = \frac{1}{2}x - 3$ is one to one. Its graph is the line shown in Figure I.26. Clearly every horizontal line meets the graph in exactly one point.

For this simple function one can calculate the point of intersection $(x, f(x))$ with horizontal lines very easily. In fact, let such a line be given by $y = u$. Then $u = f(x) = \frac{1}{2}x - 3$ can be solved for x, resulting in the unique solution $x = 2u + 6$. So the function $g : u \to 2u + 6$ describes the

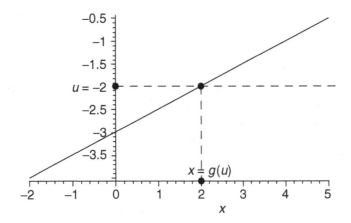

Fig. I.26 The function $f(x) = \frac{1}{2}x - 3$ is one-to-one.

inverse of f. As expected, one obtains

$$g \circ f(x) = g(f(x)) = 2(\frac{1}{2}x - 3) + 6 = (x - 6) + 6 = x.$$

Notice that one also has $f \circ g(u) = u$, i.e., f is the inverse of g. On the other hand, $(1/f)(x) = 1/(\frac{1}{2}x - 3)$ is defined for $x \neq 6$, and clearly this function differs from the inverse g of f.

More generally, if $m \neq 0$, then the linear function $f(x) = mx + b$ is one-to-one, with inverse $g(u) = \frac{1}{m}(u - b)$. Notice that the latter formula does not make sense if $m = 0$; in fact, the function $f(x) = 0x + b = b$ is constant and certainly not one-to-one.

Periodic functions are definitely not one-to-one. In particular, *sine* and *cosine* functions are not one-to-one on their natural domain $\mathbb{R}$. For example, $\sin(x) = \sin(x + 2\pi) = \sin(x + 4\pi)$, and so on. For every value $y \in [-1, 1]$ there are infinitely many solutions of $\sin x = y$, so the horizontal line test fails. However, as we shall discuss later, by suitably shrinking the domain, one can extract portions of these functions that are one-to-one and introduce appropriate inverse functions.

I.5.3 Logarithm Functions

We had seen that all exponential functions

$$E_b(x) = b^x \text{ with } 0 < b \neq 1$$

are one-to-one. The inverse function of E_b is called the **logarithm function to the base** b, and it is denoted by $\log_b$. More precisely, if

$E_b(x) = u = b^x$, then $x = \log_b u$, and vice-versa. So one has the equations

$$u = b^{\log_b u} \text{ and } \log_b(b^x) = x.$$

Recall that the values of an exponential function E_b are always positive. Moreover, it is evident from the shape of the graph of an exponential function that every positive real number $u > 0$ arises as the image $E_b(x)$ of some $x \in \mathbb{R}$.[5] Therefore the domain of the function $\log_b$ is the set $\mathbb{R}^+$ of positive real numbers. Usually, when considering exponential and logarithm functions one assumes that the base b is greater than 1, so that both functions are (strictly) increasing.

In order to visualize the graph of logarithm functions, it is best to first examine the relationship between the graphs of a function f and of its inverse g in general.

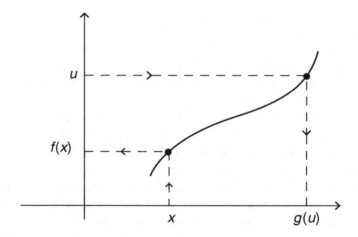

Fig. I.27   Graph of f and the reverse relation.

According to Figure I.27, the graph of the function $y = f(x)$ can also be used to describe the inverse function g of f. Just start with a point u on the vertical axis that lies in $f(\Omega)$. The horizontal line $y = u$ intersects the graph of f in precisely one point (x, u) (*horizontal* line test!); then $u = f(x)$ and $g(u) = x$. In order to be consistent with the convention to mark the input value of a function on the horizontal coordinate axis, usually labeled

---

[5]To be precise, the completeness of $\mathbb{R}$ is needed to turn this geometrically "evident" fact into a correct statement. The result used here is a consequence of the so-called *Intermediate Value Theorem* that will be discussed in the next chapter.

x-axis, we need to interchange the coordinate axis and switch notation, i.e., we interchange x with u and write $g(x) = u$. Geometrically, interchanging the coordinates of a point (a, b) leads to the point (b, a) that is visualized by reflecting (a, b) on the line $y = x$. (See Figure I.28.)

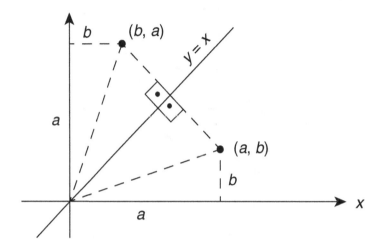

Fig. I.28 Reflection of (a, b) on the line $y = x$.

Applying this reflection to each point $(x, f(x))$ on the graph of f gives the set of points $\{(f(x), x), x \in \Omega\} = \{(u, g(u)), u \in f(\Omega)\}$. This set is evidently the graph of g displayed in the usual form, i.e., the input variable is displayed on the horizontal axis and labeled u instead of the commonly used x. Lastly one may replace u with x to make the notation consistent.

To summarize:

The graph of the inverse function g of a one-to-one function f is obtained by reflecting the graph of f on the line $y = x$.

By applying this result to the graph of $E_2(x) = 2^x$ one obtains the graph of the inverse function $y = \log_2 x$ as shown in Figure I.29.

We summarize the basic properties of logarithm functions $\log_b$, where we assume that the base $b > 1$.

i) $\log_b$ *is defined on* $\mathbb{R}^+ = (0, \infty)$ *and is strictly increasing and one-to-one;*

ii) $\log_b(1) = 0$;

iii) $\log_b(uv) = \log_b(u) + \log_b(v)$;

iv) $\log_b(u^a) = a \log_b(u)$.

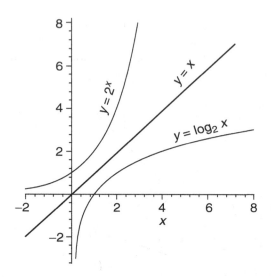

Fig. I.29   Graph of $y = \log_2 x$.

The properties iii) and iv) are the functional equations of the logarithm that follow from the corresponding equations for the exponential function. For example, in order to verify iii), set $x_1 = \log_b(u)$ and $x_2 = \log_b(v)$. Then $u = b^{x_1}$ and $v = b^{x_2}$, and therefore

$$uv = b^{x_1}b^{x_2} = b^{x_1+x_2}$$

by the functional equation for the exponential function E_b. This latter equation implies that

$$\log_b(uv) = x_1 + x_2 = \log_b(u) + \log_b(v).$$

Property iv) is left as an exercise. Properties iii) and iv) were used before the era of computers to facilitate large scale computations. Extensive tables of logarithms had been compiled, so that large numbers could be readily replaced by their logarithms, and vice-versa. By using logarithms, *multiplication* of two large numbers could be replaced by the simpler operation of *adding* the corresponding logarithms. Because our number system is based on powers of 10, the logarithms most widely used were the ones to base 10. For example, $\log_{10} 100 = 2$, $\log_{10} 1000 = 3$, and so on. Given the importance of the binary number system in today's digital world, one would think that the logarithm function to base 2 should be the more useful one today. However, it turns out that the so-called *natural* logarithm, whose base is the special transcendental number $e = 2.71828...$, is the one most widely used. We shall discuss this more in detail in Section II.2.4 when we examine the tangent problem for exponential functions.

I.5.4 *Inverting Functions on Smaller Domains*

The function $y = S(x) = x^2$, whose natural domain consists of all real numbers, is a simple function that is not one-to-one, since $S(-x) = S(x)$, and $x \neq -x$ except when $x = 0$. Note that its graph (see Figure I.30) clearly does not satisfy the horizontal line test. On the other hand, Figure I.30 suggests that the right half of the graph, taken alone, does indeed satisfy the horizontal line test. Therefore the function S^+ with graph $\{(x, x^2) : x \geq 0\}$ is one-to-one on its domain $\Omega = \{x \in \mathbb{R} : x \geq 0\} = [0, \infty)$, and hence has an inverse g that is given by $g(u) = \sqrt{u}$,[6] whose domain is also the interval $[0, \infty) = S^+(\Omega)$. After reflection on the line $y = x$ we obtain the graph of g in standard form as shown in Figure I.30.

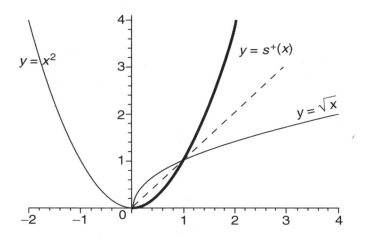

Fig. I.30   Inverse of $y = x^2$ for $x \geq 0$.

Similarly, one may consider the left half of the graph of $y = x^2$, which corresponds to the function S^- defined on $(-\infty, 0]$ by $S^-(x) = x^2$. This function is also one-to-one, and its inverse is given by $y = -\sqrt{x}$.

An analogous procedure can be done with other functions that are not one-to-one on their given domain, such as the *sine* function. We just restrict

---

[6]Recall that the symbol $\sqrt{u}$ denotes the unique non-negative number x that satisfies $x^2 = u$.

the domain to an appropriate interval on which the function *is* one-to-one. For example, the function f with domain $I = [-\frac{\pi}{2}, \frac{\pi}{2}]$ defined by $f(x) = \sin x$, whose graph is shown in Figure I.31, is clearly one-to-one.

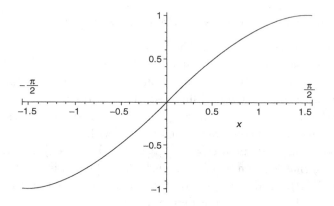

Fig. I.31   Graph of $y = \sin x$ restricted to $[-\frac{\pi}{2}, \frac{\pi}{2}]$.

Its inverse function is called the **inverse sine**, or **arc sine** function, and it is denoted by $y = \arcsin x$. Its graph is shown in Figure I.32; it is obtained by reflecting the graph of the sine function in Figure I.31.

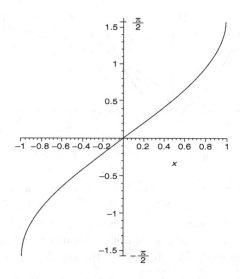

Fig. I.32   Graph of $y = \arcsin x$ on $[-1, 1]$.

Just as for the function $y = x^2$ one could obtain *different* inverse functions by restricting the function to either the domain $[0, \infty)$ or the domain $(-\infty, 0]$, one can obtain many different inverses for the sine function, for example, by restricting $y = \sin x$ to the interval $I = [\frac{\pi}{2}, \frac{3\pi}{2}]$ or, more generally, to any interval $I_k = [-\frac{\pi}{2} + k\pi, \frac{\pi}{2} + k\pi]$, where k is some integer. Our initial choice $I_0 = [-\frac{\pi}{2}, \frac{\pi}{2}]$ is singled out by being the largest interval that is symmetric about 0 on which the sine function is one-to-one; the inverse function corresponding to it is referred to as the *principal branch* of the inverse sine function.

I.5.5 Exercises

1. Find a formula for the composition $f \circ g$, where $f(u) = \sin u$ and $g(x) = \sqrt{9 - x^2}$. What is the domain of $f \circ g$?

2. With f and g as in Problem 1, show that the domain of the composition $g \circ f$ is the set of all real numbers.

3. Which of the function(s) whose graphs are shown in Figure I.33 are one-to-one? Explain!

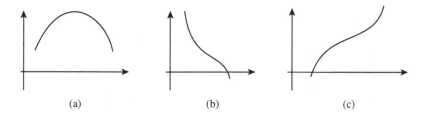

(a)   (b)   (c)

Fig. I.33   Graphs of functions for Problem 3.

4. a) Sketch the graph of the inverse function g of the function f whose graph is shown in Figure I.34.

   b) Use the graph to estimate $g(-1)$ and $g(2)$.

5. a) Find the explicit formula for the inverse function of $f(x) = 4x - 5$.

   b) Use a geometric argument to show that the product of the slopes of a non-constant linear function and of its inverse equals 1. (Hint: The corresponding graphs are the reflections of each other on the line $y = x$. Express the relevant slopes by using the point of intersection (c, c) of the two lines as one of the points. Compare with Figure I.28).

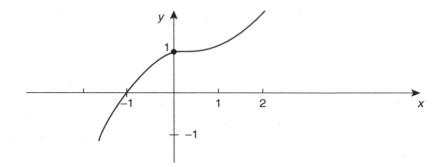

Fig. I.34   Graph of f for Problem 4.

c) Verify by algebra that for $m \neq 0$ the inverse of $y = mx + b$ is a linear function with slope $1/m$.

6. Use properties of exponential functions to verify property iv) of logarithm functions, i.e., $\log_b(u^a) = a \log_b(u)$.

7. Without using any calculators, determine the values

    i) $\log_3 9$,   ii) $\log_4(\frac{1}{2})$,   iii) $\log_{10} 10,000$,   iv) $\log_2(8^5)$.

8. Write the following functions as compositions of two simpler functions.
    i) $F(t) = 4(3^t)^2 - 5(3^t)^4$.
    ii) $F(x) = 2^{(\sin x)}$.
    iii) $H(s) = 5\cos(\log_{10} s)$
    iv) $L(x) = \sqrt{(\cos x)^4 + 1}$.

9. Write the function $G(x) = \log_2(\cos^2 x + 1)$ as a composition of three simple functions.

10. a) Show that for any base $b > 1$ one has the equation
$$\log_b x = \log_{10} x \, \frac{1}{\log_{10} b}$$
for all $x > 0$.

    b) Use a) and a graphing calculator that has the logarithm function to the base 10 (often denoted just by log) to plot the function $y = \log_4 x$.

11. Let g denote the inverse function of $f(x) = \sin x$ restricted to the interval $[-\frac{3\pi}{2}, -\frac{\pi}{2}]$.
    a) What is the domain of g?

b) Determine $g(0)$ and $g(1)$.

c) Determine $g(1/2)$. (Hint: Recall $\sin(\pi/6) = 1/2$.)

d) Sketch the graph of g.

12. Verify that the function $\cos x$ is one-to-one on the interval $[0, \pi]$ and sketch the graph of its inverse in standard form, i.e., with input on the horizontal x-axis. This function is called the *principal branch of the inverse cosine*.

I.6 Algebraic Operations and Functions

Composition and inverse of functions that we discussed in the last section are natural operations on functions that are meaningful in very general settings, not just in the case of real valued functions of real variables that we have been considering. However, for functions $f : \Omega \to \mathbb{R}$ that are *real* valued, the arithmetic properties of $\mathbb{R}$ readily lead to additional operations on functions that the reader should be familiar with. We shall briefly review the basic ideas.

I.6.1 *Sums and Products of Functions*

Given two real valued functions f, g with domain Ω, one defines the sum $f + g$ and the product fg by the formulas

$$(f + g)(x) = f(x) + g(x),$$

and

$$(fg)(x) = f(x) \cdot g(x)$$

for any $x \in \Omega$. Note that the product defined here is very different from the *composition* $f \circ g$ we considered earlier. To define quotients f/g requires that one avoids inputs x for which $g(x) = 0$. If $\Omega_g^* = \{x \in \Omega : g(x) \neq 0\}$, one defines

$$\frac{f}{g}(x) = \frac{f(x)}{g(x)} \text{ for } x \in \Omega_g^*.$$

For example, if $f(x) = 2^x$ and $g(x) = x^2 - 4$, then

$$(f + g)(x) = 2^x + x^2 - 4 \text{ for all } x \in \mathbb{R},$$
$$(fg)(x) = 2^x(x^2 - 4) \text{ for all } x \in \mathbb{R}, \text{ and}$$
$$\frac{f}{g}(x) = \frac{2^x}{x^2 - 4} \text{ for all } x \in \mathbb{R} \text{ with } x \neq -2, 2.$$

I.6.2 *Simple Algebraic Functions*

Polynomial functions are obtained by successively taking sums and products of constant functions and the identity function $id(x) = x$. Examples of such functions are described by the formulas $x^3 - 2x + 3$, $5x^4 + 2x^3 - 4x^2 + 6$, $\pi x^{52} - 2x^6$, and so on. In general, a polynomial P is a function described by a formula

$$P(x) = a_n x^n + a_{n-1} x^{n-1} + \ldots + a_2 x^2 + a_1 x + a_0$$
$$= \sum_{j=0}^{n} a_j x^j,$$

where n is a positive integer, and $a_0, a_1, \ldots, a_n \in \mathbb{R}$ are the coefficients of P. If in the above representation of P one has $a_n \neq 0$, the integer n is called the **degree of the polynomial** P; the degree is the highest power of the variable x that occurs with a non-zero coefficient. Non-zero constant functions are polynomials of degree 0, while the degree of the zero polynomial $P(x) = 0$ for all x is not defined. Note that the domain of polynomials is the set of all real numbers.

Rational functions $R(x)$ are quotients of two polynomials P, Q :

$$R(x) = \frac{P(x)}{Q(x)}, \text{ defined for all } x \text{ with } Q(x) \neq 0.$$

For example, the functions

$$y = \frac{1}{x - 1} \text{ and } y = \frac{3x^4 - 2x + 4}{x^3 - x^2 + 4x - 1}$$

are rational functions defined at all points where the denominator is different from 0.

Note that every polynomial is also a rational function (with denominator 1). By applying the familiar rules of algebra, one sees that sums, products, and quotients of rational functions are again rational.

The special polynomials $P_n(x) = x^n$, $n = 1, 2, 3, \ldots$, are also called **power functions**. These functions are strictly increasing and hence one-to-one on the set $\mathbb{R}^+ = \{x \geq 0\}$, with range $P_n(\mathbb{R}^+) = \mathbb{R}^+$. The inverse R_n of P_n is the nth **root function**.

$$R_n(x) = \sqrt[n]{x} = x^{1/n} \text{ for } x \geq 0.$$

By considering the composition of $R_n(x) = x^{1/n}$ with the power function $P_m(x) = x^m$, where $m \in \mathbb{Z}$, one obtains the power function $(x^{1/n})^m = x^{m/n}$ with rational exponent m/n.

The simple functions we just considered illustrate how the basic operations lead to the definition of new functions. Starting with the constant functions and the function $P_1(x) = x$, polynomials are obtained by repeatedly applying multiplication and addition of functions. Rational functions are obtained by dividing polynomials. Root functions are obtained by taking inverse functions of simple power functions, and compositions then lead to power functions with rational exponents. These various operations can, in turn, be applied to the functions so obtained, and this process can be continued, leading to increasingly more complicated functions defined by algebraic expressions. One must be careful with specifying the domains of the relevant functions, so that the newly obtained functions are defined correctly. For example, when taking quotients f/g of functions one must exclude points a where $g(a) = 0$. We also saw that in order to define the inverse of a function one must restrict its domain so that the function is one-to-one on the new domain. If $f_1 : \Omega_1 \to \mathbb{R}$ and $f_2 : \Omega_2 \to \mathbb{R}$ are two functions, their sum or product are defined on the domain $\Omega_1 \cap \Omega_2$, provided this set is not empty. If $\Omega_1 \cap \Omega_2 = \varnothing$, then $f_1 + f_2$ and $f_1 \cdot f_2$ are not defined. For example, $f_1 = \sqrt{4 - x^2}$ has domain $\Omega_1 = [-2, 2]$ and $f_2 = \sqrt{x^2 - 9}$ has domain $\Omega_2 = (-\infty, -3] \cup [3, \infty)$, and clearly $\Omega_1 \cap \Omega_2 = \varnothing$; so it does not make sense to consider $f_1 + f_2$. Proceeding with such operations a finite number of times one obtains the collection $\mathcal{A}$ of *algebraic functions* that we had already considered in the Prelude.

Recall that in the Prelude we recognized the importance of the Factorization Lemma for identifying tangents and defining the derivative of algebraic functions. In particular, when one considers the corresponding factorization for the inverse of such functions, an additional restriction on the domains of the functions becomes necessary. More in detail, consider $f \in \mathcal{A}$ and $a \in dom(f)$, and the factorization

$$f(x) - f(a) = q(x)(x - a),$$

where the factor q is again in $\mathcal{A}$ with the same domain as f. If one assumes that f is one-to-one (perhaps after restricting the domain appropriately), in order to obtain the corresponding factorization for the inverse function f^{-1}, one needs to divide by $q(x)$, i.e., it is necessary to require that $q(x) \neq 0$. In particular, if $q(a) = 0$, then $b = f(a)$ has to be excluded from the domain of the inverse function. For example, the function $P(x) = x^3$ is one-to-one on $\mathbb{R}$; however, the factorization

$$P(x) - P(0) = x^3 = x^2(x - 0)$$

shows that the factor $q(x) = x^2$ is zero at $x = 0$, and hence there is no corresponding factorization for the inverse $g(x) = \sqrt[3]{x}$ at the point 0. In fact, if $\sqrt[3]{x} - \sqrt[3]{0} = \sqrt[3]{x} = q_{\sqrt[3]{x}}(x)x$, then $q_{\sqrt[3]{x}}(x) = \sqrt[3]{x}/x = 1/\sqrt[2/3]{x}$ for $x \neq 0$, and $1/\sqrt[2/3]{x}$ is not defined at 0. Therefore, when viewed as an algebraic function in $\mathcal{A}$ with the relevant factorization at all points of its domain, the domain of $g(x) = \sqrt[2/3]{x}$ excludes 0, i.e., $dom(\sqrt[2/3]{x}) = \mathbb{R} - \{0\}$. Similarly, the domain of $k(x) = \sqrt{x} \in \mathcal{A}$ is the open interval $(0, \infty)$.

Let us emphasize once again that for functions $f \in A$ the factorization $f(x) - f(a) = q(x)(x - a)$ is the essential technical feature that makes such a function algebraically differentiable (see Prelude, Sections 5 and 7). In particular, the value $q(a)$ is, by definition, the derivative $D(f)(a) = f'(a)$ at the point a. If f is one-to-one, its inverse f^{-1} is (algebraically) differentiable only at points $b = f(a)$ where $D(f)(a) \neq 0$. Consequently, we shall assume that the domains of functions in $\mathcal{A}$ are restricted accordingly, so that the Factorization Lemma holds at all points in the domain.

I.6.3   *Local Boundedness of Algebraic Functions*

The following intuitively obvious property will be quite useful for establishing important properties of algebraic functions.

Theorem 6.1. *Suppose $f \in \mathcal{A}$ and let a be in the domain of f. Then there exists $\delta > 0$ such that f is bounded on $[a - \delta, a + \delta]$.*

The proof of this result is somewhat tedious and repetitive, and we shall not present all the details. Instead, we discuss a few special cases to illustrate the techniques and in order to identify the critical special property of polynomials that needs to be preserved as one performs the various operations on polynomials resulting in more and more general functions in the class $\mathcal{A}$.

As already noted in Section 7 of the Prelude, the result is essentially obvious if f is a polynomial. In fact, let I be *any* bounded interval and suppose $P(x) = \sum_{j=0}^{n} c_j x^j$. We may assume that I is contained in an interval $[-M, M]$ for some integer M. Since for $x \in I$ one then has $|x| \leq M$, standard estimations imply that

$$|P(x)| \leq \left| \sum_{j=0}^{n} c_j x^j \right| \leq \sum_{j=0}^{n} |c_j| \, |x|^j \leq \sum_{j=0}^{n} |c_j| \, |M|^j = K \text{ for } x \in I.$$

In order to prove the theorem for a rational function $R = Q/P$, where

P and Q are polynomials, we need the following immediate consequence of the local boundedness. Recall from Section 1.5 that $I_\delta(a) = (a - \delta, a + \delta)$.

Lemma 6.2. *If P is a polynomial and $a \in \mathbb{R}$, then for any $\delta > 0$ there exists a constant K such that*

$$|P(x) - P(a)| \le K\,|x - a| \quad \text{for all } x \in I_\delta(a). \tag{I.5}$$

Proof. Note that $P(x) - P(a) = q(x)(x - a)$, where q is a polynomial. By the preceding result, given $\delta > 0$, $|q(x)|$ is bounded by a constant K for all $x \in I_\delta(a)$. The result then follows by a standard estimation. ∎

We note that the above estimate, in turn, implies the local boundedness of P, since $P(x) = P(a) + [P(x) - P(a)]$ implies that $|P(x)| \le |P(a)| + |P(x) - P(a)|$, and hence $|P(x)| \le |P(a)| + K\,|x - a| \le |P(a)| + K\delta$ for all $x \in I_\delta(a)$. Furthermore, the estimate (I.5) also implies the following bound from below.

Lemma 6.3. *If P is a polynomial and $P(a) \ne 0$, there exists $\delta > 0$ such that $|P(x)| \ge |P(a)|/2 > 0$ for all $x \in I_\delta(a)$.*

Proof. By Lemma 6.2, there exists a constant K so that

$$|P(x) - P(a)| \le K\,|x - a| \quad \text{for all } x \in I_1(a).$$

Choose $0 < \delta \le 1$ so that $\delta \le |P(a)|/(2K)$. For $|x - a| \le \delta$ it then follows that $|P(x) - P(a)| \le K\delta \le |P(a)|/2$. The triangle inequality then implies that

$$|P(x)| = |P(a) + [P(x) - P(a)]| \ge |P(a)| - |P(x) - P(a)|$$
$$\ge |P(a)| - |P(a)|/2 = |P(a)|/2$$

for all $x \in I_\delta(a)$. ∎

Returning to the rational function $R = Q/P$, if $a \in dom(R)$, then $P(a) \ne 0$. Choose $\delta > 0$ according to Lemma 6.3. If K_Q is a constant so that $|Q(x)| \le K_Q$ for $x \in I_\delta(a)$, it then follows that

$$|R(x)| = \frac{|Q(x)|}{|P(x)|} \le \frac{K_Q}{|P(a)|/2} \quad \text{if } |x - a| \le \delta, \tag{I.6}$$

so that R is indeed bounded in a neighborhood of a.

In particular, the function R is defined for all $x \in I_\delta(a)$. Consequently, in the factorization $R(x) - R(a) = q_R(x)(x - a)$ the factor q_R is a rational function defined on $I_\delta(a)$ as well. Therefore, by the preceding argument, q_R is bounded as well on a suitable interval I of positive length centered at a. One thus obtains the estimate

$$|R(x) - R(a)| \le K_R\,|x - a|$$

for all $x \in I$, that is, any rational function satisfies the analogue of (I.5). Consequently, the proof of Lemma 6.3 applies to rational functions as well, so that the conclusion in that Lemma holds for rational functions and, more generally, for any other function that satisfies (I.5).

As for taking inverses, suppose $a \in dom(R)$, $b = R(a)$, and that the derivative $D(R)(a) = q_R(a) \neq 0$. As we just showed, Lemma 6.3 applies to q_R, so that $|q_R(x)| \geq |q_R(a)|/2$ for all x in some interval $I_\delta(a)$, and hence $1/q_R$ is bounded on $I_\delta(a)$. By using the completeness of the real numbers, one can show that $J = R(I_\delta(a))$ contains an interval $I_\gamma(b)$ for some $\gamma > 0$. (See Problems 7 and 8 of Exercise I.6.5 for details.) Suppose, in addition, that R is one-to-one on $I_\delta(a)$ (perhaps after choosing a smaller δ), so that the inverse g of R is defined on J, with $g(J) = I_\delta(a)$. As we saw in Section 6.3 of the Prelude, it follows that the inverse g has the factorization

$$g(y) - g(b) = \frac{1}{q_R}(g(y))(y - b)$$

for $y \in J$. The estimate for $|q_R(x)|$ then implies that $|(1/q_R)(g(y))| \leq 2/|q_R(a)|$ for $y \in I_\gamma(b) \subset J$. Therefore the inverse g also satisfies the estimate (I.5) for an appropriate constant K and interval I.

We thus see from these examples how the estimate (I.5) continues to hold as we apply various operations on rational functions that are involved in building up functions in the class $\mathcal{A}$.

In order to prove the theorem for arbitrary $f \in \mathcal{A}$ one needs to verify, more generally, that the various operations such as products, quotients, compositions, and inverses, when applied to functions that satisfy (I.5), result in functions that still satisfy such an estimate. For example, suppose it is known that $f \in \mathcal{A}$ satisfies (I.5), and we want to prove the analogous estimate for $1/f$. If $a \in dom(1/f)$, then $f(a) \neq 0$; the proof of Lemma 6.3 applies, and therefore there is $\delta > 0$ so small that $|f(x)| \geq |f(a)|/2$ for $x \in I_\delta(a)$. Since

$$\frac{1}{f(x)} - \frac{1}{f(a)} = -\frac{f(x) - f(a)}{f(x)f(a)},$$

it follows that

$$\left| \frac{1}{f(x)} - \frac{1}{f(a)} \right| \leq \frac{K}{\frac{1}{2}|f(a)||f(a)|}|x - a| \text{ for } x \in I_\delta(a).$$

We conclude this discussion by explicitly stating the two relevant properties for functions in $\mathcal{A}$ that have been used in the preceding arguments. The proofs follow by techniques analogous to those we just used for rational functions.

Corollary 6.4. *Let f be a function in $\mathcal{A}$ with domain Ω. Given $a \in \Omega$,*
i) there exist $\delta > 0$ and K so that $I_\delta(a) \subset \Omega$ and

$$|f(x) - f(a)| \leq K |x - a| \text{ for all } x \in I_\delta(a), \text{ and}$$

ii) if $f(a) \neq 0$, there exists $\delta > 0$, such that

$$|f(x)| \geq |f(a)| /2 > 0 \text{ for all } x \in I_\delta(a).$$

I.6.4 Global Boundedness

The proofs of the *local* estimates in the preceding section only required basic algebraic tools and estimations. By using the completeness of $\mathbb{R}$ one can prove a corresponding *global* result, as follows.

Theorem 6.5. *Suppose $f \in \mathcal{A}$ with domain Ω, and let J be any closed and bounded interval contained in Ω. Then f is bounded on J.*

It is noteworthy that *both* conditions on the interval J are critical. If one of the conditions is dropped, then the theorem is no longer true. For example, the interval $I = (0, 1]$ is bounded but not closed, and it is contained in the domain of $f(x) = 1/x$; note that f is NOT bounded on I. Similarly, the interval $J = [0, \infty)$ is closed (it contains its only boundary point 0) but not bounded, and the function $g(x) = x$ is NOT bounded on J.

Proof. We shall prove the result by contradiction, that is, let us assume that f is NOT bounded on J. So for each natural number n one can find a point $x_n \in J$ such that $|f(x_n)| \geq n$. Set $J = J_0$, divide the interval J_0 in half, and denote the two closed bounded intervals so obtained by J' and J''. Then at least one of the intervals J' and J'' must contain points x_n for infinitely many $n \in \mathbb{N}$. Label that half interval by J_1. By repeating this process over and over one obtains a nested sequence of closed bounded intervals $J = J_0 \supset J_1 \supset \ldots \supset J_k \supset \ldots$ such that each interval J_k contains points x_n for infinitely many n, and so that $length(J_k) = length(J_0)/2^k$. In particular, f is NOT bounded on any interval J_k. By the nested interval theorem (Theorem 1.3), there exists a point $a \in \mathbb{R}$ with $a \in J_k$ for all $k = 1, 2, 3, \ldots$
By Theorem 6.1 there exist $\delta > 0$ and K, such that $|f(x)| \leq K$ for all $x \in I_\delta(a) = \{x : |x - a| < \delta\}$. Choose an integer N such that $length(J_N) = length(J_0)/2^N < \delta$. Since $a \in J_N$, it follows that $J_N \subset I_\delta(a)$, and therefore

$|f(x)| \leq K$ for all $x \in J_N$. On the other hand, by the construction of J_N, the function f is NOT bounded on J_N! This contradiction shows that our assumption that f is *not* bounded on the interval J is incompatible with the local boundedness property of algebraic functions. Therefore it follows that f must indeed be bounded on the interval J.   ■

Corollary 6.6. *Suppose* $f \in A$ *with domain* Ω, *and let* J *be any closed and bounded interval contained in* Ω. *Given* $a \in J$, *there exists a constant* $K = K(J, a)$, *such that*

$$|f(x) - f(a)| \leq K |x - a| \text{ for all } x \in J.$$

Proof. Consider the factorization $f(x) - f(a) = q(x)(x - a)$, where $q \in A$ also has domain Ω. By the theorem, q is bounded over J, so that $|q(x)| \leq K$ for some constant K and all $x \in J$, and the desired estimate follows.   ■

Note that the factor q, and hence also the bound K depends on the fixed point a. We shall see later in Chapter III that it is possible to choose the constant K independently of $a \in J$, although K will still depend on the interval J.

I.6.5   *Exercises*

1. Determine the largest possible domain of the following functions:

    i) $f(x) = \sqrt{x^2 + 2x - 8}$;

    ii) $y = \frac{1}{1+x^4}$;

    iii) $g(s) = \frac{s-1}{s+1}$.

2. a) Find the domain of the function $g(x) = \frac{1}{1-x^2}$.

    b) Use a graphing calculator to display the graph of g.

    c) Describe the behavior of $g(x)$ as x approaches the points that are not in the domain.

    d) Determine $\lim_{x \to \infty} g(x)$.

3. Show that $(2^x + 1)(2^x - 1) = 2^{2x} - 1$ for all $x \in \mathbb{R}$.

4. Show that $(\sin x / \cos x)^2 + 1 = (1 / \cos x)^2$ for all x with $\cos x \neq 0$.

5. a) Define $Ch(x) = (2^x + 2^{-x})/2$ and $Sh = (2^x - 2^{-x})/2$. Show that $[Ch(x)]^2 - [Sh(x)]^2 = 1$ for all $x \in \mathbb{R}$.

    b) More generally, show that $[\frac{E_b(x)+E_b(-x)}{2}]^2 - [\frac{E_b(x)-E_b(-x)}{2}]^2$ is a constant and determine its value.

6. Assume that f and g are algebraic functions in the class $\mathcal{A}$ which satisfy the estimates i) and ii) in Corollary 6.4 at the point $x = a$. Show in detail that $f + g$ and fg satisfy the corresponding estimates as well.

7. Suppose $f \in \mathcal{A}$ and that a is in the domain of f. Then $f(x) - f(a) = q(x)(x - a)$, with $q \in \mathcal{A}$.

   a) Suppose that $q(a) > 0$. Show that there exists $\delta > 0$ such that $q(x) > 0$ for all x with $|x - a| \leq \delta$. (Hint: Apply Corollary 6.4 to q.)

   b) With δ as in a), show that $c = f(a - \delta) < f(a) < f(a + \delta) = d$.

   c) Formulate and prove an estimate corresponding to b) in the case $q(a) < 0$.

8. Suppose $f \in \mathcal{A}$ with domain Ω.

   a) Assume that the interval $[a, b]$ is contained in Ω and that $f(a) < 0$ and $f(b) > 0$. Use the completeness of $\mathbb{R}$ to show that there exists $x_0 \in (a, b)$ such that $f(x_0) = 0$. (Hint: Let x_0 be the Least Upper Bound of $S = \{x \in [a, b] \text{ and } f(x) \leq 0\}$; use Corollary 6.4 to show that $f(x_0) = 0$.)

   b) Use a) to show that if $f(a) < f(b)$ and λ is any number with $f(a) < \lambda < f(b)$, then there exists $x_\lambda \in (a, b)$ such that $f(x_\lambda) = \lambda$. (Hint: Apply a) to $f(x) - \lambda$.)

   c) Use b) and the notations and results from Problem 7 to show that if $q(a) \neq 0$, then the image $f(I_\delta(a))$ contains an open interval centered at $f(a)$.

Chapter II

Derivatives: How to Measure Change

As seen in the Prelude, derivatives—which were introduced by algebraic techniques based on double points and multiplicities—provide the solution to the ancient problem of finding tangent lines for large classes of curves. These elementary methods, however, fail in the case of non-algebraic functions, in particular for the important case of exponential functions. Motivated by the application to velocity that we considered in Section 4 of the Prelude, we recognized that derivatives can also be captured by an approximation process, e.g., the velocity at time t_0 is approximated by *average* velocities over decreasing time intervals containing t_0. Such *average* velocities, or more generally, average rates of change, are very easy to define. The main new difficulty thus concerns understanding the approximation process and developing a general setting where it leads to meaningful results. Starting with the elementary case of algebraic functions and guided by the example of exponential functions, we shall now investigate this approximation process in detail, culminating with a notion of differentiability that generalizes the algebraic formulation and that is equivalent to the classical version used in analysis. Along the way we shall highlight the interpretation of derivatives as *instantaneous rate of change,* a concept that is the foundation for the numerous applications of calculus to the natural sciences over several centuries and to many other disciplines in more recent times.

Before proceeding with this chapter, the reader is urged to review Sections 4 and 8 of the Prelude.

II.1   Algebraic Derivatives by Approximation

II.1.1   *From Factorization to Average Rates of Change*

As we saw in the Prelude, given a polynomial f and a point $a \in \mathbb{R}$, the elementary algebraic factorization

$$f(x) - f(a) = q(x)(x - a),$$

where q is a uniquely determined polynomial, provides the critical information to solve the tangent problem for the curve that is defined by f. In fact, the value $q(a)$ is the slope of that unique line through the point $P = (a, f(a))$ that intersects the graph of f with multiplicity 2 or higher. This is the special property that singles out the *tangent* line to the graph at P, and the value $q(a)$ is called the derivative $D(f)(a) = f'(a)$ of f at a. We then showed how this factorization, and consequently the solution of the tangent problem, extends by simple algebraic techniques to all functions that are built up from polynomials by standard algebraic operations and the natural operations of composition and taking inverses of functions. In particular, that includes the familiar rational functions and root functions.

While the calculation of an explicit formula for the factor q in concrete cases typically involves lengthy computations, we showed in the Prelude that its value at a, that is, the derivative $D(f)(a)$, can readily be found for all algebraic functions by a routine application of specific rules. We shall now consider in detail the values of q at inputs x that are *different* from a. Regardless of the nature of the function f, or whether a particular explicit expression for the factor q is available, the value $q(x)$ for $x \neq a$ is uniquely determined by

$$q(x) = \frac{f(x) - f(a)}{x - a} \text{ for } x \neq a. \tag{II.1}$$

The quotient on the right side, which is NOT defined for $x = a$ (plugging in $x = a$ leads to the expression $0/0$, which is meaningless), contains important information that not only is most useful in applications, but that is critical in order to solve the tangent problem for more general non-algebraic functions, where the value $q(a)$, i.e., the derivative, is not accessible by any elementary methods.

In order to illustrate the significance of the quotient (II.1), we shall now consider several concrete situations.

II.1.1.1   *Average Velocity*

Let us begin with the concept of "average velocity" that we had already considered in Section 4 of the Prelude. We are all familiar with the basic idea as it arises, for example, with a moving automobile. A car that travels a distance of 15 km between 12:10 p.m. and 12:22 p.m. is said to have traveled with a velocity of $\frac{15}{12}$ km/min over that time period. Note that this number does not take into account any changes that may occur during the time interval considered, such as slowing down to avoid an obstacle, stopping for a traffic light, or accelerating to pass another car. Instead, what has been measured is the *average* velocity between 12:10 p.m. and 12:22 p.m. Formally, for two distinct moments in time $t_1 < t_2$, one defines

$$\textbf{average velocity } \textit{between } t_1 \textit{ and } t_2 = \frac{\textit{distance traveled between } t_1 \textit{ and } t_2}{t_2 - t_1},$$

or, more briefly,

$$\textit{average velocity} = \frac{\textit{distance}}{\textit{time}}.$$

Note that the numerical value of the average velocity depends on the units chosen to measure distance and time. For example, since 12 minutes are 0.2 hours, the velocity of $\frac{15}{12}$ km/min corresponds to a velocity of $\frac{15}{0.2} = 75$ km/hour. Converting 15 km into 9.32 miles results in an average velocity of $\frac{9.32}{0.2} = 46.61$ miles/hour. Commonly used units for velocity are m/sec = meters/second, ft/sec = feet/second, km/h = kilometer/hour, and mi/h = miles/hour.

The distance traveled between two points in time t_1 and t_2 relates to the change in position of the automobile. Suppose the car moves along a highway, and let $s(t)$ measure its position at time t as given, for example, by the km-markers along the road. Between the times t_1 and t_2 the car will have traveled a distance $\Delta s = s(t_2) - s(t_1)$. (See Figure II.1.)

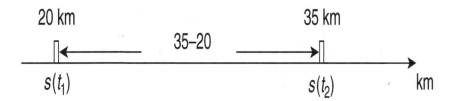

Fig. II.1   Distance traveled on a road.

The average velocity between t_1 and t_2 is thus given by the quotient

$$\frac{\Delta s}{\Delta t} = \frac{s(t_2) - s(t_1)}{t_2 - t_1},$$

where the symbol Δ (= delta = capital Greek "D") is generally used to indicate a difference of relevant quantities. Note that this latter expression has exactly the structure of the general quotient (II.1), with the function f replaced by the position function s.

A car is said to move with *constant* velocity (during a particular time period) if the average velocity between any two points in time during that period is always the same number. In that case the position of the car in dependence of time can easily be described precisely. Suppose that the car travels with *constant* (average) velocity v. Then

$$\frac{s(t_2) - s(t_1)}{t_2 - t_1} = v \text{ for any } t_1 \neq t_2.$$

Therefore, if we fix the initial time t_1, and let $t_2 = t$ be arbitrary, one obtains

$$\frac{s(t) - s(t_1)}{t - t_1} = v \text{ for any } t \neq t_1.$$

(This formula holds both when $t > t_1$ and when $t < t_1$.) This equation can be solved for $s(t)$, resulting in

$$s(t) = v(t - t_1) + s(t_1).$$

Note that the latter formula is valid also for $t = t_1$. We see that constant velocity implies that the position $s(t)$ is described by a *linear* function of time t, i.e., by a polynomial of degree 1.

II.1.1.2   *Lines and Slopes*

An analogous discussion applies when one considers the inclination or steepness of a highway, as shown in Figure II.2.

A measure of the "steepness" is given by the change in height $\Delta h = h(x_2) - h(x_1)$ that occurs over the (horizontal) distance $\Delta x = x_2 - x_1 > 0$. What matters is not the value of Δh itself, but rather the value of the ratio

$$\frac{\Delta h}{\Delta x} = \frac{h(x_2) - h(x_1)}{x_2 - x_1}.$$

Note that if $x_1 < x_2$, then $\Delta h/\Delta x < 0$ is equivalent to $\Delta h < 0$, so a negative quotient indicates that the height is decreasing, i.e., that the

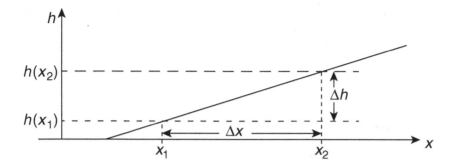

Fig. II.2   Inclination of a straight line.

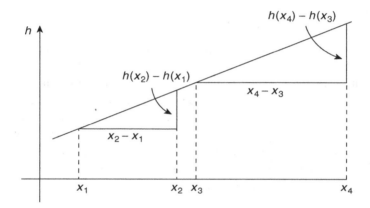

Fig. II.3   Measure of steepness is independent of position.

road goes downhill. In the case where the road follows an inclined *straight* line, the quotient $\frac{\Delta h}{\Delta x}$ is a constant m independent of x_1 and x_2, since the corresponding right triangles shown in Figure II.3 are similar, and hence the ratios of corresponding sides are equal.

This constant number

$$m = \frac{\Delta h}{\Delta x} = \frac{\text{``}rise\text{''}}{\text{``}run\text{''}},$$

which depends on the units chosen to measure length, is called the *slope* of the line. For example, if the road rises by 40 meters over a horizontal distance of 1 kilometer, $m = 40/1 = 40$ m/km. Using the same units for the height and the distance results in a slope of $40/1000 = 0.04$ m/m, which is a quantity without any "dimension". One often then writes the result as a percentage, i.e., in the example just considered one says that the road rises with a slope of 4% or at a rate of 4%.

Let us fix x_1 and consider $x_2 = x \neq x_1$ as variable. We then solve the equation

$$m = \frac{h(x) - h(x_1)}{x - x_1}$$

for $h(x)$, resulting in

$$h(x) = m(x - x_1) + h(x_1) = mx + (h(x_1) - mx_1).$$

Again, the relevant function—the height h in this case—is given by a linear function.

II.1.1.3   *Average Rate of Climb*

More generally, suppose the steepness of the road changes along the way, perhaps alternating between climbing and descending. Again, denote by x the position in the horizontal direction and let $h(x)$ be the height of the road above sea level, measured in meters, at the position x. In this case the quotient $\Delta h / \Delta x = [h(x_2) - h(x_1)]/(x_2 - x_1)$ depends on the particular locations x_1 and x_2 chosen, i.e., it is not constant along the way. This quotient thus describes the *average* rate of change in altitude between the two points x_1 and x_2, or also the *average rate of climb of the road* between the points identified by x_1 and x_2.

II.1.1.4   *Average Rates of Change*

Returning to the case of a general function f, for a fixed value $x_1 \neq a$, the numerator $f(x_1) - f(a)$ of the quotient in (II.1) simply measures the difference Δf, or change in the values of f, between the two input values a and x_1. As in the examples we just considered, division by the change $\Delta x = x_1 - a$ in the input thus provides a measure of the "rate of change of f" between the two points a and x_1. Since the quotient $[f(x_1) - f(a)]/(x_1 - a)$ contains no information whatsoever about the function f at any point x between a and x_1, this rate of change again is just an *average* rate of change of f between a and x_1.

Example. The average rate of change of $f(x) = \sin x$ over the interval $[0, \pi/2]$ is

$$\frac{\sin \frac{\pi}{2} - \sin 0}{\frac{\pi}{2} - 0} = \frac{1}{\frac{\pi}{2}} = \frac{2}{\pi} \approx 0.637.$$

Over the interval $[\pi/6, \pi/2]$ one obtains

$$\frac{\sin\frac{\pi}{2} - \sin\frac{\pi}{6}}{\frac{\pi}{2} - \frac{\pi}{6}} = \frac{1 - \frac{1}{2}}{\frac{\pi}{3}} = \frac{3}{2\pi} \approx 0.477,$$

which is—no surprise—a different value.

Let us consider a geometric interpretation of this abstract notion, as follows. The graph of a general (non-linear) function f describes a curve in the coordinate plane. For $x_1 \neq x_2$ the ratio

$$\frac{\Delta f}{\Delta x} = \frac{f(x_2) - f(x_1)}{x_2 - x_1} \tag{II.2}$$

can then be interpreted as the slope of the line through the two points $(x_1, f(x_1))$ and $(x_2, f(x_2))$ on the graph (such a line through two distinct points is called a *secant*, to distinguish it from a tangent). (See Figure II.4.)

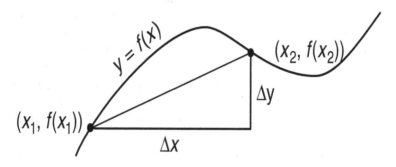

Fig. II.4 Average rate of change or slope given by $\Delta y / \Delta x$.

Clearly the graph of f varies quite a bit from that line, i.e., the secant only provides crude information about the curve. The quotient (II.2) is therefore called the *average* slope of the curve between x_1 and x_2.

Our discussion shows that (non-vertical) lines are exactly those curves whose average slopes are constant.

II.1.1.5 *Other Examples of Rates of Change*

Average rates of change occur in numerous applications. Here are some additional examples.

Suppose the function T measures the temperature in Celsius degrees at the current location in dependence of the time of day t, e.g., $T(8)$ is the temperature at 8 a.m. If $T(8) = 16^0$, and $T(11) = 25^0$ the quotient

$\Delta T/\Delta t = (25 - 16)/(11 - 8) = 3$ measures the average rate of change in temperature, i.e., the temperature is increasing between 8 and 11 a.m. at the average rate of 3^0 per hour. Again, the numerical value depends on the chosen temperature scale and units of time.

Volume V, pressure p, and temperature T of an ideal gas are related by the formula $Vp = kT$, where k is a numerical constant. Keeping T fixed, we can view V as a function of pressure that is explicity given by $V(p) = kT/p$. The ratio

$$\frac{\Delta V}{\Delta p} = \frac{V(p_2) - V(p_1)}{p_2 - p_1}$$

describes the average rate of change of volume between the pressure points p_1 and p_2.

Finally, suppose $P = P(t)$ describes the size of the population in a town in year t. If $P(2007) = 80,000$ and $P(2015) = 92,000$, the ratio $\Delta P/\Delta t = (92,000 - 80,000)/(2015 - 2007) = 1,500$ gives the average rate of growth of the population between the years 2007 and 2015, i.e., during this period the population grew at an *average* rate of $1,500$ people per year. In order to compare rates of growth of cities of different sizes, it is more useful to consider the average *relative* rate of growth, defined by $[\Delta P/\Delta t]/P$, where usually the value of the population at the beginning of the time period is chosen. In the case at hand, the relative growth rate between the years 2007 and 2015 is thus given by $1,500/P(2007) = 1,500/80,000 = 0.01875$. The relative growth rate is most commonly expressed as a percentage, i.e., one says that in the relevant period the population grew at an average rate of $1.875\% \approx 1.9\%$ per year. The use of "percentages" signals that one considers the *relative* growth rate of the population.

To summarize, we see that "average rates of change" occur in many different settings. The reader is encouraged to add additional examples related to her/his own experience and interest.

Returning to the basic factorization

$$f(x) - f(a) = q(x)(x - a),$$

we thus see that the values of the factor $q(x) = [f(x) - f(a)]/(x - a)$, $x \neq a$, contain important information about the particular process that is modeled by the function f.

We conclude this section with a basic relationship between average rates of change over *adjacent* intervals that will turn out to be significant in later discussions. To illustrate the simple idea, let us consider average velocities over two consecutive time intervals $T_1 = [t_0, t_1]$ and $T_2 = [t_1, t_2]$. For example, if the average velocity over T_1 is 40 km/h and over T_2 it is 55 km/h, then surely the average velocity over the *combined* time interval $T = T_1 \cup T_2$ must be between 40 and 55 km/h. We expect that this relationship holds in general, i.e., that the average velocity over the *combined* time interval is at least as large as the *smaller* of the average velocities over each of the two intervals, and that, similarly, it cannot exceed the larger one of these two velocities.

Indeed, this remains correct for general average rates of change, as follows. Let us assume that the function f is defined on the two adjacent intervals $[x_0, x_1]$ and $[x_1, x_2]$, where $x_0 < x_1 < x_2$. For $j = 1, 2$ we denote the average rate of change of f over the interval $[x_{j-1}, x_j]$ by $A_j = A([x_{j-1}, x_j]) = [f(x_j) - f(x_{j-1})]/(x_j - x_{j-1})$.

Lemma 1.1. *With the notations introduced above, the average rate of change $A([x_0, x_2])$ over the combined interval $[x_0, x_2]$ satisfies the estimate*

$$\min\{A_1, A_2\} \leq A([x_0, x_2]) \leq \max\{A_1, A_2\}.$$

Furthermore, if $A_1 \neq A_2$, then both inequalities are strict.

Proof. For completeness' sake we give the simple proof. By writing the definition of average rate of change in product form, one obtains

$$f(x_j) - f(x_{j-1}) = A_j(x_j - x_{j-1}) \text{ for } j = 1, 2.$$

It follows that

$$\min\{A_1, A_2\}(x_1 - x_0) \leq f(x_1) - f(x_0) \leq \max\{A_1, A_2\}(x_1 - x_0)$$

and

$$\min\{A_1, A_2\}(x_2 - x_1) \leq f(x_2) - f(x_1) \leq \max\{A_1, A_2\}(x_2 - x_1) .$$

After adding the two inequalities and rearranging, the terms involving x_1 and $f(x_1)$ cancel, and one is left with

$$\min\{A_1, A_2\}(x_2 - x_0) \leq f(x_2) - f(x_0) \leq \max\{A_1, A_2\}(x_2 - x_0) .$$

The desired result follows by dividing the last inequality by the positive number $(x_2 - x_0)$. If $A_1 \neq A_2$, we may assume that $A_1 < A_2$ after perhaps renumbering. Then $\min\{A_1, A_2\} = A_1 < A_2$ and $A_1 < A_2 = \max\{A_1, A_2\}$, so that $\min\{A_1, A_2\}(x_2 - x_1) < f(x_2) - f(x_1)$ and $f(x_1) - f(x_0) < \max\{A_1, A_2\}(x_1 - x_0)$; proceeding as before, it follows that the inequalities are now strict. ∎

The result clearly generalizes to any finite collection of adjacent intervals $[x_{j-1}, x_j]$, $j = 1, ..., n$, with $n \geq 2$. Furthermore, when the average rates of change are strictly increasing, one has the following easy consequences.

Corollary 1.2. *Let $n \geq 2$ and $x_0 < x_1 < ... < x_{n-1} < x_n$, and assume that the average rates of change of the function f defined on $[x_0, x_n]$ satisfy*

$$A([x_{j-1}, x_j]) < A([x_j, x_{j+1}]) \text{ for } j = 1, ..., n-1.$$

Then

(i) $\qquad A([x_0, x_1]) < A([x_0, x_n]) < A([x_{n-1}, x_n]) \qquad$ *and*

(ii) $\qquad A([x_0, x_j]) < A([x_0, x_{j+1}]) \qquad$ *for $j = 1, ..., n-1$.*

Proof. The proof of (i) is an immediate consequence of the Lemma, generalized to n adjacent intervals, since the hypothesis implies that $A([x_0, x_1])$ and $A([x_{n-1}, x_n])$ are the minimal, resp. maximal of the rates of change over the n intervals. As for (ii), replacing n with j in (i), one obtains $A([x_0, x_j]) < A([x_{j-1}, x_j])$. If $j < n$, the hypothesis gives $A([x_{j-1}, x_j]) < A([x_j, x_{j+1}])$, and it then follows that $A([x_0, x_j]) < A([x_j, x_{j+1}])$. We now apply the lower estimate in Lemma 1.1 to the two adjacent intervals $[x_0, x_j]$ and $[x_j, x_{j+1}]$, resulting in $A([x_0, x_j]) < A([x_0, x_{j+1}])$. ∎

Of course, corresponding results with the inequalities reversed are true as well. See Problem 7 of Exercise II.1.4 for details.

II.1.2 *From Average to Instantaneous Rates of Change*

As we saw in the preceding section, the average slopes of the graph of a function f over different intervals provide only limited information about the function. Many details are simply not captured by such averages. It therefore seems desirable to introduce more refined ways to describe the behavior of the function or of its graph. Recall that in the Prelude we had considered the classical problem of finding *tangents* to curves. In analogy to the (average) rate of change or slope between two distinct points P_1 and P_2 on the graph of a function, the tangents at points P or Q on the graph capture the rates of change or slopes at the single point P, respectively Q, thereby defining the rate of change of a function at single points. (See Figure II.5.)

The situation is particularly simple in the case where the function f is a polynomial, or more generally of algebraic type, i.e., when f is in the class $\mathcal{A}$ introduced in Section 7 of the Prelude (see also Section I.6). Given the factorization $f(x) - f(a) = q(x)(x - a)$, we saw in the Prelude that the value $q(a)$ gives the slope of the tangent line to the graph of f at the point $(a, f(a))$. On the other hand, for $x \neq a$, the value $q(x) = [f(x) - f(a)]/(x-a)$ is the average slope of f between the two distinct points $P_a = (a, f(a))$ and $P_x = (x, f(x))$. (See Figure II.6.)

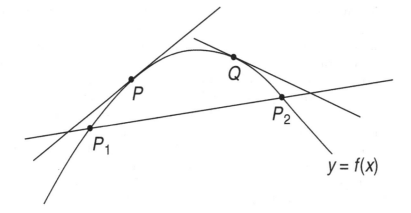

Fig. II.5   Average slopes and tangent slopes.

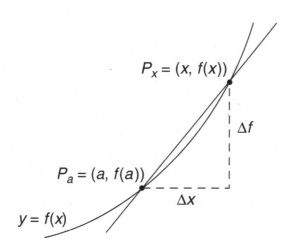

Fig. II.6 $q(x)$ gives the slope $\Delta f/\Delta x$ of the secant.

From this perspective it thus seems natural to interpret $q(a) = D(f)(a)$ in exactly the same way as $q(x) = \Delta f/\Delta x$ for $x \neq a$, that is, $q(a)$ is the (average) slope of the graph of f between the two (now identical) points P_a and P_a. The two *distinct* points of intersection of the secant with the graph of f when $x \neq a$ coincide when $x = a$, i.e., we have only *one* point P_a of intersection, which however is *counted twice*, that is, P_a is now a "double point" of intersection between the secant and the curve. According to the discussion in Section 2 of the Prelude, this is exactly the defining geometric property of the tangent line to the curve at P_a. When $x = a$ the secant has become the tangent line, and $q(a)$ still measures its slope. Since only **one** point is involved (which however is counted twice), the value $q(a)$, i.e., the slope of the tangent, measures the rate of change of f at the single point $P_a = (a, f(a))$! In more general settings, we say that the derivative $D(f)(a) = q(a)$ of f at a measures the "instantaneous" rate of change of f at the point a. In order to highlight this interpretation, in applications one often denotes the derivative by the formal quotient $\frac{df}{dx}$, i.e., $D(f)(a) = \frac{df}{dx}(a)$. This notation—which is analogous to $\frac{\Delta f}{\Delta x}$—helps to remind us that the derivative measures a "rate of change".

An analogous interpretation remains valid in all other applications. For example, if $s(t)$ describes the position of an automobile at time t, the derivative $D(s)(t_0)$ at time t_0 measures the velocity at that moment, i.e., the instantaneous velocity at time t_0, and we also write $D(s)(t_0) = ds/dt(t_0)$. Similarly, if $V(p) = kT/p$ is the volume of a fixed amount of gas in dependence of the pressure p, the derivative $D(V)(p_0) = dV/dp(p_0) = -kT/p_0^2$ describes the instantaneous rate of change of volume with respect to pressure when the pressure has value p_0.[1] Regarding the application to population growth, where $P = P(t)$ measures the size of the population in a town at time t, it turns out that in all "natural" situations the relevant function P is not of algebraic type, so the preceding discussion does not apply directly. As we shall see later, models of population growth involve exponential functions. Of course, once we have developed the appropriate concept of derivative, we shall say that the derivative $D(P)(t_0) = dP/dt(t_0)$ measures the instantaneous rate of change of the population at time t_0. Correspondingly, $D(P)(t_0)/P(t_0)$ measures the (relative) rate of growth of the population at time t_0.

II.1.3   *Approximation of Algebraic Derivatives*

The discussion of derivatives for algebraic functions in the Prelude, combined with the results of the preceding two sections, provide a detailed analysis of the critical factor q in the basic factorization $f(x) - f(a) = q(x)(x - a)$. In order to avoid any possible misunderstanding, in this section we shall indicate explicitly that the factor q, which is a function of x, depends on the fixed point a in the domain of f: changing the point a requires changing the factor q. We shall write q_a for the factor that appears

---

[1]The derivative $D(V)$ is calculated by applying the power rule for negative integer exponents discussed in Section 6 of the Prelude.

in

$$f(x) - f(a) = q_a(x)(x - a). \qquad \text{(II.3)}$$

In the preceding sections we recognized that the values $q_a(x)$ can be interpreted in a uniform and consistent manner as average rates of change of f between the two points a and x. When the two points coincide we have a special situation, but the same conceptual interpretation applies. Instead of the slope of a secant through two *distinct* points, the value $q_a(a)$, i.e., the derivative of f at a, measures the slope of the secant through the *double point* corresponding to a, that is, the slope of the tangent. Tangents are just special cases of secants, and the basic algebraic technique, that is, the factorization (II.3), treats the two in a unified way. The single formula for q_a in the algebraic case suggests that there is a strong bond between the derivative $D(f)(a) = q_a(a)$, i.e., the slope of the tangent, and the average rate of change $q_a(x) = \Delta f / \Delta x$ for $x \neq a$. This is justified also at an intuitive level. For example, regardless of how we actually define and calculate the instantaneous velocity at a single moment in time t_0, we expect that this value is very close to the average velocity over very short time intervals surrounding t_0. Furthermore, the average velocity should get closer and closer to the velocity at t_0 as the length of the time intervals shrinks to 0. In other words, the instantaneous velocity at t_0 is approximated by average velocities over shorter and shorter time intervals. Similarly, Figure II.6 clearly suggests that as the input x gets closer and closer to a, i.e., when the point P_x moves towards the point P_a on the graph of f, the secant through P_a and P_x turns towards the position of the tangent at P_a. In other words, the slope of the tangent at P_a is approximated by the average slope between P_a and P_x as x approaches a.

Conceptually, we write:

[*average slope between P_a and P_x*] $\rightarrow$ [*slope of tangent at P_a*] *as* $P_x \rightarrow P_a$,

or even more briefly,

$$q_a(x) \rightarrow q_a(a) \quad as \ x \rightarrow a.$$

What seems quite obvious to the eye is, in fact, easily justified precisely in the case of algebraic functions. In Section 4 of the Prelude we had already examined this approximation process in some detail in the case of the velocity of a freely falling object, subject only to the gravitational force of the earth. Before considering the general case, let us look at another example to illustrate the idea as concretely as possible.

Example. Let us consider the function $f(x) = x^3$ at the point $a = 1$. The relevant factorization is

$$f(x) - f(1) = x^3 - 1 = q_1(x)(x - 1),$$

where the factor q_1 is given by $q_1(x) = x^2 + x + 1$. Then $q_1(1) = f'(1) = 3$ is the slope of the tangent at the point $(1, 1)$, while for $x \neq 1$,

$$q_1(x) = \frac{f(x) - f(1)}{x - 1}$$

measures the *average* slope of f over the interval with endpoints 1 and x. Note that

$$q_1(x) - q_1(1) = x^2 + x + 1 - 3 = x^2 + x - 2$$
$$= (x + 2)(x - 1).$$

Let us consider the interval $I = [0, 2]$ centered at $a = 1$. Clearly $|x + 2| \leq 4$ for all $x \in I$, so that

$$|q_1(x) - q_1(1)| \leq 4 |x - 1| \text{ for } x \in I.$$

This explicit estimate clearly shows that $q_1(x) \to q_1(1)$ as $x \to 1$. For example, if we want to approximate the slope of the tangent within $10^{-5} = 0.00001$ by *average* slopes $q_1(x)$, it is enough to choose x within a distance of $10^{-5}/4$ from the point 1, i.e., choose any $x \neq 1$ that satisfies $|x - 1| \leq 10^{-5}/4$.

The corresponding estimate in the case of an arbitrary algebraic function is obtained by the same method. In fact, the critical result has already been discussed in detail in Section I.6. Recall from the Prelude that if f is a function in the class $\mathcal{A}$ and a is in the domain of f, the factor q_a in the basic factorization $f(x) - f(a) = q_a(x)(x - a)$ is again in the class $\mathcal{A}$, and a is in the domain of q_a as well. This is the (algebraic) generalization of the standard fact that if f is a polynomial, then the factor q_a is a polynomial as well. One therefore can apply Theorem I.6.1. It follows that there exist an interval $I_\delta(a)$ and a constant K, so that

$$|q_a(x) - q_a(a)| \leq K |x - a| \text{ for all } x \in I_\delta(a). \tag{II.4}$$

This is the critical estimate that gives precise meaning to the approximation property captured by the statement that $q_a(x) \to q_a(a)$ as $x \to a$. For example, suppose we want to approximate $q_a(a)$ within 10^{-10}. Formula II.4 obviously implies that $|q_a(x) - q_a(a)| < 10^{-10}$ for all $x \in I_\delta(a)$ that satisfy $|x - a| < 10^{-10}/K$. Clearly the same argument works if 10^{-10} is replaced by the much smaller number 10^{-100}, or for that matter, by any arbitrarily small number $\varepsilon > 0$. Just choose $|x - a| < \varepsilon/K$ to ensure that $|q_a(x) - q_a(a)| < \varepsilon$. To summarize:

The closer x is to the point a, the closer the average slope $q_a(x)$ will be to the slope of the tangent $q_a(a)$.

More generally, the preceding discussion establishes the following abstract result.

Theorem 1.3. *Let a be a point in the domain of the function $f \in \mathcal{A}$, and denote by $A(J_{x,a})$ the average rate of change of f over the interval $J_{x,a}$ with endpoints a and x. Then there exist an interval I centered at a and a constant K, such that*

$$|A(J_{x,a}) - D(f)(a)| \leq K |x - a| \text{ for } x \in I \text{ and } x \neq a.$$

In a less formal way we can say that the average rate of change of f over small intervals (with one of the endpoints at a) approaches the derivative $D(f)(a) = f'(a)$ of f at the point a as the lengths of the intervals go to zero. Symbolically we may write

$$\frac{\Delta f}{\Delta x} \to D(f)(a) = \frac{df}{dx}(a) \text{ as } \Delta x \to 0, \text{ or}$$

$$\lim_{\Delta x \to 0} \frac{\Delta f}{\Delta x} = D(f)(a) = \frac{df}{dx}(a).$$

This fundamental approximation process involving rates of change for algebraic functions is the precursor of the general concept of "limit" that needs to be considered when one studies functions that are not algebraic. We shall formalize the appropriate notions after we have examined in detail the case of exponential functions in the next section. Recall that the Prelude already culminated with a preliminary investigation of the tangent problem for such functions. In particular, we had recognized that the approximation property that we just identified and verified for all algebraic functions is the critical ingredient that suggests how to overcome the new difficulties that appear in the non-algebraic case.

In essence, we have discovered that the algebraic definition of derivative based on the identification of double points can be replaced by a new *non-algebraic* **approximation process** that views a double point as the "limiting" position of two *distinct* points that move towards each other.

Historical Remark. This approximation process is the crux of the new ideas developed by Leibniz and Newton. The history of Calculus and

Analysis in the 17th century shows that Descartes' algebraic method based on double points was never fully implemented as we have done in the Prelude. Instead, Leibniz and Newton started directly with the much deeper and more powerful approximation process. Given that apparently they were not aware of the elementary algebraic approach and of the explicit estimates that provide a direct motivation for the approximation process, the discovery of the approximation process by Leibniz and Newton to solve the tangent problem is particularly remarkable and a lasting testimony to their creativity.

II.1.4 *Exercises*

1. An airplane departed Albany, NY, at 1:50 p.m. and it landed at Newark Airport at 2:35 p.m. Determine the average velocity (in miles/hour) of the airplane on this trip. (You will need to look up a relevant piece of information that is not given here. Alternatively, use your best estimate.)

2. A motorcycle travels along a highway from 9 to 10 a.m. with a constant speed of 70 km/h. Determine the function $s(t)$ that measures the distance (in km) at time t from the position at 9 a.m.

3. Determine the average rates of change on the interval $[1, 5]$ of the functions $F(x) = x^2$ and $E_2(x) = 2^x$.

4. a) Find the average rate of change of the function $f(x) = x^3$ on the interval $[0, t]$, where $t > 0$. (Note: the answer depends on t.)

   b) Show that for any fixed number c the average rate of change of f (as in a) on the interval $[c, t]$ for $t \neq c$ can be expressed by a polynomial of degree 2 in t.

   c) Find the polynomial given in b). What is its value at $t = c$?

5. The Department of Fisheries estimates that the population of trouts in a mountain lake grew from about 50,000 to 80,000 from March 1 to July 31.

   a) Determine the average rate of growth per month of the fish population.

   b) Determine the average relative rate of growth per month of the fish population. Give the answer in percentage form.

6. a) A train traveled at an average velocity of 70 mph between 10 a.m. and 12 p.m. Thereafter, because of the poor condition of the tracks, the train had to slow down and traveled at an average velocity of only 35 mph between 12 and 1 p.m. Determine the average velocity of the train over the whole trip, i.e., between 10 a.m. and 1 p.m.

b) Verify that your answer in a) is consistent with Lemma 1.1.

7. With the same notations as in Corollary 1.2, assume that $A([x_{j-1}, x_j]) > A([x_j, x_{j+1}])$ for $j = 1, ..., n-1$. Prove that

  (i) $A([x_0, x_1]) > A([x_0, x_n]) > A([x_{n-1}, x_n])$, and
  (ii) $A([x_0, x_j]) > A([x_0, x_{j+1}])$ for $j = 1, ..., n-1$.

8. Let $f(x) = \frac{1}{2} x^4 - 3x$.

  a) Determine the difference between the slope of the tangent line to the graph of f at the point $(2, 2)$ and the slope of the line through the points $(2, 2)$ and $(2.1, f(2.1))$.

  b) Determine a precise estimate for the difference between the slope of the tangent line to the graph of f at the point $(2, 2)$ and the slope of the line through the points $(2, 2)$ and $(x, f(x))$ that is valid for *all* x between 1 and 3.

9. a) Determine the instantaneous rate of change df/dt of the function f given by $f(t) = 4\sqrt{t}$ at points $t > 0$. (Hint: Use the power rule - see Prelude, Section 6.)

  b) What happens to $df/dt(t)$ in part a) as $t \to 0$?

  c) Interprete the result in b) geometrically by looking at the graph of f and its tangents.

II.2 Derivatives of Exponential Functions

Exponential functions arise in numerous central applications. However, as we already recognized at the end of the Prelude, the study of their tangent lines and the related notion of instantaneous rate of change leads to new phenomena that transcend the elementary algebraic methods used so far. The discussion in the Prelude and in Section 1.2 showed that in the algebraic case we were able to define and calculate derivatives by an easy application of a basic algebraic factorization property. The relevant approximation property we identified was then a simple by-product of our analysis that helped us to better understand the concept of derivative of a function, i.e., the instantaneous rate of change. In contrast, for exponential functions, as well as for all other non-algebraic functions, we need to turn this process around, since no appropriate factorization is available to us at the beginning. Therefore we will have to take average rates of change as our *starting* point and use them to capture and define derivatives by a limit process that is motivated by the discussion of the algebraic case in the previous section. Given the importance of exponential functions, we shall investigate the existence and properties of their derivatives in great detail. In the subsequent section we shall then use the insights gained along the way to formulate the general concepts of limits and derivatives that are the core of analysis.

II.2.1   *Tangents for $y = 2^x$*

Let us begin with the concrete exponential function $E_2(x) = 2^x$ that we already considered in Section 8 of the Prelude. (The reader may wish to review Sections I.4.4 and I.4.5 before proceeding.) Inspection of the graph of this function suggests that it has a (non-vertical) tangent line at every point. (See Figure II.7.) Again, the point P of intersection is a double point (just rotate the tangent slightly to reveal the two points), but there is no (algebraic) technique to identify that particular slope for which the corresponding line through P intersects the graph with multiplicity at least two.

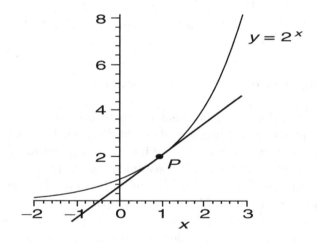

Fig. II.7   Graph of $y = 2^x$ with a tangent.

The basic problem thus is to find the slope of such tangents. Given a fixed number a, we consider the factorization

$$E_2(x) - E_2(a) = q_a(x)(x - a)$$

where $q_a(x)$ is uniquely defined for all $x \neq a$ by the average rate of change

$$q_a(x) = \frac{E_2(x) - E_2(a)}{x - a} \, .$$

In contrast to the algebraic case, there is no explicit formula for q_a that is defined for $x = a$, so there is no obvious way to capture the derivative of E_2 at $x = a$. However, motivated by the results in the algebraic case we obtained in Section 1, we consider the double point as the limiting position of two distinct points that approach each other. For $x \neq a$, the average slope between the two corresponding points on the graph of E_2 is given by $q_a(x)$. We are thus led to investigate whether there is a value L that arises as the "limit" of $q_a(x)$ as $x \to a$. In analogy to the algebraic case, that limit L would give us the desired slope of the tangent to the graph of E_2 at the (double) point $(a, E_2(a))$. Note that this problem is quite a bit more complicated than the approximation in the algebraic case. Since we do not know any value of L a priori—note that we cannot evaluate q_a at $x = a$—we cannot attempt to estimate $|q_a(x) - L|$ directly. Instead, we need to determine the existence of the "limit" by carefully studying the behavior of $q_a(x)$ when $x \neq a$ gets closer and closer to a.

Before proceeding with the analysis, let us simplify the problem as follows. Set $h = x - a$, so that $x = a + h$ and $x \to a$ corresponds to $h \to 0$. Then notice that by the functional equation of the exponential function one has

$$\begin{aligned} E_2(a + h) - E_2(a) &= 2^{a+h} - 2^a = 2^a 2^h - 2^a \\ &= 2^a[2^h - 1] = E_2(a)[E_2(h) - E_2(0)] \\ &= E_2(a)[q_0(h) \cdot h], \end{aligned}$$

where in the last step we have used the factorization $E_2(h) - E_2(0) = q_0(h)(h - 0)$ at the point $a = 0$ for $h \neq 0$. Since $E_2(a + h) - E_2(a) = q_a(a + h) \cdot h$, it follows that

$$q_a(a + h) = E_2(a)q_0(h) \quad \text{for all } h \neq 0. \tag{II.5}$$

This shows how finding the derivative of E_2 at the arbitrary point a is reduced to finding the derivative for the case $a = 0$. More precisely, assume for the moment that the limiting process for $q_0(h)$ as $h \to 0$ indeed leads to a meaningful result that we denote by the expression

$$\lim_{h \to 0} q_0(h).$$

Equation (II.5) then shows that the corresponding limit process for q_a at the arbitrary point a also is meaningful, and that it results in

$$\lim_{h \to 0} q_a(a + h) = E_2(a) \lim_{h \to 0} q_0(h).$$

Geometrically, this means that the slope of the tangent to the graph of E_2 at the point $x = a$ equals $E_2(a) \cdot c_2 = 2^a \cdot c_2$, where $c_2 = \lim_{h \to 0} q_0(h)$ is the slope of the tangent at $x = 0$. As in the algebraic case, let us denote the slope of the tangent line at $(a, E_2(a))$, that is, the derivative of the function E_2 at the point $x = a$, by $D(E_2)(a)$, or also by $E_2'(a)$. Thus

$$D(E_2)(a) = \lim_{h \to 0} q_a(a + h).$$

This is consistent with the corresponding results in the algebraic case discussed in the Prelude and reviewed in Section 1.3, where $D(f)(a) = q_a(a) = \lim_{h \to 0} q_a(a + h)$ for $f \in \mathcal{A}$. The preceding arguments show that one has the formula

$$D(E_2)(a) = E_2(a)D(E_2)(0), \text{ or}$$
$$(2^x)' = 2^x \cdot derivative \ at \ 0.$$

So, in order to determine the derivative of E_2 at arbitrary points, it is enough to study in detail the derivative at $a = 0$.

II.2.2   *The Tangent to $y = 2^x$ at $x = 0$*

The relevant factorization for E_2 at the point 0 is given by $2^x - 1 = q_0(x)x$. Since the point $a = 0$ is fixed in this section, we shall simplify notation by using q instead of q_0.

Figure II.8 shows that for $x \neq 0$ the slopes $q(x) = (2^x - 1)/x$ of the secant lines through the points $(0, 1)$ and $(x, 2^x)$ increase as x increases. In particular, $q(x)$ decreases as $x \to 0$ from the right side.   Let us recall the explicit numerical data for the values of $q\left(10^{-k}\right)$ that we had already considered in Section 8 of the Prelude.

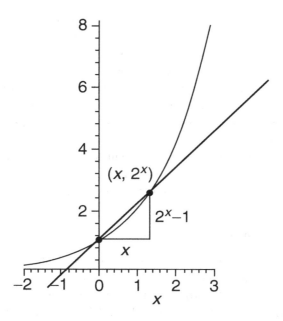

Fig. II.8   Secant to $y = 2^x$ of slope $(2^x - 1)/x$ for $x > 0$.

$$x_k \qquad q(x_k) = (2^{-x_k} - 1)/x_k$$

x_k	$q(x_k) = (2^{-x_k} - 1)/x_k$
10^{-1}	0.7177346253
10^{-2}	0.6955550056
10^{-3}	0.6933874625
10^{-4}	0.6931712037
10^{-5}	0.6931495828
10^{-6}	0.6931474207
10^{-7}	0.6931472045
10^{-8}	0.6931471829
10^{-9}	0.6931471808
10^{-10}	0.6931471805

Table II.1.

Table II.1 confirms that the numbers $q(10^{-k})$ decrease as k gets larger and larger, i.e., as $10^{-k} \to 0$, and that they seem to approximate a number whose decimal expansion begins with 0.69314... . We will now verify precisely that $q(x)$ is indeed an increasing function on $\mathbb{R} - \{0\}$, and that the limiting process suggested by Table II.1 is meaningful, that is, we shall identify a specific real number c_2 that is the "limit" of $q(x)$ as $x \to 0$ in an appropriate sense. (The subscript 2 is in reference to the base 2 of the exponential function that is considered here.) In contrast to the algebraic case, there is no explicit formula for c_2; instead, the existence of the limit must be verified by a more abstract argument that utilizes the axioms of the real numbers $\mathbb{R}$, including the critical "completeness" property introduced in Section I.1.4.

For the remainder of this section we will continue to think of limits and use their basic natural properties in an informal and intuitive way. We shall formulate the notion of "limit" and the basic rules more precisely in the next section.

Lemma 2.1. *Suppose $x_1, x_2 \neq 0$, and $x_1 < x_2$. Then $q(x_1) < q(x_2)$.*

As mentioned earlier, this statement is geometrically evident from the graph of E_2, as seen in Figure II.8. Also, Table II.1 provides numerical evidence. However, as the property stated in the Lemma is critical in order to make precise the existence of the desired limit for $q(x)$ as $x \to 0$, we shall present the somewhat technical details of the proof. The reader may skip these details on first reading and just accept the geometric argument.

Proof. Note that for $x > 0$ $q(x) = \frac{E_2(x) - E_2(0)}{x - 0}$ is the average rate of change $A([0, x])$ of the function E_2 over the interval $[0, x]$. For fixed $\delta > 0$ and any $x \in \mathbb{R}$ one has

$$A([x, x + \delta]) = \frac{E_2(x + \delta) - E_2(x)}{\delta} = E_2(x)\frac{E_2(\delta) - 1}{\delta}.$$

Since E_2 is a (strictly) increasing function, one sees that $A([x, x + \delta])$ is a strictly increasing function of x. We shall first consider the case of two non-zero rational numbers $r_1 < r_2$ and use Lemma 1.1 and its Corollary to relate the average rates of change over appropriate intervals, as follows. By choosing a sufficiently large common denominator $n \in \mathbb{N}$ for r_1 and r_2, we may assume that $r_l = m_l/n$ for $l = 1, 2$, where $m_1, m_2 \in \mathbb{Z}$, and $m_1 < m_2$. We consider the intervals $[\frac{j}{n}, \frac{j}{n} + \frac{1}{n}] = [\frac{j}{n}, \frac{j+1}{n}]$ for $j \in \mathbb{Z}$. By what we just observed, $A([x, x + 1/n])$ is strictly increasing in x, and therefore $A([\frac{j}{n}, \frac{j+1}{n}]) < A([\frac{j+1}{n}, \frac{j+2}{n}])$ for all j, so that the hypotheses of Corollary 1.2, with $x_j = j/n$, are satisfied for any finite collection of successive intervals. Note that $x_0 = 0$. If $r_1 > 0$, and hence also $m_1 > 0$, part (ii) of that Corollary for $0 \le j \le m_2$ then implies that

$$A([0, x_j]) < A([0, x_{j+1}]) \le A([0, x_{m_2}])$$

for any j with $0 < j < m_2$. In particular, for $j = m_1$ one obtains $q(r_1) = A([0, x_{m_1}]) < A([0, x_{m_2}]) = q(r_2)$. In the case $r_1 < 0$, note that $q(r_1) = A([x_{m_1}, 0])$ with $m_1 < 0$. For any j with $m_1 < j < 0$, part (i) of the Corollary implies that

$$A([x_{m_1}, x_j]) \leq A([x_{j-1}, x_j]) < A([x_j, x_{j+1}]) \leq A([x_j, x_0]).\ ^2$$

Lemma 1.1 applied to the intervals $[x_{m_1}, x_j]$ and $[x_j, 0]$ then implies that

$$A([x_{m_1}, 0]) < A([x_j, 0]). \tag{II.6}$$

In particular, if $m_2 < 0$ we can take $j = m_2$ in (II.6), resulting in $q(r_1) = A([x_{m_1}, 0]) < A([x_{m_2}, 0]) = q(r_2)$. Finally, if $m_2 > 0$, use $j = -1$ in (II.6) to obtain $A([x_{m_1}, 0]) \leq A([x_{-1}, 0])$ (the case $m_1 = -1$ being trivial). It then follows that

$$q(r_1) = A([x_{m_1}, 0]) \leq A([x_{-1}, 0]) < A([0, x_1]) \leq A([0, x_{m_2}]) = q(r_2),$$

thereby completing the proof of Lemma 2.1 in the case of rational numbers.

For the general case, recall from Section I.4 that $E_2(x)$ for x real is approximated by $E_2(r)$ for rational numbers r with $r \to x$. Given $x \neq 0$, it then follows from general properties of limits—made precise in the next section—that one has

$$q(r) = \frac{E_2(r) - 1}{r} \to \frac{E_2(x) - 1}{x} = q(x)$$

as $r \to x$. Therefore, if $x_1 < x_2$, approximate x_1 and x_2 sufficiently closely by non-zero rational numbers r_1 and r_2 with $x_1 < r_1 < r_2 < x_2$. Since $q(r_1) < q(r_2)$ by the first part of the proof, it then readily follows that $q(x_1) < q(x_2)$ as well. (See Problem 4 of Exercise II.2.8 for more details.)∎

We now want to identify the limit of $q(h)$ as $h > 0$ approaches 0. Clearly

$$q(h) = \frac{2^h - 1}{h} > 0 \text{ for } h > 0,$$

so that the set $S^+ = \{q(h) : h > 0\}$ is bounded from below. Therefore, by the completeness axiom of the real numbers, S^+ has a *greatest lower bound* in $\mathbb{R}$ that we denote by c_2. We will now explain why this number $c_2 = \inf S^+$ is the desired limit of $q(h)$ as $h \to 0$. Note that in particular c_2 is a lower bound for S^+, i.e., $q(h) \geq c_2$ for $h > 0$. More significantly, c_2 is the *greatest* lower bound, that is, for any natural number n the number $c_2 + 1/n > c_2$ is NOT a lower bound for S^+. So there exists $h_n > 0$ such

---

^{2}In the Corollary, the counter j began at 0, while here the counter begins at a negative integer, but that detail is irrelevant for the conclusions.

that $c_2 \leq q(h_n) < c_2 + 1/n$. Since by Lemma 2.1 $q(h)$ is getting smaller as h decreases, it then follows that one even has

$$c_2 \leq q(h) < q(h_n) < c_2 + 1/n \text{ for all } h \text{ with } 0 < h < h_n.$$

Thus, no matter how small $1/n > 0$ has been chosen, all the values $q(h)$ will eventually be at a distance $< 1/n$ from c_2 provided $h > 0$ is sufficiently small. Surely these estimates make precise that *the numbers $q(h)$ have the* ***limit*** *c_2 as $h > 0$ goes to zero* . We thus write

$$\lim_{h \to 0^+} q(h) = c_2, \tag{II.7}$$

where we use the notation $h \to 0^+$ (instead of $h \to 0$) to encode that h approaches 0 from the right only, i.e., from the positive side. The numerical data in Table II.1 suggests that $c_2 = 0.69314...$.

In order to complete the discussion and obtain a more precise numerical estimate for c_2 we need to also consider the values $q(h)$ for *negative h* that approach 0. Note that since for $h < 0$ one has $q(h) < q(s)$ for all $s > 0$ by Lemma 2.1, it follows that $q(h) \leq c_2$ for $h < 0$. Here is some relevant numerical data.

x_k	$q(x_k) = (2^{x_k} - 1)/x_k$
-10^{-1}	0.6696700846
-10^{-2}	0.6907504562
-10^{-3}	0.6929070095
-10^{-4}	0.6931231584
-10^{-5}	0.6931447783
-10^{-6}	0.6931469403
-10^{-7}	0.6931471565
-10^{-8}	0.6931471781
-10^{-9}	0.6931471803
-10^{-10}	0.6931471805

Table II.2.

According to the data in Tables II.1 and II.2 we have the estimate

$$q(-10^{-9}) = 0.6931471803... \leq c_2 \leq 0.6931471808... = q(10^{-9}).$$

This estimate determines the first 9 digits of c_2. More digits of c_2 can be captured by increasing the computing technology, i.e., by increasing the number of significant digits in the calculations, and evaluating $q(-x_k)$ and $q(x_k)$ for numbers x_k closer and closer to zero.

The numerical evidence in Table II.2 clearly suggests that

$$\lim_{h \to 0^-} q(h) = c_2, \tag{II.8}$$

where the notation $h \to 0^-$ encodes that h approaches 0 from the left, i.e., through *negative* numbers. We will now show that this last equation is indeed correct. Note that for $h \neq 0$ one has

$$q(-h) = \frac{2^{-h} - 1}{-h} = 2^{-h}\frac{1 - 2^h}{-h} \tag{II.9}$$

$$= 2^{-h}\frac{2^h - 1}{h} = 2^{-h}q(h).$$

Now recall from Section I.4.4, that $2^h \to 1$ as $h \to 0$, i.e., $\lim_{h \to 0} 2^h = 1$, and consequently also $\lim_{h \to 0^+} 2^{-h} = 1$. It then follows from (II.9) that

$$\lim_{h \to 0^+} q(-h) = 1 \cdot \lim_{h \to 0^+} q(h) = c_2. \; ^3 \tag{II.10}$$

Since for $h > 0$ one has $-h < 0$, equation (II.10) surely proves equation (II.8). Since the limits from the two sides coincide, we can combine the equations (II.7) and (II.8) into the single statement

$$\lim_{h \to 0} q(h) = \lim_{h \to 0} \frac{2^h - 1}{h} = c_2 = 0.6931471805.... \tag{II.11}$$

We want to emphasize that the limit c_2 in the preceding equation is identified precisely as $c_2 = \inf\{q(h) : h > 0\}$, and that the proof is theoretical and does not depend on any numerical data. Computing technology is only used in order to obtain the decimal expansion of c_2 to any desired level of accuracy, subject to the limitations of the technology.

Note that while the *decimal expansion* of c_2 looks quite mysterious, the *geometric meaning* of c_2 is very simple and not at all mysterious:

c_2 *is the slope of the tangent to the graph of* $y = 2^x$ *at the point* $(0,1)$!

The number c_2 can thus be readily visualized by the length of the short vertical line segment shown in Figure II.9.

This limit c_2 identified in equation (II.11) represents the "missing value $q(0)$" for q at $h = 0$: it is that particular number that is approximated by slopes of secants, just as in the case of algebraic functions discussed in Section 1.3. In analogy to the familiar case of algebraic functions we thus say that the function $y = E_2(x)$ is *differentiable at* $x = 0$ (i.e., there

---

^{3}We have used natural properties of limits, for example, that the limit of a product equals the product of the limits. This and other rules will be formalized in the next section.

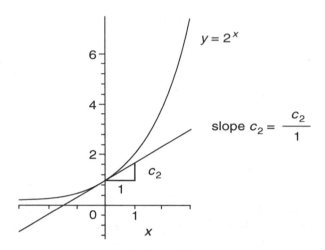

Fig. II.9   Visualization of c_2, the slope of the tangent at $(0, 1)$.

exists a well-defined tangent at the point $(0, 1)$), and that the value $c_2 = \lim_{h \to 0} q(h)$ is the *derivative* $D(E_2)(0) = (E_2)'(0)$ of E_2 at 0. As we already observed at the beginning of this section, the functional equation for E_2 then implies that E_2 is differentiable at each point $a \in \mathbb{R}$, and that

$$(E_2)'(a) = \lim_{h \to 0} q_a(a + h) = E_2(a) \cdot c_2,$$

where c_2 is defined by (II.11). We have thus solved the tangent problem at every point of the graph of $E_2(x) = 2^x$.

II.2.3   *Other Exponential Functions*

The discussion in the preceding section readily generalizes to exponential functions $E_b(x) = b^x$ with an arbitrary base $b > 0$. In particular, for each such b there exists a real number c_b defined by

$$c_b = \lim_{h \to 0} \frac{b^h - 1}{h} = \lim_{h \to 0} \frac{E_b(h) - E_b(0)}{h}$$

that is the slope of the tangent to the graph of $y = E_b(x)$ at the point $(0, 1)$. We call c_b the derivative $E_b'(0)$ of the exponential function E_b at 0. Furthermore, for any other point $a \in \mathbb{R}$ one has

$$E_b'(a) = \lim_{h \to 0} \frac{E_b(a + h) - E_b(a)}{h} = E_b(a) \cdot c_b.$$

Note that for $b = 1$ one trivially has $c_1 = 0$, but otherwise the numbers c_b are quite unexpected and intriguing. For example, with the help of a computer one readily obtains the following numerical approximations:

$$c_3 = 1.098612...$$
$$c_4 = 1.386294...$$
$$c_5 = 1.609437...$$
$$c_{10} = 2.302585...\,.$$

These numbers c_b with mysterious decimal representations thus appear very naturally as soon as one considers the tangent problem for exponential functions. We repeat that their geometric meaning is clear:

c_b *equals the slope of the tangent to the graph of* $y = b^x$ *at the point* $(0, 1)$.

We shall soon find another interpretation of these numbers that gives further insight than just the (approximate) decimal expansion.

II.2.4 The "Natural" Exponential Function

Surely the strange values c_b that we identified prompt the question whether there exists a base b for which the slope of the tangent to $y = b^x$ at $x = 0$ is some simple natural number, say the number 1. Since $c_2 = 0.693... < 1$ and $c_3 = 1.098... > 1$, it appears reasonable that there is a base $b^\#$ somewhat smaller than 3 for which $c_{b^\#} = 1$ exactly. A simple "rescaling" argument will allow us to verify this precisely and to determine this particular base $b^\#$ in terms of the number c_2, i.e., in terms of the derivative of E_2 at 0.

In the discussion that follows, we shall continue to use the intuitive idea of *limit* and relevant natural rules as they occurred in the previous section. A more detailed discussion of limits will be given in Section 3.1.

Starting with $c_2 = \lim_{h \to 0}(2^h - 1)/h$, we can use logarithms to express the values of c_b for arbitrary bases $b > 0$ in terms of c_2, as follows. Since $\log_2$ is the inverse function of $E_2(x) = 2^x$, one has $b = E_2(\log_2 b) = 2^{\log_2 b}$, and hence, by the properties of exponentials,

$$b^h = (2^{\log_2 b})^h = 2^{h \log_2 b} \text{ for any } h.$$

Therefore,

$$c_b = \lim_{h \to 0} \frac{b^h - 1}{h} = \lim_{h \to 0} \frac{2^{h \log_2 b} - 1}{h}.$$

We now rescale: instead of considering $h \to 0$ we consider $t = h \log_2 b \to 0$. This is analogous to a change in units in measurements of physical quantities, for example, changing from meters to centimeters (or feet) to measure a distance that goes to 0. Let us assume that $b \neq 1$, so that $\log_2 b \neq 0$. Substituting $h \log_2 b = t$, so that $h = t/\log_2 b$, one obtains

$$\frac{2^{h \log_2 b} - 1}{h} = \frac{2^t - 1}{t/\log_2 b} = \log_2 b \, \frac{2^t - 1}{t}.$$

Since $t \to 0$ precisely when $h \to 0$, it now follows that

$$c_b = \lim_{h \to 0} \frac{2^{h \log_2 b} - 1}{h} = (\log_2 b) \lim_{t \to 0} \frac{2^t - 1}{t} = \log_2 b \cdot c_2.$$

Note that $c_1 = 0$ and $\log_2 1 = 0$, so this last equation holds also for $b = 1$. We thus see that

$$c_b = c_2 \log_2 b \text{ for any } b > 0.$$

Rather than starting with a familiar base b for an exponential function, such as $b = 2$, thereby ending up with the mysterious and awkward value $c_2 = 0.69314718...$, we can now turn matters around and prescribe a convenient value for the slope c_b. The preceding equation then allows us to determine the corresponding—and possibly quite strange—base b. Of particular interest is to find the base $b^{\#}$ that satisfies $c_{b^{\#}} = 1$—clearly as simple as it gets—so that the exponential function with the corresponding base $b^{\#}$ will satisfy the differentiation formula

$$(E_{b^{\#}})' = c_{b^{\#}} E_{b^{\#}} = 1 \, E_{b^{\#}} = E_{b^{\#}}.$$

This base $b^{\#}$ is uniquely determined by

$$1 = c_{b^{\#}} = c_2 \log_2 b^{\#}, \text{ i.e., } \log_2 b^{\#} = \frac{1}{c_2},$$

and hence

$$b^{\#} = E_2(\log_2 b^{\#}) = 2^{1/c_2} = 2^{1/0.6931478...} = 2.7182818... \, .$$

This number $b^{\#}$ turns out to be ubiquitous in mathematics, at a par with the number π. It is most commonly denoted by the letter "e". The corresponding exponential function $y = e^x$ is called the *natural exponential function*, and we shall also denote it by the letter "E", i.e., E denotes the function that is defined by $E(x) = e^x$ for all real numbers x. It is important to clearly understand that this number

$$e = 2.7182818...$$

has been identified as that unique base for which the tangent at the point $(0,1)$ to the graph of the corresponding exponential function $E(x) = e^x$ has slope 1. In particular e satisfies

$$\lim_{h \to 0} \frac{e^h - 1}{h} = 1.$$

Later on we shall discover other formulas that directly express the number e as a limit. Let us mention in passing that e—just as π—is **not** an *algebraic* number, and hence, in particular, e is not rational.

II.2.5   *The Natural Logarithm*

The natural exponential function $E(x) = e^x$ with base e was singled out among all possible exponential functions so as to satisfy the property

$$E'(x) = E(x) \text{ for all } x \in \mathbb{R}.$$

Just as any other exponential function with base $b \neq 1$, $E(x)$ is one-to-one on its whole domain $\mathbb{R}$, and hence is invertible. Its inverse function is the logarithm $\log_e$ to the base e. This particular function is called the *natural* logarithm, and it is denoted by $\ln x$, or $\log x$ (no indication of base). Accordingly, the number e is also referred to as the *base of the natural logarithm*. Its domain is the set of all positive real numbers. The inverse relationship between $E(x) = e^x$ and $\ln x$ is captured by the formulas

$$e^{\ln y} = y \text{ for } y > 0, \text{ and}$$
$$\ln(e^x) = x \text{ for } x \in \mathbb{R}.$$

We now revisit the expression for the number c_b in terms of $\log_2 b$ that we had obtained earlier by using the *natural* logarithm instead. Just as before, by replacing $b = e^{\ln b}$ and then $h \ln b = t$, it follows that

$$c_b = \lim_{h \to 0} \frac{b^h - 1}{h} = \lim_{h \to 0} \frac{e^{h \ln b} - 1}{h}$$
$$= \lim_{t \to 0} \frac{e^t - 1}{t / \ln b} = \ln b \cdot \lim_{t \to 0} \frac{e^t - 1}{t}$$
$$= \ln b \cdot 1.$$

We have therefore identified the number c_b as the natural logarithm $\ln b$ of b, i.e., c_b is that unique number that satisfies $b = e^{c_b}$. In particular,

$$c_2 = 0.693147... = \ln 2, \text{ and } e^{0.693147...} = 2.$$

More generally, the above formula shows how to calculate approximations of $\ln b$ for any $b > 0$ by considering the limit

$$\ln b = \lim_{h \to 0} \frac{b^h - 1}{h}.$$

The formula for the derivatives of exponential functions now takes the form

$$D(E_b)(x) = (b^x)' = \ln b \ b^x,$$

or

$$E_b'(x) = \ln b \ E_b(x).$$

The natural exponential function $E(x) = e^x$ and its inverse $y = \ln x$ occur so often in mathematics and in many applications that most *scientific* calculators have special function keys for them.[4] The reader should get familiar with evaluating these functions by practicing with a suitable calculator. Most numerical work involving these functions is now handled with the aid of such scientific calculators, which have replaced the use of tables or slide rules from decades ago.

II.2.6   *The Derivative of* $\ln x$

Recall from Section I.5.3 that the graph of the inverse of the function E is obtained by reflection of the graph of E on the line $y = x$. Since the tangent to the graph of E at $(0, 1)$ has slope 1 and hence is parallel to the line $y = x$ (see Figure II.10), reflection of that tangent on the line $y = x$ gives the line through $(1, 0)$ with that same slope 1. Clearly this line is the tangent to the graph of $y = \ln x$ at that point. Again, we say that $y = \ln x$ is differentiable at $x = 1$, i.e., the graph has a tangent line at that point, and we call the slope of that tangent the derivative of $\ln$ at $x = 1$. This geometric argument thus suggests that $D(\ln)(1) = (\ln)'(1) = 1$.

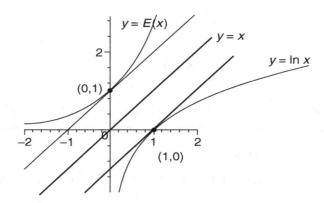

Fig. II.10   Reflection of the tangent of $y = e^x$ at $(0, 1)$.

A similar simple relationship holds true at all other points (a, b) on the graph of E, where $b = e^a$. The tangent at that point with slope $e^a = \Delta_2/\Delta_1$ is reflected to the tangent to the graph of the natural logarithm at the point

---

[4]Sometimes one of the functions is accessed by entering the inverse key before the other function key, for example e^x is obtained by entering a number for x followed by [inv] + [ln].

$(b, a) = (b, \ln b)$. (See Figure II.11.)

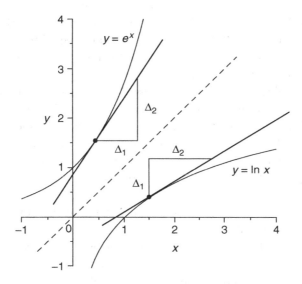

Fig. II.11   Slope of tangent to $y = \ln x$ at $b = e^a$.

A look at the triangles shown in Figure II.11 reveals that the slope of the reflected tangent at $(b, \ln b)$ is $\Delta_1/\Delta_2 = 1/e^a = 1/b$.

We have thus verified by an intuitive geometric argument that the function given by $y = \ln x$ has a tangent, i.e., is differentiable, at all points $x > 0$, and that

$$D(\ln)(x) = (\ln x)' = \frac{1}{x} \text{ for all } x > 0.$$

This differentiation formula is a special case of a general relationship between the derivative of a function and that of its inverse that we had already established in the case of algebraic functions in Section 6.2 of the Prelude. As we shall see in Section 6.4 later in this chapter, this relationship remains valid in general.

II.2.7   *The Differential Equation of Exponential Functions*

We conclude this section by focusing on a relationship that characterizes the family of exponential functions E_b. Each such function $y = E_b(x)$ satisfies the simple "differential equation"

$$y' = k y$$

for some constant k. Note that the trivial case $b = 1$, with $E_1(x) = 1^x = 1$, is covered as well, with the corresponding constant $k = 0$. Conversely, we shall eventually see in Section III.2.4 that *any* function f that satisfies such an equation, i.e., $f'(x) = k f(x)$, is necessarily equal to some constant multiple $C E_b(x)$ of a particular exponential function. The base b must satisfy $\ln b = k$, so that $b = e^k$.

More generally, a relationship expressed by some equation that involves an (unknown) function and its derivative is called a (first order) differential equation. Many phenomena in nature are modeled by such differential equations, which often involve "higher order derivatives" and/or more than one variable. It is one of the principal tasks of mathematical analysis to determine specific properties of the functions that satisfy a given differential equation.

To illustrate how differential equations and properties of their solutions can be used to estimate and make predictions about processes whose behavior can be modeled by such equations, let us consider a population of bacteria whose size at time t is given by the function $P = P(t)$, which we shall assume to be differentiable. Under suitable external conditions, the instantaneous rate of change $dP/dt(t) = P'(t)$ of the population at time t is proportional to the size of the population $P(t)$ at that time. This property is consistent with common sense and can be verified experimentally in numerous settings. This means that there exists a constant k so that

$$P'(t) = kP(t).$$

It then follows from the preceding discussion that $P(t)$ must be described by an exponential function. This same process applies to many other "populations" that are growing according to an "exponential model". This information can be used to answer questions about the particular population under consideration, and to make predictions about the future.

Example. The population of a city is projected to grow at a rate of 2% annually over the next five years. Assuming the population is 450,000 in 2013, estimate the size of the population in 2018.

We measure time t in years and denote the population at time t by $P(t)$. The statement about the growth rate means that the *relative* instantaneous rate of change $P'(t)/P(t)$ at time t equals 2% per year. This translates into the equation

$$P'(t) = 0.02P(t).$$

As we mentioned earlier, solutions of this equation are functions of the form $P(t) = CE_b(t)$ for suitable constants C and b. In the present case we must have $\ln b = 0.02$, or $b = e^{0.02}$. If we count the years so that $t = 0$ corresponds to the year 2013, then $C = P(0) = 450,000$. Therefore, according to this model, the population after t years will be

$$P(t) = 450,000 \, (e^{0.02})^t = 450,000 \, e^{0.02t}.$$

Hence the population in 2018 is estimated by $P(5) = 450,000 \, e^{0.02 \cdot 5} = 450,000 \, e^{0.1} = 450,000 \cdot 1.1052 \approx 497,000$.

Other questions can readily be answered based on this formula for the population. For example, suppose the city continues to grow at the same rate. After how many years will the population reach about 1 million? The number of years t asked for must satisfy $1,000,000 = 450,000 \, e^{0.02t}$, or $e^{0.02t} = 1/.45 = 2.2222$. Therefore $0.02t = \ln 2.2222 = .7985$, and consequently $t = .7985/0.02 \approx 40$. It follows that assuming the same growth rate in future years, the population will reach the one million mark in approximately 40 years, that is, in 2053.

We shall study further applications of this sort in Chapter III.

II.2.8 *Exercises*

1. a) Find the derivative of $y = 3^x$.
   b) Determine the equation of the tangent lines to the graph of $y = 3^x$ at the points where $x = 0$ and where $x = 1$.
2. Find the equation of the tangent line to the graph of $y = 2^x$ that goes through the point $(2,0)$. (Note that the given point is NOT on the graph. Make a sketch including the (unknown) point of tangency.)
3. At what point does the tangent to the graph of $y = 2^x$ have slope 1?
4. Complete the details in the final step of the proof of Lemma 2.1 (Hint: Given non-zero $x_1 < x_2$, choose non-zero rational numbers $r_1^{\#}, r_2^{\#}$ with $x_1 < r_1^{\#} < r_2^{\#} < x_2$. Then $d = q(r_2^{\#}) - q(r_1^{\#}) > 0$. Approximate $q(x_l)$ by $q(r_l)$, $r_l \neq 0$, $l = 1, 2$, within $d/4$, where $r_1 < r_1^{\#} < r_2^{\#} < r_2$. Verify that $q(x_2) - q(x_1) > d/2 > 0$.)
5. Use a scientific calculator to evaluate e^2 and $\sqrt{e}$. (If your calculator does not have a separate key for the exponential function, try [inv] + [ln], or check the website for your calculator.)
6. Simplify the following expressions *without using any calculators*!

   a) $\ln(e^3)$, b) $\ln(\frac{1}{e})$, c) $5\ln 4 + 7\ln\frac{1}{2}$, d) $e^{1/\ln(e^2)}$.

7. a) Evaluate $\ln 10$, $\log_{10} e$, and the product $\ln 10 \cdot \log_{10} e$ by using a calculator.

   b) Show by using the properties of logarithms that for any $b > 0$ with $b \neq 1$ one has

   $$\ln b \cdot \log_b e = 1.$$

   c) More generally, verify the equation $\log_a b \cdot \log_b a = 1$ for any $a, b > 0$ and $\neq 1$.

8. Find the equation of the tangent to the graph of $y = \ln x$ at the point where $x = 2$.

9. Use numerical approximations to find the value of c_6, the derivative of $y = 6^x$ at $x = 0$.

10. Suppose f is an exponential function, and let $k > 0$ be a constant. Introduce the rescaled function f_k by $f_k(x) = f(kx)$. Show that $f_k'(a) = kf'(ka)$ at the point $x = a$. (Hint: Replace $kh = t$, and hence $1/h = k/t$, in the average rates of change that approximate $f_k'(a)$, as in the proof of $c_b = \log_2 b \cdot c_2$.)

11. Suppose a population of bacteria of size $P(t)$ at time t (in hours) grows according to the differential equation $\frac{dP}{dt} = kP$. Determine k if the population doubles in 12 hours.

II.3 Differentiability and Local Linear Approximation

The discussion of the tangent problem for exponential functions in the preceding section shows that approximations and limits are the critical new ideas that allow us to capture the slope of the tangent, i.e., the derivative, for non-algebraic functions. We clearly need a solid understanding of these new "transcendental" concepts in order to formulate and understand tangents and derivatives in the most general setting.

II.3.1 *Limits*

We had seen in Section 1.2 that an algebraic function f in the class $\mathcal{A}$ satisfies the following important approximation property at each point a in the domain.

$$f(x) \to f(a) \text{ as } x \to a.$$

More precisely, we had seen that there exist an interval I centered at a and a constant K, such that

$$|f(x) - f(a)| \leq K|x - a| \quad \text{for all } x \in I. \tag{II.12}$$

The estimate (II.12) implies, in particular, that for any natural number n one has

$$|f(x) - f(a)| < \frac{1}{n} \text{ for all } x \in I \text{ that satisfy } 0 < |x - a| < \frac{1}{nK}.$$

While studying the tangent problem for exponential functions in the preceding section we saw that a more general abstract version of this relationship between $|f(x) - f(a)|$ and $|x - a|$ needs to be considered, as follows.

Definition 3.1. *A function F defined at all points near a, but not necessarily at a, has a **limit** at a if there exists a number L such that the values $F(x)$ for $x \neq a$ approach L arbitrarily close for all x that are sufficiently close to a. The number L is determined uniquely by this property. It is called the limit of F at a, and one writes*

$$L = \lim_{x \to a} F(x),$$

or also

$$F(x) \to L \text{ as } x \to a.$$

While the intuitive idea is quite clear, the precise relationship is rather subtle. It is most easily understood in the case of algebraic functions, where we can rely on the explicit estimation (II.12) given above, as follows. We interpret "arbitrarily close" to mean that $|F(x) - L| < 1/n$ for an arbitrarily chosen natural number n. Of course a fixed value x cannot satisfy this estimate for *all* n unless $F(x) = L$. Therefore, given a "measure of closeness" $1/n$ to the limit L, one has to specify the corresponding meaning of "all x that are sufficiently close to a", which will depend on the choice of $1/n$. In the algebraic case described above one sees that an appropriate "closeness condition" between x and a is specified by requiring x ($\in I$ and $\neq a$) to satisfy $|x - a| < 1/(nK)$. In Problem 1 of Exercise II.3.5 we will formulate a more general and precise formulation of the above definition that forms the foundation for all technical proofs involving limits. For our purposes, it will be quite sufficient to have a firm understanding of the concept of limit as formulated in the definition above.

Let us verify that if a function has a limit, then that limit is determined uniquely. Suppose that L_1 and L_2 are limits of F at a. Let n be any natural number. Then for all $x \neq a$ sufficiently close to a one has both

$$|F(x) - L_1| < 1/n \text{ and } |F(x) - L_2| < 1/n.$$

In particular, we can choose x_n in the domain of F so close to a that

$$|F(x_n) - L_1| < 1/n \text{ and } |F(x_n) - L_2| < 1/n.$$

This implies that

$$|L_1 - L_2| \le |L_1 - F(x_n)| + |F(x_n) - L_2| < 2/n. \qquad \text{(II.13)}$$

Therefore $|L_1 - L_2| < 2/n$ for all $n \in \mathbb{N}$, and this clearly implies that $|L_1 - L_2| = 0$, i.e., $L_1 = L_2$.[5]

The basic estimate for algebraic functions shows that $f \in \mathcal{A}$ has a limit at every point $a \in dom(f)$, and that, in fact,

$$\lim_{x \to a} f(x) = f(a).$$

At points that are not in the domain, algebraic function may or may not have limits. For example, $g(x) = 1/x$ defined for $x \ne 0$ has *no* limit as $x \to 0$, since the values of $1/x$ become arbitrarily large as $x \to 0$ from the right side, so that they cannot approach any fixed number L. On the other hand, consider the rational function $k(x) = (x^2 - 4)/(x + 2)$ that is defined for $x \ne -2$; by canceling the common factor $x + 2 \ne 0$ from the numerator and denominator one obtains $k(x) = x - 2$ for $x \ne -2$, which clearly implies $\lim_{x \to -2} k(x) = -4$.

Another situation occurs for the function

$$A(x) = \frac{|x|}{x}, \text{ defined for } x \ne 0.$$

Note that $A(x) = x/x = 1$ for $x > 0$ and $A(x) = (-x)/x = -1$ for $x < 0$, so the values of $A(x)$ cannot approach a single number L as $x \to 0$. Therefore the function $A(x)$ has no limit as $x \to 0$. This example suggests that sometimes it might be useful to consider *one-sided* limits. The notation $x \to a^+$ means that one only considers $x > a$, with the corresponding meaning for $x \to a^-$. With $A(x)$ as above, one then has the correct statements $\lim_{x \to 0^+} A(x) = +1$, and $\lim_{x \to 0^-} A(x) = -1$, while $\lim_{x \to 0} A(x)$ does NOT exist. One can easily show the following result.

A function F has a limit at a if and only if the two one-sided limits $\lim_{x \to a^+} F(x)$ *and* $\lim_{x \to a^-} F(x)$ *exist and have equal value.*

We emphasize once more that F need not be defined at a in order to have a limit at a. This allows us to consider functions that are more general than the algebraic ones, such as the function

$$q(x) = \frac{2^x - 1}{x}, \text{ defined for } x \ne 0,$$

---

[5]To be precise, if $|L_1 - L_2| > 0$, by Lemma I.1.2 there exists a positive integer l, such that $0 < 1/l < |L_1 - L_2|$, that is $\frac{2}{2l} < |L_1 - L_2|$. The integer $n = 2l$ then violates the estimate (II.13).

that we studied in Section 2.2. Recall that we had established that $\lim_{x \to 0} q(x) = c_2$, where $c_2 = \inf\{q(x) : x > 0\}$. In fact, the details of the proof for the one-sided limit $\lim_{x \to 0^+} q(x) = c_2$ given there fit exactly the somewhat more precise formulation given in the definition above, as follows. The (arbitrary) measure of closenesss is given by $1/n$ for an arbitrary $n \in \mathbb{N}$. We then saw that there exists a number $h_n > 0$, such that $c_2 \leq q(x) < c_2 + 1/n$ for all x with $0 < x < h_n$. In other words, given n, there exists h_n such that

$$|q(x) - c_2| < \frac{1}{n} \text{ for all } x > 0 \text{ that satisfy } |x - 0| < h_n.$$

Based on the argument given in Section 2.2, one can verify—after suitably changing h_n—that the estimate above holds for negative x as well, although we shall not go through the details. What matters is that given the fixed, but arbitrarily small bound $1/n$, in this case it is the estimate $|x - 0| < h_n$ that gives precise meaning to "all x sufficiently close to 0" in the definition of limit. Notice that in contrast to the algebraic case we now do **not** have any simple formula that expresses h_n explicitly in terms of n. Instead, we only determined the *existence* of a suitable h_n by appropriate abstract arguments that relied on the properties of the greatest lower bound (i.e., on the completeness of $\mathbb{R}$) and of the exponential function E_2.

II.3.2 Continuous Functions

Guided by the case of algebraic functions, we now highlight the relationship between the *limit* of a function f as $x \to a$ and its *value* $f(a)$ as follows.

Definition 3.2. *The function f defined on an interval I centered at a is said to be **continuous** at a if f has a limit as $x \to a$ and*

$$\lim_{x \to a} f(x) = f(a).$$

Corollary 3.3. *Every algebraic function in the class $\mathcal{A}$ is continuous at every point of its domain.*

A function is said to be *continuous* if it is continuous at every point of its domain.

The geometric interpretation of continuity means that the graph of a continuous function has no holes or tears (though corners are possible). Figures II.12 and II.13 show some examples.

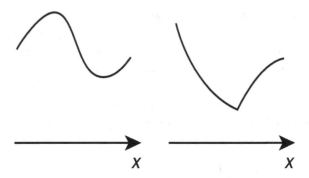

Fig. II.12   Graphs of continuous functions.

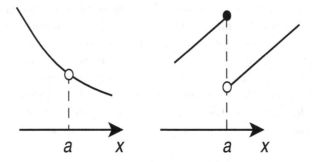

Fig. II.13   Functions that are NOT continuous at $x = a$.

The x-axis, i.e., the real number line, which has no holes by the completeness (or *continuity*) axiom, is the prototype of a graph of a continuous function, namely of the function f defined by $f(x) = 0$ for all x. The graphs of continuous functions are precisely those that are obtained by reshaping an interval of the number line into some "curve" that satisfies the vertical line test, without making any holes, cuts, or tears, although "kinks" are allowed.

Aside from the algebraic functions, most other functions that occur "naturally" are continuous. This statement makes precise the ancient Latin saying *"natura non facit saltum"*, i.e., "nature does not make any jumps". In particular, all exponential functions E_b and logarithm functions $\log_b$ are continuous on their domains. At an intuitive level this should be clear from a look at their graphs.

In order to gain more experience with these new concepts, let us consider an exponential function E_b in more detail. The verification of the continuity of E_b uses ideas that were already introduced as we defined E_b for non-rational inputs in Chapter I. Recall from Section I.4 that if a is a real (non-rational) number, $E_b(a)$ was defined by

$$E_b(a) = \sup\{E_b(r) : r \text{ rational and } r < a\}. \tag{II.14}$$

Furthermore, we saw that if a is rational, the above equation still remains correct. Let us first show that E_b is continuous at $a = 0$, i.e., that $E_b(x) \to 1 = E_b(0)$ as $x \to 0$. In fact, this property was already identified in statement (I.3) in Section I.4.5. For completeness' sake, let us review the essential arguments in the more precise language of limits that is now available. Let us assume that $b > 1$, so that E_b is increasing. Given any natural number n, we must show that

$$|b^x - 1| < 1/n$$

for all x that are sufficiently close to 0. Let $S_0 = \{b^r : r < 0\}$. By applying (II.14) with $a = 0$, one sees that $1 = b^0 = \sup S_0$. So the number $1 - 1/(n+1)$ is not an upper bound for S_0, and therefore there exists $r_n > 0$, so that $1 - 1/(n + 1) < b^{-r_n} < 1$. By taking reciprocals (note $1 - 1/(n + 1) = n/(n + 1)$, and careful with the inequalities), one obtains

$$1 < b^{r_n} < \frac{n+1}{n} = 1 + 1/n.$$

Since E_b is increasing, it follows that for all x with $-r_n < x < r_n$ one has $1 - 1/n < b^{-r_n} < b^x < b^{r_n} < 1 + 1/n$, and hence

$$|b^x - 1| < 1/n \text{ for all } x \text{ with } |x - 0| < r_n.$$

Since $1/n$ can be chosen arbitrarily small, this last statement verifies that

$$\lim_{x \to 0} b^x = 1 = b^0,$$

i.e., that E_b is continuous at 0.

The functional equation then implies for an arbitrary number a that

$$\lim_{x \to a} E_b(x) = \lim_{x \to a} E_b(a)E_b(x - a) = E_b(a) \lim_{x \to a} E_b(x - a).$$

Since $(x - a) \to 0$ as $x \to a$, one has $\lim_{x \to a} E_b(x - a) = 1$. Therefore $\lim_{x \to a} E_b(x) = E_b(a)$, which means that E_b is continuous at a as well.

Note that a function that has a limit at $x = a$ and that is also defined at a is not necessarily continuous at a. For example, let

$$k(x) = \begin{cases} (x^2 - 4)/(x + 2) & \text{for } x \neq -2 \\ 0 & \text{for } x = -2 \end{cases}.$$

We had seen earlier that $\lim_{x \to -2} k(x) = \lim_{x \to -2}(x-2) = -4 \neq k(-2)$, so k is not continuous at $x = -2$. If you think that this example is somewhat artificial you are right. Since all algebraic functions in the class $\mathcal{A}$ are continuous at each point in their domain, a function that is discontinuous at some point needs to be built up in some "artificial" way, that is, it cannot just be given by a single basic algebraic formula. Note that our example fails to be continuous because we made a poor choice for $k(-2)$. Clearly the discontinuity disappears if one changes that value, i.e., if one *defines* $k(-2) = \lim_{x \to -2} k(x) = -4$.

II.3.3   *Differentiable Functions*

The example at the end of the previous section illustrates the following important procedure. Suppose the function f has a limit $\lim_{x \to a} f(x) = L$. Then f may fail to be continuous at $x = a$ either because f is not defined at a, i.e., if there is *no value* $f(a)$, or because $f(a) \neq L$. In either case, the discontinuity of f is said to be **removable**: by (re)defining $f(a) = L$ one readily transforms f into a function that is continuous at a. Thus the existence of a limit as $x \to a$ is the essential property that allows to extend the domain of function defined for $x \neq a$ to $x = a$, so that it becomes a function that is continuous at a.

We now apply the principle of removing a discontinuity to the factor $q_a(x)$ in the formula $2^x - 2^a = q_a(x)(x-a)$ that we studied in Section 2.2. Recall that $q_a(x)$ is not defined at $x = a$. On the other hand, we had seen that

$$\lim_{x \to a} q_a(x) = 2^a \cdot \ln 2.$$

We therefore add a to the domain of q_a by defining $q_a(a) = 2^a \cdot \ln 2$. The function so extended is then continuous at $x = a$, that is, we have verified the existence of a factorization

$$2^x - 2^a = q_a(x)(x-a),$$

where the factor q_a is *continuous* at $x = a$. We are thus led to the following definition.

Definition 3.4. *Suppose f is defined on an interval containing the point a. We say that f is differentiable at a if there exists a factorization*

$$f(x) - f(a) = q_a(x)(x-a),$$

where the function q_a is continuous at $x = a$. The value $q_a(a)$ is then called the derivative of f at a and it is denoted by $D(f)(a)$ or $f'(a)$.

Note that since the values $q_a(x)$ are completely determined by f for $x \neq a$, the requirement of continuity at $x = a$ and the uniqueness of limits imply that the value $q(a)$ is determined uniquely by the differentiable function f.

Remark. We had seen in Section 7 of the Prelude, that if $f \in \mathcal{A}$ is an algebraic function and a is in the domain of f, then there exists a factorization as above with $q_a \in \mathcal{A}$ and a in the domain of q_a as well. We also verified (Corollary 3.3) that every function in $\mathcal{A}$ is continuous. Consequently $q_a \in \mathcal{A}$ is continuous at a, and therefore f is differentiable at a according to the new definition, with $D(f)(a) = q_a(a)$. The definition we just gave thus generalizes in a natural way the elementary definition of "algebraically differentiable" that we had introduced in the Prelude.

Note that if f is differentiable at a, then its derivative $D(f)(a) = q_a(a) = \lim_{x \to a} q_a(x)$. Since for $x \neq a$ the value $q_a(x)$ is the average rate of change of f between a and x, the derivative is approximated by average rates of change, and hence it measures the instantaneous rate of change of f at a, just as in the algebraic case.

Corollary 3.5. *If f is differentiable at $x = a$, then f is continuous at $x = a$.*

Proof. Note that $f(x) = f(a) + q_a(x)(x - a)$. Since the constant function $f(a)$ and the factor $(x - a)$ are clearly continuous at a, and q_a is continuous at a by the hypothesis, this representation of f readily implies that f is continuous at a as well. (See Section 4.2 below for an explicit formulation of relevant natural rules for continuous functions.) ∎

On the other hand, there are continuous functions that are NOT differentiable. For example, consider the function $f(x) = |x|$ that is continuous at $x = 0$. (See Figure II.14.)

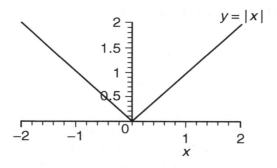

Fig. II.14 $f(x) = |x|$ is not differentiable at $x = 0$.

The factorization $f(x) = q(x)x$ implies $q(x) = f(x)/x = |x|/x$ for $x \neq 0$; we had seen earlier in Section 3.1 that the function $A(x) = |x|/x$ does not have a limit as $x \to 0$, so that there is no way to define $A(0)$ to make A continuous at 0. Consequently, f is not differentiable at 0.

The following result is an immediate consequence of the results in Sections 2.2 and 2.3, and of the discussion that led up to the definition of differentiability formulated above. We state it as a separate theorem because of its central importance.

Theorem 3.6. *For each $b > 0$ the exponential function $y = E_b(x) = b^x$ is differentiable at all points $x \in \mathbb{R}$, with derivative*

$$D(E_b)(x) = \ln b \; E_b(x).$$

Finally we state the following characterization of differentiability that is the one that has traditionally been used in most calculus texts.

Theorem 3.7. *f is differentiable at the point a if and only if*

$$\lim_{x \to a} \frac{f(x) - f(a)}{x - a}$$

exists. The limit is then equal to the derivative $D(f)(a) = f'(a)$.

Proof. Just notice that $f(x) - f(a) = q_a(x)(x - a)$ is equivalent to $q_a(x) = [f(x) - f(a)]/(x - a)$ for $x \neq a$. Then q_a extends to a continuous function at $x = a$ if and only if $\lim_{x \to a} q_a(x)$ exists. ∎

II.3.4   *Local Linear Approximation*

Suppose the function $y = f(x)$ is differentiable at $x = a$. In particular this implies that the graph of f has a well defined tangent at $(a, f(a))$ whose slope is given by $f'(a)$. The equation of the tangent line is given by the linear function

$$L_{f,a}(x) = f(a) + f'(a)(x - a).$$

In the algebraic case, the tangent line is distinguished among all possible lines through the point $(a, f(a))$ by the fact that it intersects the graph of f at the point $(a, f(a))$ with multiplicity at least two. In particular, the error $\mathcal{E}_a(x) = f(x) - L_{f,a}(x)$ between the function and its tangent satisfies $\mathcal{E}_a(x) = k(x)(x - a)^2$, and therefore it goes to 0 *much* faster than $(x - a)$. In less explicit form this remains true in the general *differentiable* case, i.e., the rate of decrease of the error $\mathcal{E}_a(x)$ is *faster* than $(x - a)$, although it may, in general, be slower than $(x - a)^2$. To understand this better, note that

$$\mathcal{E}_a(x) = f(x) - L_{f,a}(x) = f(x) - f(a) - f'(a)(x - a).$$

Since f is differentiable, one has $f(x) - f(a) = q_a(x)(x - a)$, with q_a continuous at $x = a$ and $q_a(a) = f'(a)$. Therefore

$$\mathcal{E}_a(x) = q_a(x)(x - a) - f'(a)(x - a)$$
$$= [q_a(x) - q_a(a)](x - a).$$

Since $[q_a(x) - q_a(a)] \to 0$ as $x \to a$, when x is sufficiently close to a the error $\mathcal{E}_a(x)$ is considerably smaller than $x - a$, in the sense that the *relative* error $\mathcal{E}_a(x)/(x - a)$ goes to zero as well, i.e.,

$$\lim_{x \to a} \frac{\mathcal{E}_a(x)}{x - a} = \lim_{x \to a} [q_a(x) - q_a(a)] = 0.$$

Figure II.15 shows the relationship geometrically.

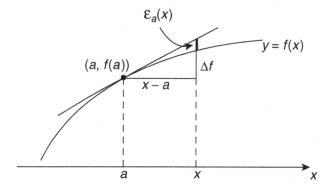

Fig. II.15   Error $\mathcal{E}_a(x)$ between the function and its tangent.

So the graph of f and its tangent are very close indeed near the point $(a, f(a))$, and the approximation gets better as $x \to a$. The intuitive geometric interpretation of this approximation suggests that one can say that a function is differentiable at $x = a$ precisely when its graph looks essentially like a line (that is, the tangent) near that point. (See Figure II.16.) In fact, if one looks with a magnifying glass into a very small neighborhood of the point $(a, f(a))$, the graph of f and the line become de facto indistinguishable.

This discussion verifies one part of the following characterization of differentiability.

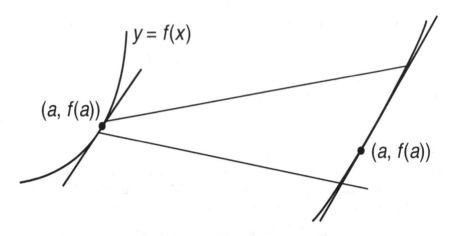

Fig. II.16   Graph of function and tangent near $(a, f(a))$.

Theorem 3.8. *The function f is differentiable at a if and only if f is well approximated near a by a linear function $L_{f,a}(x) = f(a) + m(x - a)$, in the sense that the "error" $\mathcal{E}_a(x) = f(x) - L_{f,a}(x)$ is of the form*

$$\mathcal{E}_a(x) = g(x)(x - a), \text{ where } g \text{ is continuous at } x = a \text{ with } g(a) = 0.$$

If this condition is satisfied, the slope m of the approximating line is uniquely determined and agrees with the derivative $D(f)(a)$.

As noted earlier, the crux of the condition on g is the requirement that

$$\lim_{x \to a} g(x) = 0.$$

In essence, this Theorem gives a precise quantitative description of the ancient notion that the tangent "touches" the curve but does not cut it. (See Section 2 in the Prelude.)

Proof. To complete the proof of the second part of the theorem, note that if the approximation of f by $L_{f,a}$ holds for some m, with the error $\mathcal{E}_a(x)$ satisfying the condition stated in the theorem, then

$$f(x) - f(a) = m(x - a) + \mathcal{E}_a(x)$$
$$= [m + g(x)](x - a).$$

The factor $q_a(x) = m + g(x)$ is then continuous at $x = a$ with $\lim_{x \to a} q_a(x) = m + g(0) = m$, so that f is indeed differentiable at a, with $f'(a) = q_a(a) = m$ as required.   ∎

In the days before hand-held calculators became widely available, the linear approximation of a differentiable function was often used to approximate (unknown) values of a function near a particular point at which the value is known, or in order to estimate errors in experimental work. (See Problem 7 of Exercise II.3.5.) For example, for the exponential function $E_2(x) = 2^x$ with $E_2'(x) = \ln 2 \, 2^x$, one knows that $E_2(2) = 4$, and one can then estimate

$$2^{2.1} \approx E_2(2) + E_2'(2)(2.1 - 2)$$
$$= 4 + 4 \ln 2 \cdot (0.1)$$
$$= 4.2773,$$

or, more generally,

$$2^{2+h} \approx 4 + 4(\ln 2) \cdot h \quad \text{for small } h.$$

Since the derivative of $y = e^x$ at $x = 0$ is equal to 1, the linear approximation for this function at $x = 0$ is particularly simple and gives the practical estimate

$$e^x \approx 1 + x \text{ for small } x.$$

Aside from such practical estimations, which are of less interest today, the most important application of the linear approximation is the conceptual understanding that is summarized in the following statement.

Differentiability is equivalent to good local linear approximation.

Stated differently, one can say:

Locally the graph of a differentiable function

looks like a (non-vertical) line.

This property is the foundation for the important principle that whatever is correct for *linear* functions should remain correct—locally—for differentiable functions in general. As we shall see later, this principle turns out to be useful for understanding key properties of differentiable functions. Furthermore, it is this idea of good local linear approximation that is critical for understanding differentiability in more general contexts, for example in the case of functions of *more than one* variable.

Based on the property stated above, it is easy to recognize from the graph of a continuous function where that function fails to be differentiable. For example, if the graph has a corner at a point P, as in the case of $f(x) = |x|$ at $(0,0)$, the graph surely does not look like a line near P, and consequently the function represented by that graph is NOT differentiable at that point. (See Figure II.14 above.) Another problem occurs if the graph of a function does have a tangent that is *vertical*, so that no slope is defined. Since derivatives evaluate the slope of tangents, a function cannot be differentiable at such points.

Example. Consider the function $g(x) = x^{1/3}$. Since $|g(x)| \leq |x|^{1/3}$, one surely has $\lim_{x \to 0} g(x) = 0$, so that g is continuous at $x = 0$. (g is continuous at all $x \neq 0$ since g is algebraic.) The tangent to the graph of g at $(0,0)$ is given by the y-axis (see Figure II.17), i.e., it is vertical. By factoring

$$x^{1/3} = q(x)x \text{ for } x \neq 0, \text{ with } q(x) = x^{-2/3},$$

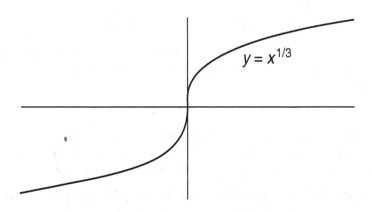

Fig. II.17   Graph of $y = x^{1/3}$ with vertical tangent at $(0,0)$.

one sees that $q(x)$ has no limit as $x \to 0$. Consequently, $q(x)$ cannot be extended as a continuous function to $x = 0$, that is, $g(x) = x^{1/3}$ is **not** differentiable at $x = 0$. On the other hand, relating this example to the general rule about inverse functions (rule III in Prelude, Section 6), we note that g is the inverse of $f(u) = u^3$. Consequently, by that rule, one knows that g is differentiable at all points $b = u^3$ with $f'(u) = 3u^2 \neq 0$, that is, at all points $b = u^3 \neq 0$, and that $g'(b) = 1/f'(u) = 1/(3u^2) = 1/(3b^{2/3}) = \frac{1}{3}b^{-2/3}$. However, the rule cannot be applied at the point 0, since $f'(0) = 0$. The function f has a horizontal tangent at $(0,0)$, and consequently, by reflection, its inverse g has a *vertical* tangent at that point.

To summarize, if near a point $P = (a, f(a))$ the graph of a function f looks very much like a **non**-vertical line (which is then part of the tangent line), one may safely conclude that the function f is differentiable at $x = a$.

Finally, we use the function g just considered to define the function $F = g^4$. By the chain rule, or simply because $F(x) = x^{4/3}$ is algebraic, F is (algebraically) differentiable at all points different from 0. In contrast to g, however, it turns out that F is differentiable at 0 as well. Just note that $F(x) - F(0) = x^{4/3} = x^{1/3} \cdot x$, where the factor $g(x) = x^{1/3}$ is continuous at 0. So F is indeed differentiable at 0, with $F'(0) = g(0) = 0$. However, F is not *algebraically* differentiable at 0, since the error term $\mathcal{E}_0(x) = F(x) - 0 = g(x)x$ does not have a zero of multiplicity 2 at 0.

II.3.5 *Exercises*

1. The technical formulation used to make the definition of limit more precise, and that needs to be used in order to rigorously prove statements about limits, is the following. The function f defined on an interval I centered at a, but not necessarily defined at a itself, is said to have a limit L at a, written as $\lim_{x \to a} f(x) = L$, if for every positive $\varepsilon > 0$ one can find a positive $\delta = \delta(\varepsilon) > 0$ such that
   $$|f(x) - L| < \varepsilon \text{ for all } x \in I \text{ that satisfy } 0 < |x - a| < \delta.$$
   a) Use the definition above to verify that if $\lim_{x \to a} f(x) = L_1$ and $\lim_{x \to a} f(x) = L_2$, then $L_1 = L_2$. This shows that limits are unique. (Hint: Look at the intuitive proof given in the text.)
   b) Show that if $\lim_{x \to a} f(x) = L$ and $L > 0$, then there exists an interval $J \subset I$ centered at a such that $f(x) > L/2$ for all $x \in J$ with $x \neq a$.
   c) Formulate and prove a statement corresponding to b) if $L \neq 0$.

2. Use the technical definition of limit in Problem 1 to carefully prove that if f_1 and f_2 have a limit at $x = a$, then $f_1 + f_2$ has a limit as well, and $\lim_{x \to a}(f_1 + f_2)(x) = \lim_{x \to a} f_1(x) + \lim_{x \to a} f_2(x)$.

3. Let $c_2^- = \sup\{ q(h) : h < 0 \}$ and $c_2^+ = \inf\{ q(h) : h > 0 \}$, where $q(h) = (2^h - 1)/h$ for $h \neq 0$.
   a) Show that $c_2^- \leq c_2^+$.
   b) Show that for $h > 0$ one has $q(-h) = (1/2^h)q(h)$.
   c) Use a) and b) to show that $0 \leq c_2^+ - c_2^- \leq q(h)(1 - 1/2^h)$ for all $h > 0$.
   d) Let $h \to 0$ and conclude that $c_2^+ - c_2^- = 0$. (You may use that $2^h \to 1$ as $h \to 0$.)

4. Prove that if f is continuous at $x = a$, then there exists an interval I centered at a and a constant K such that $|f(x)| \leq K$ for all $x \in I$. Note

that this generalizes the corresponding result given in Section I.6.3 for functions in $\mathcal{A}$. (Hint: Use the definition of limit stated in Problem 1 above.)

5. a) Use techniques of linear approximation to justify the approximation $e^h \approx 1 + h$ for small values of h.

   b) Estimate the error $\mathcal{E}_0(h) = k(h) \cdot h$ made by the above approximation in the case $h = 0.1$, 0.01 and 0.001 with the help of a calculator.

   c) Evaluate $k(h) = \mathcal{E}_0(h)/h$ for the values of h in b). Do the results appear to support the statement that $\lim_{h \to 0} k(h) = 0$?

   d) Use additional smaller values of h, if necessary, to estimate $\lim_{h \to 0} k(h)/h$.

6. Let $f(x) = x^2$ and denote its linear approximation at $x = a$ by $L_{f,a}$.

   a) Determine $L_{f,a}(x)$ explicitly. (Recall $f'(x) = 2x$).

   b) Evaluate the error $\mathcal{E}_a(x) = f(x) - L_{f,a}(x)$ by algebra and simplify as much as possible.

   c) Use the result in b) to verify $\lim_{x \to a} \frac{\mathcal{E}_a(x)}{x-a} = 0$.

7. In order to estimate the area of a circular platform one measures its radius with a measuring tape. The result is 5 m ± 0.001. Use the linear approximation of the area function $A(r) = \pi r^2$ at $r = 5$ to estimate that the area of the platform equals $25\pi \pm \pi(0.01)\text{m}^2$.

II.4   Properties of Continuous Functions

The differentiation rules that we discussed in the Prelude in the context of algebraic functions extend naturally to the more general class of differentiable functions that was introduced in the last section. All that is needed in addition are some basic natural properties of continuous functions that we shall make explicit in this section. We shall discuss the differentiation rules in Sections 6 and 7 later on.

II.4.1   *Rules for Limits*

We begin by summarizing basic properties of limits. We had already verified the following fundamental property of the limits of algebraic functions f in the class $\mathcal{A}$.

Rule i) *If a is a point in the domain of the function $f \in \mathcal{A}$, then* $\lim_{x \to a} f(x) = f(a)$.

Furthermore, the following somewhat more abstract rules do appear to be intuitively quite reasonable.

Rule ii) *If* $\lim_{x \to a} f(x)$ *and* $\lim_{x \to a} g(x)$ *exist, then the following limits exist as well, with their values as indicated:*

ii-a) *If* c, d *are constants, then* $\lim_{x \to a}(cf(x) + dg(x)) = c \lim_{x \to a} f(x) + d \lim_{x \to a} g(x);$

ii-b) $\lim_{x \to a}[f(x)g(x)] = \lim_{x \to a} f(x) \lim_{x \to a} g(x);$
ii-c)

$$\lim_{x \to a} \frac{f(x)}{g(x)} = \frac{\lim_{x \to a} f(x)}{\lim_{x \to a} g(x)} \quad provided \ that \ \lim_{x \to a} g(x) \neq 0;$$

Note that by Problem 1 b) and c) of Exercise II.3.5, the assumption on g implies that $g(x) \neq 0$ for all $x \neq a$ in some interval I centered at a. Hence the quotient f/g is defined on $I - \{a\}$.

Rule iii) *If* $f(x) \leq h(x) \leq g(x)$ *for all* $x \neq a$ *near* a *and* $\lim_{x \to a} f(x) = \lim_{x \to a} g(x) = L$, *then* $\lim_{x \to a} h(x)$ *exists and equals* L *as well.*

Rule iv) $\lim_{x \to a} f(x) = f(a)$ *if* f *is continuous at* a.

We note that iv) is not really a "rule", but is just a restatement of the definition of the continuity of f at the point a. However, once one knows large classes of continuous functions, this property is often quite useful. In fact, a fundamental technique to analyze a limit $\lim_{x \to a} q(x)$ when q is not defined at a or it is not readily identified as a function continuous at a, involves transforming the expression that defines $q(x)$ (without changing its values at points $x \neq a$ near a!) by whatever (algebraic or other) means available into some other expression $F(x)$ known or recognized to be a function that is continuous at a. Then

$$q(x) = F(x) \ \text{for} \ x \neq a \ \text{implies that} \ \lim_{x \to a} q(x) = \lim_{x \to a} F(x) = F(a).$$

Here is a typical simple example of this technique. Suppose we want to determine whether $g(x) = (\sqrt{x} - 3)/(x - 9)$ has a limit at $x = 9$, which is the one point where the algebraic function g is not defined, so that rule i) does **not** apply. Observe that for $x \neq 9$ one has

$$\frac{\sqrt{x} - 3}{x - 9} = \frac{\sqrt{x} - 3}{x - 9} \frac{\sqrt{x} + 3}{\sqrt{x} + 3} = \frac{x - 9}{(x - 9)(\sqrt{x} + 3)}$$

$$= \frac{1}{\sqrt{x} + 3}.$$

The final expression is algebraic and defined at $x = 9$, so
$\lim_{x \to 9} g(x) = \lim_{x \to 9} 1/(\sqrt{x} + 3) = 1/(\sqrt{9} + 3) = 1/6$.

The formal verification of these rules requires a precise technical definition of limit. (See Problems 1 and 2 of Exercise II.3.5.) We shall not go into these details. The basic message to remember is that limit statements that appear "reasonable" and that can be written in a meaningful way are indeed correct. Rule ii-c) illustrates the message clearly; the additional assumption that $\lim_{x \to a} g(x) \neq 0$ is necessary, since otherwise the answer would be meaningless.

II.4.2   *Rules for Continuous Functions*

We already stated that most natural functions are "continuous" at all points in their domains. Familiar examples include all algebraic functions in the class $\mathcal{A}$, and the exponential and logarithm functions. The rules for limits stated above readily imply corresponding results for continuity, as follows.

Theorem 4.1. *If f and g are continuous at the point a, then $f \pm g$, fg, and f/g are continuous at a, provided one assumes that $g(a) \neq 0$, i.e., the denominator is non-zero, in the case of a quotient.*

Next we consider composition of continuous functions. Suppose g is defined on the interval I and assume that $g(I)$ is contained in the domain of f, so that the composition $f \circ g$ is defined on I.

Theorem 4.2. *If g is continuous at $a \in I$ and f is continuous at $b = g(a)$, then $f \circ g$ is continuous at a.*

Proof. Let $u = g(x)$. Then $b = g(a) = \lim_{x \to a} g(x)$ implies that $u \to b$ as $x \to a$. Since $\lim_{u \to b} f(u) = f(b)$, it follows that

$$\lim_{x \to a} (f \circ g)(x) = \lim_{x \to a} f(g(x)) = \lim_{u \to b} f(u) = f(b),$$

which equals $(f \circ g)(a)$. This shows that $f \circ g$ is continuous at a. ∎

Finally we consider inverse functions. Note that the hypotheses require more than continuity at just one point.

Theorem 4.3. *Suppose f is one-to-one on the open interval I and that it is continuous on I, that is, at each point $a \in I$. Then $J = f(I)$ is an interval, and the inverse $g : J \to I$ of f is continuous on J.*

The rigorous proof of this result is somewhat more delicate, as it does require explicit use of the completeness of the real numbers, e.g., the Intermediate Value Theorem, discussed in the next section, as well as a rather precise formulation of continuity. (See Problem 1 of Exercise II.3.5.) However, at the intuitive geometric level the result is clear, as follows. The hypotheses imply that the graph of f is a "continuous curve" without any cuts or holes that satisfies both the vertical and horizontal line tests. Recall from Section I.5.2 that the graph of the inverse g is obtained by reflecting the graph of f on the line $y = x$, a process that does not make any changes to the graph but just places it in a different way in the coordinate system. Hence the graph of the inverse g is also a "continuous curve" without any cuts or holes, i.e., g is continuous as well. We shall discuss a more rigorous proof in the next subsection.

II.4.3 *The Intermediate Value Theorem*

One of the conclusions in the last theorem was the claim that the image $J = f(I)$ of the interval I is also an interval. This result remains correct as long as f is continuous on the interval I, and not necessarily also one-to-one. This fact is an easy consequence of the following important theorem, known as the **Intermediate Value Theorem.**

Theorem 4.4. *Suppose the function f is continuous on the interval I and that for two points $a < b$ in I one has $f(a) \neq f(b)$. Let λ be any number between $f(a)$ and $f(b)$. Then there exists a point $x_0 \in (a, b)$ with $f(x_0) = \lambda$.*

Note that in the case where f is in the class $\mathcal{A}$ a version of this result was already stated in Problem 8 of Exercise I.6.5.

Proof. Since the graphs of continuous functions have no holes or tears, this result is geometrically clear. Its precise verification requires the completeness of the real numbers. (See Problem 9 of Exercise II.4.5 for details.)

Let us see how this result implies that $J = f(I)$ is an interval if f is continuous at all points in I. Suppose $A < B$ are any two distinct points in J. It is enough to show that $(A, B) \subset J$. Choose $a, b \in I$ with $f(a) = A$ and $f(b) = B$, and let $\lambda \in (A, B)$. By the theorem (interchange a and b if $a > b$) there exists $x_0 \in (a, b)$ with $f(x_0) = \lambda$. Thus $\lambda \in J$ and we are done.

We shall now use this result to prove Theorem 4.3. So assume that f is continuous and one-to-one on the open interval I. It then follows that f is either strictly *increasing* or f is strictly *decreasing* on I. This is usually known directly for any specific function, so we shall skip the somewhat technical verification of this fact. So let us assume that f is strictly increasing. (The case of a decreasing function follows by an analogous argument.) As shown above, $J = f(I)$ is an interval, and since f is one-to-one, its inverse g is defined on J, with $g(y) \in I$ for $y \in J$. Let $b \in J$. We want to prove that g is continuous at b, i.e., $\lim_{y \to b} g(y) = g(b)$. Note that $g(b) = a \in I$. Since I is assumed open, a is not a boundary point of I. We prescribe the (arbitrary) closeness of $g(y)$ to the expected limit $g(b) = a \in I$ by fixing an arbitrarily small closed interval $[a - \varepsilon, a + \varepsilon] \subset I$ with $\varepsilon > 0$. Let us show that for all y that are sufficiently close to b one has $g(y) \in (a - \varepsilon, a + \varepsilon)$. Since f is increasing, $f(I_\varepsilon(a)) = (f(a - \varepsilon), f(a + \varepsilon))$, and this latter interval is open and contains the point b, since $f(a - \varepsilon) < f(a) = b < f(a + \varepsilon)$. Now choose $\delta > 0$ sufficiently small, so that the interval $J_\delta(b)$ is contained in $f(I_\varepsilon(a))$. Then $g(J_\delta(b)) \subset I_\varepsilon(a)$, i.e.,

$$|g(y) - g(b)| < \varepsilon \text{ for all } y \text{ with } |y - b| < \delta,$$

as required. ∎

Additional applications of the intermediate value theorem are given in the Exercises.

II.4.4 *Continuity and Boundedness*

It follows readily from the definition of continuity that a function that is continuous at a point a is bounded in some neighborhood of a, thereby generalizing a property we had established in Section I.6.3 for algebraic functions $f \in \mathcal{A}$. In fact, since $\lim_{x \to a} f(x) = f(a)$, the estimate $|f(x) - f(a)| \leq 1$ must hold for all x in the domain of f that are sufficiently close to a, that is, there must exist a number $\delta > 0$ such that the estimate holds for all x with $|x - a| < \delta$. Then

$$|f(x)| = |[f(x) - f(a)] + f(a)| \leq |f(a) - f(x)| + |f(a)|$$
$$\leq 1 + |f(a)| \text{ for all } x \in I_\delta(a).$$

The proof of Theorem I.6.5 can therefore be applied to prove the following important result.

Theorem 4.5. *Suppose the function f is continuous on the closed and bounded interval I. Then f is bounded on I.*

As we had already noticed in the case of algebraic functions, the conditions on the interval are indeed necessary for this result to hold. In fact, $f(x) = x$ is not bounded on $[0, \infty)$, and $g(x) = 1/x$ is not bounded on $(0, 1]$.

Suppose the function f is continuous on the closed and bounded interval I. By the preceding Theorem, the set $\{f(x) : x \in I\}$ is bounded. Consequently, by the completeness of $\mathbb{R}$, $\sup_I f = \sup\{f(x) : x \in I\}$ and $\inf_I f = \inf\{f(x) : x \in I\}$ are well defined real numbers. In fact, it is a remarkable result that under the given hypotheses both $\sup_I f$ and $\inf_I f$ are actually elements of $\{f(x) : x \in I\}$, as follows.

Theorem 4.6. *(Existence of maximum and minimum) Suppose the function f is continuous on the closed and bounded interval I. Then there are points x_m and x_M in I, so that*

$$f(x_m) \leq f(x) \leq f(x_M) \text{ for all } x \in I.$$

We call $\max_I f = f(x_M)$ the *maximum* and $\min_I f = f(x_m)$ the *minimum* of f on the interval I. We emphasize that in general a function that is bounded on a set S does not necessarily have a maximum (or minimum) on that set. For example, $f(x) = x$ is clearly bounded on $S = (0, 1)$, with $\sup_S f = 1$, but f does not have a maximum on S. Of course f has a maximum $f(1) = 1$ on the *closed* interval $[0, 1]$. These differences may appear irrelevant in practical situations, but it is important to understand the distinction between $\max f$ and $\sup f$ made by precise mathematical language.

Proof. The proof involves another application of the completeness of $\mathbb{R}$. We shall prove the existence of the maximum of f. The existence of the minimum then follows by observing that $\min f = -\max(-f)$. Let $M = \sup_I f$. Divide the interval in half, i.e. $I = I' \cup I''$, where the lengths of I' and I'' equal $1/2$ the length of I. Let $M' = \sup_{I'} f$ and $M'' = \sup_{I''} f$. Since $\max\{M', M''\} \leq M$ is an upper bound for $\{f(x) : x \in I\}$, and M is the *least* upper bound, we must have $\max\{M', M''\} = M$. Let I_1 denote a half of I for which $\sup_{I_1} f = M$ (if both halves satisfy this, choose the one on the left). Continue this process to obtain a nested sequence $I_1 \supset I_2 \supset \ldots$ of closed and bounded intervals I_n with $\sup_{I_n} f = M$ and length $I_n = 1/2^n \times$ length I. By Theorem I.1.3 there exists a point $\xi \in \cap_{j=1}^{\infty} I_n$. We claim that $f(\xi) = M$, i.e., ξ can be chosen as the point x_M required in the theorem. Clearly $f(\xi) \leq M$. Suppose we had $f(\xi) < M$. Then $\varepsilon = M - f(\xi) > 0$. By the continuity of f at ξ there exists an interval $I_\delta(\xi)$ so that $|f(x) - f(\xi)| < \varepsilon/2$ for all $x \in I_\delta(\xi)$. It follows that $f(x) < f(\xi) + \varepsilon/2 < M - \varepsilon/2$ for all

$x \in I_\delta(\xi)$. Since length $I_n \to 0$ as $n \to \infty$, we can chose n so large that $I_n \subset I_\delta(\xi)$. Then $f(x) < M - \varepsilon/2$ for all $x \in I_n$, so that $M - \varepsilon/2$ would be an upper bound for $\{f(x) : x \in I_n\}$ that is smaller than M, contradicting the fact that $\sup_{I_n} f = M$.   ∎

II.4.5   *Exercises*

1. Determine the limits

  a) $\lim_{x \to 2}(x^3 - 2x^2 + 3x)$,   b) $\lim_{x \to 2}(3^x - x^3 2^{-x})$, c) $\lim_{t \to 2} \frac{2^t}{t^2}$,
  d) $\lim_{s \to 0} \sqrt[4]{2(s)^3 + 9 \cdot 3^s + 7}$, e) $\lim_{x \to 2}(2^x - 5x)$,
  f) $\lim_{x \to -1} \frac{x3^x}{x^2 + 3x - 1}$.

  (Hint: Use the continuity of the functions.)

2. Determine the limits

$$a) \quad \lim_{x \to 25} \frac{x - 25}{\sqrt{x} - 5}, \qquad b) \quad \lim_{r \to 2} \frac{4 - r^2}{r^3 - 8}.$$

3. Let the function f be differentiable at $x = 2$, and define

$$G(x) = \begin{cases} \frac{f(x) - f(2)}{x - 2} & \text{for } x \neq 2 \\ f'(2) & \text{for } x = 2 \end{cases}.$$

  Is G continuous at $x = 2$? Explain!

4. Define $g(x) = \begin{cases} \frac{x-1}{x+3} & \text{for } x \neq -3 \\ -4 & \text{for } x = -3 \end{cases}$. At which points $a \in \mathbb{R}$ is g continuous? Explain!

5. Define $P(t) = \begin{cases} 2^t + 3t^2 & \text{for } t \leq 1 \\ (t+2)^2 - 4t & \text{for } t > 1 \end{cases}$. Is the function P continuous at $t = 1$? Explain!

6. Determine $\lim_{x \to 3} \frac{2^x - 8}{x - 3}$. (Hint: Think of a particular derivative!)

7. a) Show that $x^3 - 7 = 0$ has a solution between 1.9 and 2.

  b) Use the Intermediate Value Theorem to show that every polynomial of odd degree has at least one zero in $\mathbb{R}$.

8. Suppose $g : [0, 1] \to [0, 1]$ is continuous at every point in $[0, 1]$. Show that there exists at least one point $x \in [0, 1]$ with $g(x) = x$. (Hint: Apply the Intermediate Value Theorem to $f(x) = g(x) - x$.)

9. Prove the Intermediate Value Theorem. (Hint: Suppose $f(a) < \lambda < f(b)$; let x_0 be the least upper bound of $\{x \in [a, b] : f(x) < \lambda\}$. Use the continuity of f at x_0 to prove $f(x_0) = \lambda$. (See also Problem 8 of Exercise I.6.5.)

II.5   Derivatives of Trigonometric Functions

II.5.1   *Continuity of sine and cosine Functions*

The trigonometric functions *sine* and *cosine* were introduced in Section
I.3.1. The reader should briefly review that section before proceeding. Since
sine and *cosine* are defined in terms of the unit circle in the x, y - coordinate
plane, we shall use t to denote the input variable. Values of these functions
at points $t \in \mathbb{R}$ are best obtained by means of a scientific calculator. A
look at their graphs—obtained with a graphing calculator (See Figures I.19
and I.20 in Section I.3.4)—suggests that these functions are continuous at
all real numbers t, so that $\lim_{t \to a} \sin t = \sin a$ for each a, and so on. In
particular,

$$\lim_{t \to 0} \sin t = \sin 0 = 0, \quad \lim_{t \to \pi/2} \sin t = \sin \pi/2 = 1, \text{ and}$$

$$\lim_{t \to 0} \cos t = \cos 0 = 1, \quad \lim_{t \to \pi/2} \cos t = \cos \pi/2 = 0.$$

The preceding statements follow readily from the geometric definition of
sine and cosine on the unit circle. For example, as shown in Figure II.18,
the length $2 \sin t$ of the secant spanned by the arc of length $2t$ centered at
the point $(1, 0)$ surely is shorter than the arc.

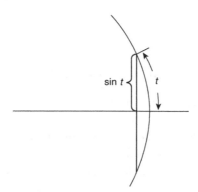

Fig. II.18   The arc t and $\sin t$.

It follows that $\sin t \leq t$ for $t > 0$, which implies that

$$|\sin t| \leq |t| \text{ for all } t \text{ near } 0.$$

This is the same kind of estimate we are familiar with for algebraic functions. It clearly shows that $\sin t \to 0 = \sin 0$ as $t \to 0$. This type

of estimate generalizes to prove continuity at an arbitrary point a. Let $t \neq a$ be close to a. Then the distance $dist(P(a), P(t))$ between the points $P(a) = (\cos a, \sin a)$ and $P(t) = (\cos t, \sin t)$ on the unit circle is less than or equal to the length $|t - a|$ of the arc on the circle connecting these two points, i.e., $dist(P(a), P(t)) \leq |t - a|$. By the formula for the distance between these two points one then has

$$|\cos t - \cos a| \leq dist(P(a), P(t)) \leq |t - a| \text{ and}$$
$$|\sin t - \sin a| \leq dist(P(a), P(t)) \leq |t - a|.$$

Just as in the case of algebraic functions, these estimates clearly imply the continuity of *sine* and *cosine* at the point a.

By general rules about continuous functions it then follows, for example, that $\tan t = \sin t / \cos t$ is continuous at all points $t \neq \pi/2 + k\pi, k \in \mathbb{Z}$. Also, since $\cos t = \sin(\pi/2 - t)$ and $\sin t = \cos(\pi/2 - t)$ for all t, it is usually enough to verify basic results just for one of the trigonometric functions, and then apply appropriate general principles to extend the results to other functions. For example, once one knows that $\sin t$ is continuous, since $h(t) = \pi/2 - t$ is clearly continuous, the composition $\cos t = \sin(h(t))$ is continuous as well at each point.

II.5.2 *The Derivative of* $\sin t$ *at* $t = 0$

The graph of $y = \sin t$ suggests that this function has a *tangent* at every point, that is, $\sin t$ is differentiable everywhere. In order to study its derivative, we shall rely on the definition of the *sine* function on the unit circle in the x, y - coordinate plane.

Let us begin by considering the point $t = 0$. As usual, we need to study the factorization $\sin t = q(t) \cdot t$. The factor $q(t)$ is uniquely determined for $t \neq 0$ by $q(t) = \sin t/t$, but just as in the case of the exponential function, the value of q at $t = 0$ is missing, and there is no obvious formula that would produce a suitable value $q(0)$ that would make q continuous at 0. So we must examine directly the behavior of $q(t)$ as $t \to 0$.

Let us first consider numerical approximations for $t_k = 10^{-k}, k = 1, 2, 3, \ldots$, as shown in the following Table II.3.

k	t_k	$q_a(t_k)$
1	10^{-1}	0.998334166468282
2	10^{-2}	0.999983333416666
3	10^{-3}	0.999999833333342
4	10^{-4}	0.999999998333333
5	10^{-5}	0.999999999983333
6	10^{-6}	0.999999999999833
7	10^{-7}	0.999999999999998
8	10^{-8}	1.00000000000000
9	10^{-9}	1.00000000000000
10	10^{-10}	1.00000000000000

Table II.3. Approximations to derivative of sine function at 0.

The data convincingly shows that

$$\lim_{t \to 0} q(t) = \lim_{t \to 0} \frac{\sin t}{t} = 1.$$

How can we recognize that this statement is indeed correct, without relying on incomplete and possibly misleading numerical "evidence"? There is no obvious method to simplify the expression $\sin t/t$, so we try to argue by using the geometric definition of the *sine* function, as shown in Figure II.19.

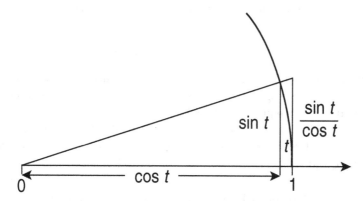

Fig. II.19   The arc t, and the values of $\sin t$ and $\cos t$.

According to this figure, if the number $t > 0$ represents the length of the arc on the unit circle, then $\sin t$ measures a line segment that is approximately of length t when t is "small". As $t \to 0$, the approximation appears to improve, so that—geometrically—it is reasonable to expect that

$$\frac{\sin t}{t} \to 1 \text{ as } t \to 0.$$

We can make this precise by comparing the areas of the right triangles shown in Figure II.19 with the area $t/2$ of the relevant circular sector. For $t > 0$ one obtains

$$\frac{1}{2}\sin t \cos t \leq \frac{t}{2} \leq \frac{1}{2}\frac{\sin t}{\cos t},$$

and hence, after dividing the inequalities by $(\sin t)/2 > 0$, it follows that

$$\cos t \leq \frac{t}{\sin t} \leq \frac{1}{\cos t}.$$

Now take reciprocals (careful with the inequalities!) to obtain

$$\frac{1}{\cos t} \geq \frac{\sin t}{t} \geq \cos t.$$

The geometric argument assumed $t > 0$, but all three expressions above do not change their values if t is replaced by $-t$, so that the latter inequalities hold for all $t \neq 0$. Since the expression in the middle remains squeezed between the numbers $1/\cos t$ and $\cos t$, both of which approach 1 as $t \to 0$, by limit rule iii) in Section II.4.1 one obtains

$$\lim_{t \to 0} \frac{\sin t}{t} = 1.$$

This is exactly the result we expected based on the numerical data.

We now define

$$q(0) = \lim_{t \to 0} \frac{\sin t}{t} = 1,$$

thereby extending $q(t)$, defined for $t \neq 0$ by $q(t) = \sin t/t$, to a function that is continuous at 0. Altogether, we have verified that $y = \sin t$ is differentiable at 0 with derivative equal to $q(0) = 1$.

Let us translate the limit result we just obtained to the setting of average slopes. Figure II.20 shows the graph of $y = \sin t$, and the line through the points $(0,0)$ and $(t, \sin t)$ (i.e., secants) for several values $t > 0$. (Replace x by t in Figure II.20.)

Clearly the slope of each secant line is the average rate of change

$$\frac{\sin t - \sin 0}{t - 0} = \frac{\sin t}{t} = q(t)$$

of $y = \sin t$ between 0 and t. As $t \to 0$, these secants turn around the point $(0,0)$ to approximate a line that is "tangential" to the graph of $y = \sin t$

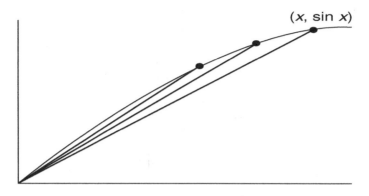

Fig. II.20   Graph of $y = \sin x$ with some secants through $(0, 0)$.

at the point $(0, 0)$. The slope of this limiting line is $\lim_{t \to 0} \frac{\sin t}{t}$, which has value 1 as we just determined. We conclude that the line of slope 1 through $(0, 0)$, i.e., the graph of

$$y = t,$$

is the tangent to the graph of $y = \sin t$ at the point $(0, 0)$.

II.5.3   *The Derivative of* $\sin t$

In order to find the derivative of $\sin t$ at other points $a \neq 0$, we shall use a geometric variation of the argument that we used for the derivative at $a = 0$. It, too, relies on the definition of the sine function in terms of the unit circle. (See Section I.3) Let us first point out a useful consequence of the important formula $\lim_{t \to 0} \sin t/t = 1$ that we established in the preceding section. By reflecting the arc of length t and the corresponding segment of length $\sin t$ on the x-axis (see Figure II.21), one obtains that the ratio of the chord $2 \sin t$ over the length $2t$ of the corresponding arc also has limit 1 as the length of the arc goes to zero. The result states a general geometric property of arcs centered at any point on the unit circle, as shown in Figure II.22.

One always has

$$\lim_{h \to 0} \frac{c(h)}{h} = 1.$$

Let us now consider the factorization $\sin(a + h) - \sin a = q_a(a + h)h$. For $h \neq 0$ the factor $q_a(a+h) = (\sin(a+h) - \sin a)/h$. The situation is visualized in Figure II.23.

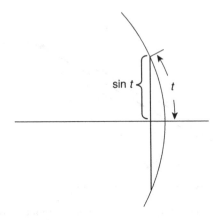

Fig. II.21   Chord spanned by arc of length $2t$.

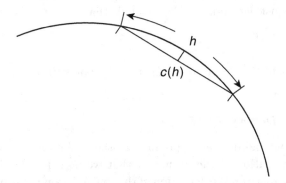

Fig. II.22   Arc of length h spans a chord of length $c(h)$.

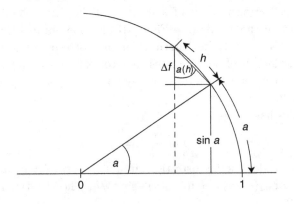

Fig. II.23   Geometric visualization of $q_a(a + h) = \Delta f / h$.

We note the small right triangle whose hypotenuse is the chord $c(h)$ spanned by the arc of length $h \neq 0$ that has the angle $a(h)$ opposite to $P(a)$. Figure II.24 shows an enlargement of that small triangle.

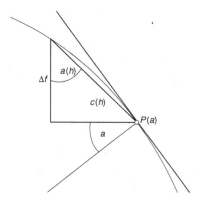

Fig. II.24 Triangle with hypothenuse $c(h)$ and leg Δf.

Since the leg Δf in this right triangle is adjacent to the angle $a(h)$, one has $\Delta f / c(h) = \cos a(h)$. Note that as $h \to 0$ the chord $c(h)$ (i.e., the secant through $P(a)$ and $P(a+h)$) will rotate into the direction of the tangent at $P(a)$, which is perpendicular to the radius at $P(a)$. Therefore the angle $\pi/2 - a(h)$ at the vertex $P(a)$ opposite to the angle $a(h)$ will have limit $\pi/2 - a$; hence the angle $a(h)$ has limit a as $h \to 0$.

It follows that

$$q_a(a+h) = \frac{\Delta f}{h} = \frac{\Delta f \cdot c(h)}{c(h) \cdot h} = \frac{c(h)}{h} \cos a(h).$$

Since $\cos(s)$ is continuous, one has $\cos a(h) \to \cos a$ as $h \to 0$. By basic limit rules one therefore obtains

$$\lim_{h \to 0} q(a+h) = \lim_{h \to 0} \frac{\Delta f}{h} = \lim_{h \to 0} \frac{c(h)}{h} \lim_{h \to 0} \cos a(h) = 1 \cdot \cos a.$$

By defining $q_a(a) = \lim_{h \to 0} q_a(a+h) = \cos a$, we extend q_a to a function that is continuous at point a as well. If one sets $t = a + h$, the factorization $\sin t - \sin a = q_a(t)(t - a)$ with q continuous at $t = a$ therefore confirms that $\sin t$ is differentiable at the arbitrary point a, and that its derivative satisfies

$$D(\sin)(a) = (\sin)'(a) = q_a(a) = \cos a \quad \text{for all } a \in \mathbb{R}.$$

Another proof of this differentiation formula, based on the functional equation of the sine function, is outlined in Problem 7 of Exercise II.5.6.

II.5.4 *The cosine Function*

A geometric argument quite similar to the one we just used for the *sine* function can be used to analyze $\cos t$. (See Problem 13 of Exercise II.5.6.) A different argument is based on the relation $\cos t = \sin(g(t))$, where $g(t) = \pi/2 - t$, as follows. Fix the point a, set $b = g(a) = \pi/2 - a$, and consider the factorization $\sin(u) - \sin b = q_b(u)(u - b)$, where q_b is continuous at b, with $q_b(b) = \cos b$, that was just established in the previous section. Substitute $u = g(t)$, and note that $u - b = g(t) - g(a) = -t - (-a) = -(t - a)$. It then follows that

$$\sin g(t) - \sin g(a) = q_b(g(t))[g(t) - g(a)] = -\,q_b(g(t))(t - a).$$

Since g is continuous at a and q is continuous at $b = g(a)$, the composition $-\,q_b \circ g$ is continuous at a; the above factorization shows that $\sin g(t)$ is differentiable at a, with derivative

$$D(\sin \circ g)(a) = -\,q_b(g(a)) = -\cos(\pi/2 - a).\ ^6$$

Since $\cos(\pi/2 - a) = \sin a$, we have therefore established that $\cos t = (\sin \circ g)(t)$ is differentiable at a and that

$$D(\cos)(a) = (\cos)'(a) = -\sin a \quad \text{for all } a \in \mathbb{R}.$$

The derivatives of the *sine* and *cosine* functions are very simple indeed. It is important *not to overlook the minus sign* that appears in the preceding formula in contrast to the earlier one for the derivative of $\sin t$.

II.5.5 *A Differential Equation for sine and cosine*

As we just verified, the *sine* and *cosine* functions satisfy the simple differentiation formulas

$$(\sin x)' = \cos x,$$
$$(\cos x)' = -\sin x.$$

By standard differentiation rules (see Prelude for the case of algebraic functions, and the next section for the general case) it then follows that a

---

[6]The reader may notice that this argument involves just a special case of the chain rule for derivatives that we had verified earlier for algebraic functions. The proof in the case at hand— as well as in the case of general differentiable functions discussed in the next section—is a simple modification of the earlier argument.

function f defined by $f(x) = A\sin x + B\cos x$, where A and B are constants, satisfies

$$D(f)(x) = f'(x) = (A\sin x + B\cos x)' = A\cos x - B\sin x.$$

Upon differentiating one more time one obtains

$$D(f')(x) = f''(x) = -A\sin x - B\cos x = -f(x).$$

We thus see that f satisfies the differential equation

$$y'' + y = 0.$$

In analogy to the differential equation $y' = ky$ that models processes involving growth and decay, the differential equation $y'' + y = 0$, or more generally

$$y'' + \omega^2 y = 0$$

for some constant ω, appears in many applications that involve periodic processes such as waves or a pendulum, to mention just a few. As we shall prove later on, it is indeed the case that *any* solution of this equation, i.e., any function $y = f(x)$ that satisfies the equation $f''(x) + \omega^2 f(x) = 0$ on some interval I is of the form $f(x) = A\sin\omega x + B\cos\omega x$ for suitable constants A, B. This result and related applications will be discussed in Chapter III.

II.5.6 *Exercises*

1. a) The basic addition formula for the sine function states that $\sin(\alpha + \beta) = \sin\alpha\cos\beta + \cos\alpha\sin\beta$. (This result is usually proved in a course on trigonometry.) Use this formula to show that $\sin x$ is continuous at every $a \in \mathbb{R}$. (Hint: Apply the addition formula to $x = (x - a) + a$, and note that $x \to a$ if and only if $(x - a) \to 0$.)

   b) Use a) and the identity $\cos x = \sin(\pi/2 - x)$ to prove that $\cos x$ is continuous at every $a \in \mathbb{R}$.

2. Use numerical methods to estimate the slope of the tangent line to the graph of $y = \sin(2x)$ at the point $(0, 0)$.

3. Given that $\lim_{x\to 0}\frac{\sin x}{x} = 1$, verify that $\lim_{x\to 0}\frac{\sin(ax)}{x} = a$ for any number a. (Hint: Replace $xa = h$, and consider $h \to 0$.)

4. The tangent function is defined by $\tan x = \frac{\sin x}{\cos x}$ for $x \ne \frac{\pi}{2} + n\pi$, n integer. Determine

$$\text{a)} \ \lim_{s\to 0}\frac{\tan s}{s}, \qquad \text{b)} \ \lim_{s\to 0}\frac{s}{\tan(3s)}.$$

5. a) The graph of the cosine function suggests that its tangent at $(0, 1)$ is the horizontal line $y = 1$. Use numerical data to confirm this geometric conclusion.

   b) Use an accurate graph of $y = \cos x$ to estimate the slope of the tangent at the point $(\frac{\pi}{3}, \frac{1}{2})$.

   c) Use a calculator to confirm your answer in b) by approximating the slope of the tangent by average rates of change on intervals $[\frac{\pi}{3}, \frac{\pi}{3} + 10^{-k}]$ for $k = 2, 3, ..., 6$.

6. Determine $\lim_{x \to \pi/2} \frac{\cos x}{x - \pi/2}$. (Hint: Use $\cos(\frac{\pi}{2} - t) = \sin t$.)

7. As stated in Problem 1, the *sine* function satisfies the functional equation (i.e., addition formula) $\sin(\alpha + \beta) = \sin \alpha \cos \beta + \sin \beta \cos \alpha$. Use this equation to give an *analytic* proof of the formula for the derivative of $\sin x$. Use the following outline.

   a) Prove that $\lim_{x \to 0} \frac{\cos x - 1}{x} = 0$.

   b) Apply the functional equation to $\sin(a + h)$ and use the result to determine a formula for $q_a(a + h) = [\sin(a + h) - \sin a]/h$ for $h \neq 0$.

   c) Take the limit as $h \to 0$ in the formula in b).

8. Determine the limits

$$\text{a) } \lim_{x \to 0} \frac{\cos x - 1}{x^2}, \qquad \text{b) } \lim_{t \to 0} \frac{t^{3/2}}{\sin t}.$$

9. Determine whether $\lim_{x \to 0} \sin \frac{1}{x}$ and $\lim_{x \to 0} |x|^{1/2} \sin \frac{1}{x}$ exist, and find the limits, if possible. Make sure to justify your conclusions.

10. a) Find the 4th derivatives of $\sin x$ and $\cos x$.

    b) Find a general formula for the nth derivative $(\sin x)^{(n)}$, $n = 1, 2, 3, ...$.

11. Find the equation of the tangent line to the graph of $y = \cos x$ at $x = \frac{\pi}{4}$.

12. Use derivatives to decide where the graph of $y = \sin x$ has horizontal tangents.

13. Work out the details of the *geometric* argument analogous to the one used for the sine function to find the derivative of $f(t) = \cos t$ directly. (Hint: Start with a sketch similar to Figure II.23; note that Δf now corresponds to a different segment than in the case of the *sine* function.)

II.6   Simple Differentiation Rules

We shall now extend the differentiation rules we had established in the Prelude for algebraic functions in the class $\mathcal{A}$ to arbitrary differentiable functions. The essential new ingredient is the application of the rules for continuous functions (see Section 4.2) at the appropriate places in the proofs that were given in the Prelude.

II.6.1   *Linearity*

Rule I. If f and g are differentiable at the point a and c, d are constants, then $cf(x) + dg(x)$ is differentiable at a and

$$D(cf + dg)(a) = (cf + dg)'(a) = cD(f)(a) + dD(g)(a).$$

As in the case of algebraic functions (see the Prelude), the factor q in the factorization $(cf + dg)(x) - (cf + dg)(a) = q(x)(x - a)$ is given by

$$q(x) = cq_f(x) + dq_g(x),$$

where q_f and q_g are the corresponding factors for f and g. Since q_f and q_g are continuous at a by assumption, q is continuous at a as well, and the rule follows.

   Examples.

$$(3 \cdot 4^x + 5x^4)' = 3 \ln 4 \cdot 4^x + 20x^3,$$

$$D(5 \sin x - 4 \cos x + 2e^x) = 5 \cos x - 4(-\sin x) + 2e^x.$$

II.6.2   *Chain Rule*

The composition $f \circ g$ of two functions f and g is a most important and natural operation on functions. Hence one would expect that the corresponding differentiation rule is particularly simple. The formula established in the algebraic case (Rule II in Section 6.1 of the Prelude) surely confirms this: *the derivative of the composition is the product of the derivatives.* It does not get any simpler than this! At an intuitive level, the extension to general differentiable functions is then clear: since the *linear* approximations L_f and L_g of f and g at the relevant points are algebraic, one has $(L_f \circ L_g)' = (L_f)' \cdot (L_g)'$, and since $L_f \circ L_g$ is again linear, it provides the linear approximation to $f \circ g$, and the result follows. In essence, since differentiability is equivalent to good local linear approximation, what is correct for linear functions remains correct (locally) for arbitrary differentiable functions.

The precise verification of the preceding intuitive argument is just as easy. Suppose that g is differentiable at a and f is differentiable at $b = g(a)$. As in the case of functions in $\mathcal{A}$, by direct substitution it then follows that

$$(f \circ g)(x) - (f \circ g)(a) = q_f(g(x)) \cdot q_g(x)(x - a),$$

where q_f and q_g are the *continuous* factors in the relevant factorizations of f and g, that is, the resulting factor q for the composition $f \circ g$ satisfies

$$q(x) = q_f(g(x)) \cdot q_g(x).$$

Since g and q_g are continuous at a and q_f is continuous at $b = g(a)$, the rules for continuous functions now imply that q is continuous at a as well. Therefore $f \circ g$ is differentiable at a, and

$$\begin{aligned}
D(f \circ g)(a) = q(a) &= q_f(g(a)) \cdot q_g(a) \\
&= D(f)(g(a)) \cdot D(g)(a) \\
&= D(f)(b) \cdot D(g)(a).
\end{aligned}$$

We have thus verified **Rule II (Chain Rule)**, that is

$$D(f \circ g) = [D(f) \circ g] \cdot D(g), \text{ or even shorter,}$$
$$(f \circ g)' = f' \cdot g',$$

for arbitrary differentiable functions. In the second formulation one of course needs to be careful to choose the input values correctly. In particular, the formula for $(f \circ g)'(a)$ cannot involve $f'(a)$, since f and hence f' need not be defined at all near the point a. Instead, for the composition to be defined, f must be defined near the point $b = g(a)$. Correspondingly, the appropriate input for the derivative f' is this point b as well. In fact, the formula $(f \circ g)'(a) = f'(b) \cdot g'(a)$ that we just obtained exhibits the only meaningful way to choose the input values in each factor.

Let us consider some examples.

i) $F(x) = \cos(4 \cdot 2^x)$. We recognize that $F = f \circ g$, where $f(u) = \cos(u)$ and $g(x) = 4 \cdot 2^x$. Hence, by the chain rule, F is differentiable and

$$F'(x) = f'(g(x)) \, g'(x) = -\sin(4 \cdot 2^x)(4 \ln 2 \cdot 2^x).$$

ii) $\left(e^{x^2}\right)' = e^{x^2}(x^2)' = e^{x^2}(2x).$

iii) $F(x) = (\sin x)^{10}$. This function is the composition of $\sin x$ with $f(u) = u^{10}$. Hence

$$D((\sin x)^{10}) = 10(\sin x)^9 D(\sin x) = 10(\sin x)^9 \cos x.$$

iv) $H(x) = \sin(2x + (4x^3 + 5x)^6)$.

Here the chain rule must be used twice. Let us denote the *inner* function by $g(x) = 2x + (4x^3 + 5x)^6$. Then $H(x) = \sin(g(x))$, so that $H'(x) = \cos(g(x)) \cdot g'(x)$. Next, differentiation of the second summand in g requires the chain rule once again (unless one wants to expand the bracket according to the binomial theorem), resulting in $g'(x) = 2 + 6(4x^3 + 5x)^5(12x^2 + 5)$. Putting everything together gives the somewhat messy formula

$$H'(x) = \cos(2x + (4x^3 + 5x)^6) \cdot [2 + 6(4x^3 + 5x)^5(12x^2 + 5)].$$

Note: Unless explicitly asked for, no further (algebraic or other) simplification should be attempted! The formula above clearly reflects the structure of the differentiation rules that have been applied; any further transformations would make it difficult to recognize the steps that have been taken.

II.6.3   *Power Functions with Real Exponents*

The chain rule and the derivative formula for the natural logarithm (see Section 2.6) allow us to easily handle power functions $p(x) = x^r$ where the exponent r is an arbitrary real number. Recall that in the Prelude we already verified the power rule $(x^r)' = rx^{r-1}$ in the case where the exponent $r = m/n$ is rational. When the exponent r is not rational, the power function x^r is **not** of algebraic type, and therefore it needs to be studied by different techniques. We also note that—except for certain special exponents—the domain of such a function is the set of *positive* real numbers.[7] For example, x^m is defined for all $x \neq 0$ if m is an integer, and also at 0 if $m \geq 0$. In order to handle arbitrary *real* exponents $r \in \mathbb{R}$, assuming that $x > 0$, we use the natural logarithm to replace $x = e^{\ln x}$. By the rules for exponentials it follows that

$$p(x) = x^r = (e^{\ln x})^r = e^{r \ln x},$$

so that $p(x)$ is transformed into the composition of e^u with $u = r \ln x$. By the chain rule it follows that $p(x)$ is differentiable for all $x > 0$, and

---

[7] Recall the discussion in Section 1.4: powers b^r with arbitrary (rational) exponent are defined only for $b > 0$!

furthermore (recall $(e^u)' = e^u$!) one obtains

$$D(x^r) = D(e^{r\ln x}) = D(e^u)(r\ln x) \cdot D(r\ln x)$$
$$= e^{r\ln x}(r\,\frac{1}{x}) = r\,(x^r\frac{1}{x})$$
$$= r\,x^{r-1}.$$

The result matches exactly the power rule that we had obtained earlier in the case of a *rational* exponent. For example, one now has the non-algebraic differentiation formula

$$(x^\pi)' = \pi x^{\pi-1} \text{ for } x > 0.$$

It is important to remember that this power rule applies only when the variable appears in the base. When the variable x appears in the exponent rather than in the base, one deals with an *exponential* function, whose differentiation formula looks quite different, namely $D(b^x) = b^x \ln b$.

II.6.4   *Inverse Functions*

Next we shall discuss the simple formula for the derivative of the inverse of a differentiable function. In particular we will justify the formula for the derivative of the natural logarithm $\ln x$—which is the inverse of the exponential function $E(x) = e^x$—that we just used in the preceding section, and that we had established in Section 2.6 by a geometric argument.

Let us first discuss the intuitive argument that is at the heart of the proof. Since differentiable functions are locally essentially linear, we consider first the case of a linear function $L(x) = mx + b$. Such a function is one-to-one precisely if $m \neq 0$, i.e., if $L' \neq 0$. In this case the inverse g is again linear, and by Rule III in the Prelude (or by direct calculation) one has $g' = 1/L'$. Turning to an arbitrary invertible function f (i.e., f must be one-to-one) that is differentiable at a, we consider its linear approximation $L_{f,a}(x)$, and we assume that $f'(a) = L'_{f,a} \neq 0$. Let $b = f(a)$. By reflecting the graphs of f and $L_{f,a}$ (note that the latter is the tangent line of f at the point $(a, f(a))$), we obtain the graph of the inverse g of f together with its tangent line at the point $(f(a), a) = (b, g(b))$. (See Figure II.25.)

As seen in Figure II.26, if the line L_1 has slope $m_1 = d/c$, the reflected line L_2 has slope $m_2 = c/d = 1/m_1$.

Since the reflection clearly preserves the geometric properties of the linear approximation, i.e., the relationship between the graph of the function and its tangent line, one obtains **Rule III**:

The inverse g of f is differentiable at $b = f(a)$, with

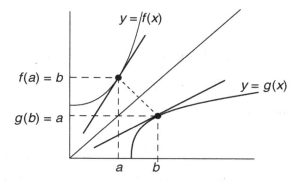

Fig. II.25   Graph of f and its inverse, with tangents.

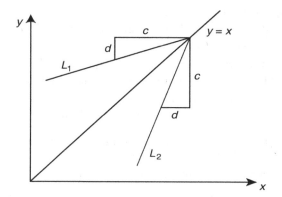

Fig. II.26   The slopes of the reflection L_2 of the line L_1.

$$D(g)(b) = \frac{1}{D(f)(a)} = \frac{1}{D(f)(g(b))}.$$

This geometric argument suggests that exactly the same formula that we had verified for algebraic functions in the class $\mathcal{A}$ (Prelude, Rule III), remains correct for general differentiable functions provided $f'(a) \neq 0$. This is easily verified precisely by considering the factorization

$$f(x) - f(a) = q(x)(x - a), \qquad \text{(II.15)}$$

with q continuous at a and $q(a) = f'(a) \neq 0$. It follows that $1/q$ is defined on a small interval I centered at a and is continuous at a as well. Since f is one-to-one and continuous on I, it follows that the inverse g is continuous as well on the interval $J = f(I)$. (See Theorem 4.3 in Section 4.2.) Consequently, assuming $x \in I$, by substituting $f(x) = y \in J$, $f(a) = b$, and $x = g(y)$, $a = g(b)$ into the above factorization for f one obtains for all $y \in J$ that

$$y - b = q(g(y))(g(y) - g(b)), \text{ i.e.,}$$

$$g(y) - g(b) = \frac{1}{q(g(y))}(y - b). \tag{II.16}$$

Since the composition $(1/q) \circ g$ of continuous functions is continuous at b, the last equation (II.16) shows that g is differentiable at b, with derivative $D(g)(b) = 1/q(g(b)) = 1/f'(a)$, as claimed.

This result is another concrete instance of the general principle that whatever is true for linear functions will usually remain correct locally for differentiable functions as well.

Note that Rule III is meaningless if $f'(a) = 0$. In this case the tangent line to the graph of f is horizontal, and hence its reflection is vertical, so that its slope is not defined. Therefore the inverse g of f is NOT differentiable at points $b = f(a)$ where $f'(a) = 0$.

Let us apply what we just learned to the exponential function $E(x) = e^x$. Here $E'(a) \neq 0$ for all a (why?), and hence the inverse function $g(x) = \ln x$ of E is differentiable at all points $b = e^a$, i.e., at all points of its domain $\{b \in \mathbb{R} : b > 0\}$. Furthermore, by Rule III one obtains

$$g'(b) = \frac{1}{E'(a)} = \frac{1}{E(a)} = \frac{1}{b} \text{ where } b = E(a) = e^a.$$

We have thus verified the differentiation formula

$$D(\ln)(x) = \frac{1}{x} \text{ for all } x > 0,$$

for the natural logarithm function that we had already obtained in Section 2.6 by a geometric argument.

Completely analogous arguments show that

$$D(\log_b)(x) = \frac{1}{x \ln b}$$

for any base $b \neq 1$ and all $x > 0$.

II.6.5   *Inverse Trigonometric Functions*

A more intriguing application of the differentiation formula for inverse functions concerns the function $y = \sin x$. Since this function is **not** one-to-one on the whole real line, we proceed as in Section I.5.4 and select a suitable smaller section of its graph. We thus consider S defined for $x \in [-\pi/2, \pi/2]$ by $S(x) = \sin x$, whose graph is the thick curve shown in Figure II.27.

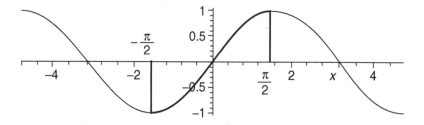

Fig. II.27 Graph of $\sin x$ for $-\pi/2 \leq x \leq \pi/2$.

Clearly this function is strictly increasing on $[-\pi/2, \pi/2]$, and so it has an inverse

$$g : [-1, 1] \to [-\pi/2, \pi/2],$$

that is continuous by Theorem 4.3. The graph of g is shown in Figure II.28.

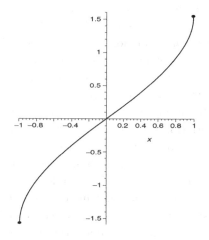

Fig. II.28 Graph of the inverse of $S(x) = \sin x$.

Since S is differentiable with $S'(x) = \cos x$, and $S'(a) \neq 0$ for all $a \in I = (\pi/2, -\pi/2)$ (no endpoints!), Rule III in the preceding section implies that the inverse g of $S(x)$ is differentiable at all points $b = \sin a \in S(I) = (-1, 1)$, and that

$$g'(b) = \frac{1}{S'(a)} = \frac{1}{\cos a}.$$

In order to write this formula in terms of $b = \sin a$, recall that $\sin^2 a + \cos^2 a = 1$ for all a. Therefore

$$\cos^2 a = 1 - \sin^2 a.$$

Since for $a \in (-\pi/2, \pi/2)$ one has $\cos a > 0$, it follows that

$$\cos a = +\sqrt{1 - \sin^2 a} = \sqrt{1 - b^2} \text{ for } a \in (-\pi/2, \pi/2),$$

and hence

$$g'(b) = \frac{1}{\sqrt{1 - b^2}} \text{ for } -1 < b < 1.$$

After introducing the name *arcsin* (or inverse sine) for the function g and replacing b with x, this formula translates into the differentiation formula

$$D(\arcsin x) = \frac{1}{\sqrt{1 - x^2}} \text{ for } -1 < x < 1$$

for the inverse of the (partial) *sine* function. Note that the inverse sine is not differentiable at the endpoints -1 and $+1$, even though $\arcsin x$ is defined and continuous at these points.

Remark. Let us conclude this discussion by pointing out a remarkable feature. While differentiation of the simple transcendental (i.e. non-algebraic) functions $E(x) = e^x$, $S(x) = \sin x$, and $C(x) = \cos x$ leads to the same type of *transcendental* functions, differentiation of the corresponding *inverse* functions (where defined) results in *algebraic* functions, i.e., functions of a completely different nature that are, in fact, more elementary. Differentiation thus establishes a surprising and deep link between transcendental and algebraic objects! As we saw in the Prelude, the tangent problem and differentiation are solved for algebraic functions by algebraic methods. This appears to be the more elementary setting, as no completeness of the real numbers and no limits need to be introduced. On the other hand, once differentiation is extended to the elementary transcendental functions—indispensable for understanding growth phenomena and periodic processes—one discovers that the differentiation formulas for these functions have a particularly simple structure, much simpler than for most algebraic functions. Even more surprising, the derivatives of the *inverses* of these functions turn out to be—in the important cases considered here—algebraic functions. In Chapter IV we will see that *reversing* the process of *differentiation* also establishes a connection between algebraic and transcendental functions.

II.6.6   *Exercises*

1. Determine the derivatives of the functions

$$\text{a)} \ 3 \cdot 4^x - 5x^4, \qquad\qquad \text{b)} \ \pi x^5 - \frac{4}{x} \text{ for } x \neq 0.$$

2. Find the derivatives of the functions

   a) $5\cos x + 3^x$,   b) $6x^4 - 2\sin x$,   c) $3\sin x + 2x^7 - 3e^x$.

3. Let $g(x) = x + \sin x$.

   a) Find the equation of the tangent line to the graph of g at the point where $x = \frac{\pi}{2}$.

   b) Are there any points on the graph of g where the tangent is horizontal? If yes, find all such points.

   c) Plot the graph of g with a calculator to verify your conclusions in b).

4. Find the derivatives of the following functions.

   a) $f(x) = \ln(x^2+1)$,   b) $g(t) = \sin(2^t)$,   c) $f(u) = 3^{\cos u}$,

   d) $p(x) = x^{\sqrt{3}}$,   e) $f(x) = x^e + e^x$,   f) $q(s) = \dfrac{1}{s^4 + 2s^2 + 3}$.

   (Hint for f): Write $q(s)$ as power with negative exponent.)

5. Determine the derivative of $F(x) = \cos(\sin(x^2 + 1))$.

6. Suppose that g is differentiable with $g(4) = 2$ and $g'(4) = 1/2$, and that f is differentiable at 2, with $f(2) = 0$ and $f'(2) = 3$. Determine the derivative of the composition $f \circ g$ at $x = 4$.

7. Let $f : \mathbb{R} \to \mathbb{R}$ be differentiable and satisfy the property $f(0) = 0$.

   a) Show that $(f \circ f)'(0) = [f'(0)]^2$ and $(f \circ f \circ f)'(0) = [f'(0)]^3$.

   b) Let n be a positive integer. Determine the derivative of $(f \circ f \circ ... \circ f)$ (n compositions) at 0 in dependence of n.

8. Let $b > 0$ and define $f_1(x) = b^{-x}$ and $f_2(x) = (b^x)^{-1}$.

   a) Evaluate the derivatives of f_1 and f_2 by using the chain rule and power rule as needed.

   b) Do your answers in a) agree? Explain!

9. Suppose f is a one-to-one differentiable function with differentiable inverse g. Apply the chain rule to the function $f \circ g$ to find the relationship between f' and g'. (Hint: Recall that $(f \circ g)(x) = x$; take derivatives on both sides!)

10. Use the formula for the derivative of the inverse function to verify in detail that

$$D(\log_b x) = \frac{1}{x \ln b}, \quad x > 0$$

for any base $b \neq 1$.

11. a) Verify that $C(x) = \cos x$ restricted to the domain $[0, \pi]$ is one-to-one with image $[-1, 1]$. (Look at the definition of $\cos x$ on the unit circle.)

    b) Let g be the inverse of the function in a). Modify the argument in the text used for the *inverse sine* to determine the derivative of g. Where does g fail to be differentiable?

12. The function $S^+(x) = x^2$ is one-to-one on the interval $[0, \infty)$, with inverse $g(x) = \sqrt{x}$ also defined on $[0, \infty)$. Use the result in this section to obtain the derivative of g. Where does g fail to be differentiable?

13. If $r = m/n$, with m and n both positive integers and n odd, the function $p(x) = x^r$ is defined in a full neighborhood of 0. Determine the values $r = \frac{m}{n} > 0$ for which p is *not* differentiable at 0, and those values for which p *is* differentiable at 0.

II.7　Product and Quotient Rules

By now the pattern is clear. All differentiation formulas established in the Prelude for algebraic functions continue to apply to arbitrary differentiable functions. As we already noticed in the Prelude, the rules for differentiating products and quotients turn out to be quite a bit more complicated than the rules we discussed so far.

II.7.1　*Statement of the Rules*

Based on the general principle that local properties of differentiable functions simply reflect the corresponding properties for the respective linear approximations, we can now understand better why things get more complicated with these last two rules. Rules I - III discussed in Section 6.2 simply reflect that sums, compositions, and inverses of the respective linear approximations L_f and L_g of two given functions f and g are again *linear*, and hence provide natural candidates for the *linear* approximations of the corresponding functions obtained by combining f and g. However, the product $L_f \cdot L_g$ and the quotient L_f / L_g are NOT linear. So neither provides us with potential *linear* approximations of product and quotient of f and g at the point a under consideration. Instead, when considering the product fg, one needs to take an additional step and determine first the linear approximation

$$L_{(L_f \cdot L_g)}$$

of the product $L_f \cdot L_g$. Application of the basic principle would then suggest that the known (algebraic) differentiation formula for the product $L_f \cdot L_g$, i.e., $D(L_f \cdot L_g) = D(L_f) \cdot L_g + L_f \cdot D(L_g)$ (recall Prelude, Section 6), remains valid for the derivative of the product $f \cdot g$ of differentiable functions in general, that is, we expect a formula

$$D(fg) = D(f)\, g + f\, D(g).$$

Similarly, applying the algebraic quotient rule to the quotient of the linearizations would lead to a corresponding quotient rule for general differentiable functions.

The precise verifications of these rules for general differentiable functions simply follow from the proofs given in the Prelude by noticing that the relevant combinations of the factors q_f and q_g obtained in the algebraic case will be continuous at the point a as a consequence of basic theorems about continuous functions.

For completeness' sake let us state the two rules one ends up with.

Assume that f and g are differentiable at the point a. Then

Product Rule : *fg is differentiable at a and*
$$D(fg) = D(f)\, g + f\, D(g) \quad \text{at } a,$$

and

Quotient Rule : *If $f(a) \neq 0$, then g/f is differentiable at a,*
$$\text{and } D\left(\frac{g}{f}\right) = \frac{D(g)\, f - g\, D(f)}{f^2} \quad \text{at } a.$$

Notice the complicated structure of the quotient rule. In particular, the numerator resembles the result of the product rule, except for the minus sign. Therefore the order of the terms in the numerator must be firmly observed. Always begin by differentiating the numerator of a quotient!

II.7.2   *Examples*

i) Find the derivative of $f(x) = \sin x \cdot 3^x$. We use the product rule to obtain

$$D(\sin x \cdot 3^x) = \cos x \cdot 3^x + \sin x \cdot (\ln 3 \cdot 3^x).$$

ii) If $G(x) = (x^3 + 4x^2 - 2x + 5) e^{5x}$, then
$$D(G)(x) = (3x^2 + 8x - 2) e^{5x} + (x^3 + 4x^2 - 2x + 5) e^{5x} 5.$$

iii)
$$D(\frac{x^3 - 2\sin x}{\cos^2 x + 1}) = \frac{(3x^2 - 2\cos x)(\cos^2 x + 1) - (x^3 - 2\sin x)(2\cos x)(-\sin x)}{(\cos^2 x + 1)^2}.$$

Do not attempt any (algebraic or other) simplification. Such changes would destroy the structure of the quotient rule, without any particular benefit.

iv) The tangent function is defined by the quotient $\tan x = \frac{\sin x}{\cos x}$ for all x with $\cos x \neq 0$. Its derivative is given by

$$D(\tan x) = D\left(\frac{\sin x}{\cos x}\right) = \frac{\cos x \cos x - \sin x(-\sin x)}{(\cos x)^2}$$

$$= \frac{\cos^2 x + \sin^2 x}{\cos^2 x} = \frac{\cos^2 x}{\cos^2 x} + \frac{\sin^2 x}{\cos^2 x} = 1 + (\tan x)^2.$$

By using the identity $\cos^2 x + \sin^2 x = 1$ before splitting up the fraction, the answer can also be written in the form

$$D(\tan x) = \frac{1}{\cos^2 x}.$$

v)
$$D\left(\frac{3x}{2^x + \cos x}\right) = \frac{3(2^x + \cos x) - 3x\, D(2^x + \cos x)}{(2^x + \cos x)^2}$$

$$= \frac{3(2^x + \cos x) - 3x\, (\ln 2 \cdot 2^x - \sin x)}{(2^x + \cos x)^2}.$$

vi)
$$D(\ln \frac{x}{x^2 + 1}) = \frac{1}{\frac{x}{x^2 + 1}} D(\frac{x}{x^2 + 1})$$

$$= \frac{x^2 + 1}{x} \frac{1 \cdot (x^2 + 1) - 2x \cdot x}{(x^2 + 1)^2}.$$

Alternatively, one may first use the equation $\ln \frac{x}{x^2 + 1} = \ln x - \ln(x^2 + 1)$; differentiation is then much simpler and leads to the answer

$$D(\ln \frac{x}{x^2 + 1}) = \frac{1}{x} - \frac{2x}{x^2 + 1}.$$

We leave it to the reader to check by algebra that the two answers indeed give the same rational function.

As we saw in examples iv) and vi) above, different correct procedures may lead to answers that may look quite different. Readers should be careful when comparing their answers to a particular problem with those of other readers.

II.7.3 Exercises

DO NOT attempt to simplify the expression you obtain after carrying out all the differentiation steps!

1. Find the derivatives of the following functions.

    a) $g(x) = x^2 3^x$,   b) $h(s) = \sqrt{s} \cos s$,   c) $f(t) = \frac{\sin t}{t^2 + 1}$,

    d) $f(x) = e^x \sin x$ e) $q(t) = \frac{2^t - 3t^4}{\cos t}$

    f) $F(x) = (x + \sin x)^5 (3x - \cos x)^6$.

2. Find the derivatives of

    a) $G(t) = \frac{4 \cos^2 t + e^t}{(\sin^2 t + 1)^5}$,   b) $H(x) = \sin(4x^2 e^x)$.

3. a) Find the derivative of $f(x) = \log_2 x \cdot (x^3 + 4x^2 - 1) \cdot \cos x$. (Hint: Apply the product rule twice!)

    b) More generally, show that

    $$(f \cdot g \cdot h)' = f' \cdot g \cdot h + f \cdot g' \cdot h + f \cdot g \cdot h'.$$

4. Consider the function $F(x) = x^x$ defined for $x > 0$. Note that 0^0 is meaningless (is it 0 or 1?), so that F is not defined at 0. It is not clear what rule of differentiation could be applied, since F combines the features of a power function with those of an exponential function. Determine the derivative of F at points $x > 0$ as follows.

    a) Verify that $\ln F(x) = x \ln x$.
    b) Explain why $\ln F$ is differentiable, and find its derivative.
    c) Explain why it follows from b) that F is differentiable.
    d) Take the derivative of $\ln F(x)$ by using the chain rule.
    e) Use b) and d) to find the derivative of F.

    The technique outlined in this problem is known as **logarithmic differentiation**.

5. The product rule can also be obtained by logarithmic differentiation (see Problem 4) as long as the factors have no zeroes, as follows.

    a) If f and g are differentiable at a, and f and g are > 0 in a neighborhood of a, apply the rules of logarithms to $\ln(fg)$.
    b) Take the derivative at a on both sides of the equation obtained in a) by using the chain rule.
    c) Solve the resulting equation in b) for $(fg)'(a)$.
    d) If $f > 0$ and $g < 0$ on some interval, apply the previous procedure to f and $(-g)$. Note that $fg = -[f(-g)]$.

e) Discuss how to handle the case when both f and g are negative.

6. Apply logarithmic differentiation as in Problem 5 to find the derivative of f/g when f and g are as in 5 a).

7. Explain why $x^x = e^{x \ln x}$ for $x > 0$. Use this result and the chain rule to obtain the derivative of $F(x) = x^x$. (This approach is somewhat more direct than the one suggested in Problem 4.)

8. Find the equation of the tangent line to the graph of $f(x) = \frac{1}{1+x^2}$ at the point where $x = 1$.

Chapter III

Some Applications of Derivatives

III.1 Exponential Models

III.1.1 *Growth and Decay Models*

We had already seen in Section II.2.7 how a simple model for the growth of a population leads to the differential equation $y' = ky$, where k is a constant. We will show in the next section that *all* solutions f of this equation (i.e., functions f that satisfy $f'(t) = kf(t)$) are functions of the form $f(t) = Ce^{kt}$. This is the result that is the basis for all further analysis of this and similar models. The basic hypothesis underlying this model states that the rate of change $\frac{dP}{dt}(t)$ of a population $P(t)$ is proportional to the size $P(t)$ of the population, i.e., that there exists a constant k so that

$$\frac{dP}{dt}(t) = kP(t) \text{ for all times } t \text{ under consideration.}$$

It is of course assumed that there is no change in any of the relevant external conditions during the time period that is studied. This is certainly not true in concrete situations. At best, one may say that conditions remain approximately stable over a limited period of time, so that the "exponential growth model" has to be applied with care.

A much more stable, and hence better, situation arises with natural *decay* processes, such as they occur with radioactive substances. For example, the isotope U_{238} of uranium emits radiation that arises from the splitting of the uranium atoms into other element. Experiments reveal that this radioactive decay process is not affected by any changes in the environment whatsoever, and hence is extremely stable over very long periods of time. The amount $A(t)$ of U_{238} present at time t decays at a rate $\frac{dA}{dt}$ that is proportional to the amount $A(t)$, so

$$\frac{dA}{dt}(t) = -\lambda A(t)$$

189

for a constant $\lambda > 0$ that is called the *decay constant* (specific to U_{238}).[1] Experimental data leads to the value $\lambda = 0.155 \times 10^{-9}$ when t is measured in years. The differential equation then implies that

$$A(t) = Ce^{-\lambda t}$$

for a constant C that is identified with the amount $A(0) = Ce^0 = C$ that is present at time $t = 0$.

Rather than describing the decay process by the decay constant λ, physicists often use the *half life* T of a radioactive element. This is that time T in which *half* of the initial amount has decayed. So T satisfies the equation

$$A(0)e^{-\lambda T} = A(T) = \frac{1}{2}A(0),$$

or

$$e^{-\lambda T} = \frac{1}{2}.$$

Notice that this last equation does not depend on the initial amount $A(0)$. It follows that

$$e^{\lambda T} = 2, \text{ or } \lambda T = \ln 2.$$

For example, the half life of uranium U_{238} equals $T = \ln 2/\lambda = 4.47 \times 10^9 = 4.47$ billion years. Uranium does indeed decay very slowly.

III.1.2 *Radiocarbon Dating*

For other radioactive substances the half life is much shorter. For example, the isotope C_{14} of carbon is radioactive (hence called *radiocarbon*) with a half life of about 5730 years. The corresponding decay constant is $\lambda = \ln 2/5730 = 0.00012$.

Radiocarbon has been used successfully to date ancient objects that have biological origins. The method is based on the fact that radiocarbon occurs naturally in the atmosphere, and that it is continuously created in the upper atmosphere from nitrogen subject to intensive cosmic radiation, at a rate that compensates the loss due to radioactive decay. Consequently, the ratio of radiocarbon to the normal carbon C_{12} in the atmosphere has remained quite stable over very long periods of time. All living organisms assimilate radiocarbon along with normal carbon, but this process stops when the organism dies, and hence the amount of radiocarbon in the

---

[1] In a decay process, dA/dt is negative; it is convenient to write the relevant equation so as to make this clearly visible.

remnants will then decrease over time. The amount of radiocarbon in an ancient object derived from living organisms can be measured, and that information can be used to estimate the age of the object.

Example. Archeologists find remnants of a skeleton in a cave. Analysis of a specimen reveals that the ratio of radiocarbon to carbon is about 82% of the ratio q found in the atmosphere. If $A(t)$ is the amount of radiocarbon in the specimen at time t (measured in years, with $t = 0$ corresponding to the time the specimen stopped living), then the decay model

$$\frac{dA}{dt} = -\lambda\, A(t)$$

implies that $A(t) = A(0)\, e^{-\lambda t}$. The value $A(t)$ measured today is $.82\, A(0)$, so that

$$.82\, A(0) = A(0)\, e^{-\lambda t}$$

implies $e^{-\lambda t} = .82$. Hence $-\lambda t = \ln(0.82)$. With $\lambda = 0.00012$, one obtains

$$t = -\ln(0.82)/\lambda \approx 1653.$$

Therefore one may conclude that the cave was inhabited about 1650 years ago.

III.1.3 *Compound Interest*

An exponential growth model arises also in finance. A principal amount of $A(t)$ dollars at time t (t measured in years) is said to grow under *continuous compounding* at an annual rate r, if the rate of growth of the capital $A(t)$ satisfies

$$\frac{dA}{dt}(t) = r A(t).$$

It then follows that

$$A(t) = A(0)e^{rt}.$$

Other compounding methods are based on dividing the year into n equal compounding periods of length $1/n$ years. At the end of each such period simple interest is added to the principal at the beginning of that period at the rate r/n obtained by dividing the annual interest rate evenly over the n periods. The resulting formula for compound interest after t years reads

$$A_n(t) = A(0)(1 + \frac{r}{n})^{nt}.$$

(See Section I.4). Commonly used methods are annual compounding ($n = 1$), quarterly compounding ($n = 4$), and daily compounding ($n = 360$).

The question arises about the relationship of these compound interest formulas with *continuous compounding*. Let us investigate what happens with $A_n(t)$ as n gets increasingly larger, thereby making the compounding periods increasingly shorter. Since

$$A_n(t) = A(0)(1 + \frac{r}{n})^{nt} = A(0)\left((1 + \frac{r}{n})^n\right)^t,$$

we are led to consider

$$\lim_{n\to\infty} (1 + r/n)^n.$$

For simplicity, consider $r = 1$ first, so we need to analyze $(1 + 1/n)^n$. Let us set $1/n = h$, so that $n = 1/h$. Then $n \to \infty$ is equivalent to $h \to 0$, and therefore

$$\lim_{n\to\infty} (1 + \frac{1}{n})^n = \lim_{h\to 0} (1 + h)^{1/h}.$$

Now, for h close to 0, yet $h \neq 0$, use the equation $(1+h) = e^{\ln(1+h)}$ to write

$$(1 + h)^{1/h} = [e^{\ln(1+h)}]^{1/h} = e^{[\ln(1+h)]/h}.$$

Recall that since $\ln 1 = 0$, one obtains

$$\lim_{h\to 0} \frac{\ln(1 + h)}{h} = \lim_{h\to 0} \frac{\ln(1 + h) - \ln 1}{h}$$
$$= \text{ derivative of } \ln x \text{ at the point } x = 1.$$

Since $(\ln x)' = 1/x$, the value of this limit is $1/1 = 1$. The continuity of $E(x) = e^x$ then implies that

$$\lim_{n\to\infty} (1 + 1/n)^n = \lim_{h\to 0} (1 + h)^{1/h}$$
$$= \lim_{h\to 0} e^{[\ln(1+h)]/h} = \lim_{u\to 1} e^u = e^1 = e.$$

We have thus discovered the representation

$$e = \lim_{n\to\infty} (1 + \frac{1}{n})^n$$

for the base e of the natural logarithm.

A minor modification of this argument shows that

$$e^r = \lim_{n\to\infty} (1 + \frac{r}{n})^n \text{ for any real number } r.$$

Consequently, we obtain that continuous compounding of interest at the annual rate r, given by the function $A_c(t) = A(0)e^{rt}$, arises as the limiting case $n \to \infty$ of compounding over n equal periods per year. Continuous compounding may be viewed as "instantaneous" compounding, i.e., interest is added to the capital at every moment.

It is of interest to compare the growth of capital under various compounding methods. Suppose the annual interest rate is 6%, i.e., $r = 0.06$, and that $A(0) = \$100,000$. After 10 years, annual compounding results in

$$A_1(10) = 100000 \times (1 + 0.06)^{10} = 1.79084\,77 \times 10^5 = \$179,848.$$

Quarterly compounding gives

$$A_4(10) = 100000 \times (1 + 0.06/4)^{40} = 1.81401\,84 \times 10^5 = \$181,402,$$

while daily compounding results in

$$A_{360}(10) = 100000 \times (1 + 0.06/360)^{3600} = 1.82202\,99 \times 10^5 = \$182,023.$$

Finally, continuous compounding gives

$$A_c(10) = 100000 \times \exp(0.06 * 10) = 1.82211\,88 \times 10^5 = \$182,212.$$

Notice that continuous compounding yields the best result. While the difference to daily compounding is perhaps insignificant, continuous compounding yields almost $\$2,500$ more than annual compounding. For this reason, banks often state the "yield" of continuous compounding, i.e., that annual rate that produces the same result by annual compounding. For example, for continuous compounding at 6%, the yield is determined by solving

$$e^{0.06} = (1 + r)$$

for r. The result is $r = e^{0.06} - 1 = 1.06183\,65 - 1 = .0\,61836\,5$, i.e., the yield is 6.18%.

III.1.4 *Exercises*

1. A bank offers a 5-year certificate of deposit which pays interest at an annual rate of 7.5% compounded continuously. Determine the effective yield of the certificate.

2. Bank ABC offers a certificate of deposit at 6.125% compounded monthly, while its competitor Bank QRS offers 6.1% compounded continuously. Which bank would you choose? Justify your choice by comparing yields.

3. A population of bacteria in a laboratory grows exponentially at the rate of 5% per day. If the initial size is 1000, after how many days will the population have grown to 2000?

4. Radon 222 is a radioactive gas that is found to be harmful to humans if they are exposed to it in excessive amounts. Its half life is about 3.8 days. Because of a leak, the basement of a factory has reached dangerously high levels of Radon 222, and the health inspector recommends that no one should enter the basement until the radioactive level has decreased to 10% of the original level. How many days should people stay out of the basement?

5. Newton's Law of Cooling states that the temperature $T(t)$ of an object placed in an environment at constant temperature A changes at a rate that is proportional to the difference $A - T(t)$, i.e., there is a constant $k > 0$ so that

$$\frac{dT}{dt} = k(A - T).$$

   a) Explain the meaning of the sign of dT/dt. Consider the cases $A > T$ and $A < T$ separately.
   b) Show that if $T(0) > A$, then $T(t) = A + (T(0) - A)e^{-kt}$. (Hint: Set $y = T(t) - A$ and show that $y' = -ky$.)
   c) Will the temperature $T(t)$ ever be equal to A? Explain.

6. At 5 p.m. police find the body of the victim of a murder in a room whose temperature was maintained at 20^0C. At that time the temperature of the body was measured to be 30^0C, and two hours later it had decreased to 25^0. Assuming that the normal body temperature of a living human is 37^0 C, determine how many hours ago the murder was committed. (Hint: Use Newton's Law of Cooling (Problem 5). Use the initial condition $T(0) = 30^0$ and $T(2) = 25^0$ to determine k. Then set $T(0) = 37^0$ to determine the time t at which $T(t) = 30^0$.)

7. An ice cube tray with water at 12^0C is placed in a freezer kept at -10^0C. An hour later the temperature of the water is measured at 6^0C. Estimate how much longer it will take until the ice cubes are ready. (Assume this will happen when the water has reached the freezing temperature 0^0C. Use Newton's Law of Cooling (Problem 5).)

III.2   The Inverse Problem and Antiderivatives

III.2.1   *Functions with Zero Derivative*

In order to apply a differential equation such as $y' = ky$ in studying exponential models in applications, as we did in Section 1, one needs to know that *all* its solutions are of the form

$$f(t) = Ce^{kt}.$$

We shall first investigate the analogous, but apparently simpler problem of determining all solutions of the differential equation $y' = 0$. Once this case is well understood, other cases can often be handled by appropriate simple modifications.

Obviously any function $f(x) = c = constant$ satisfies $f' = 0$. Are the constants the only functions with this property? We have not encountered any other function whose derivative is always zero, and—intuitively—it is hard to imagine a *non-constant* function that has zero derivative everywhere. Geometrically, the graph of a function that has always a horizontal tangent appears to necessarily be a horizontal line. Yet attempts to turn these intuitive ideas into precise form run into difficulties. Further analysis reveals that any correct verification of this apparently so "obvious" conclusion ultimately requires the *completeness* of the real numbers. In fact, if all we could "see" were just the rational numbers (realistically, is this not the case?), one could indeed build such strange non-constant functions with zero derivative "everywhere" (i.e., at all visible rational points).

We shall now discuss a process that will allow us to extract a precise verification of what seems so obvious to the eye. In order to motivate our arguments, let us translate the problem into the setting of motion and velocity. So we assume that the function $s = s(t)$ measures the position of a vehicle at time t. Suppose we know that the *instantaneous* velocity $v(t) = \frac{ds}{dt}(t)$ is zero for all t in a time interval I. We want to conclude that $s(t)$ is constant on I, i.e., that there is no motion at all. Alternatively, assume that there is some motion between two points in time t_1 and t_2, i.e., there are $t_1 < t_2 \in I$ with $s(t_1) \neq s(t_2)$; we must then be able to find a time t^* in the time interval $[t_1, t_2]$ at which the *instantaneous* velocity $v(t^*) = ds/dt(t^*) \neq 0$. Our experience tells us that this must indeed be correct, but how do we back up our experience with a solid argument? The only thing we can be absolutely certain about is that the *average* velocity $\frac{s(t_1)-s(t_2)}{t_1-t_2}$ in the time interval $[t_1, t_2]$ is non-zero. We want to use this fact to produce a point t^* with $v(t^*) \neq 0$. Since the instantaneous velocity is

approximated by average velocities over very short time intervals, we need
to find a sequence of *shrinking* intervals, so that the corresponding average
velocities will remain non-zero, and will in fact *converge to a non-zero limit*,
i.e., to a non-zero instantaneous velocity. The crux of the construction is
based on the relationship between average rates of change that we had
already established in Lemma II.1.1. Here we shall just use the following
version of that result.

*If during each of two successive time periods $[t_0, t_1]$ and $[t_1, t_2]$ the av-
erage velocity is less than or equal to v^*, then the average velocity over the
combined period $[t_0, t_2]$ is also less than or equal to v^*.*

Again, this statement is consistent with our experience and completely
"obvious". But in contrast to the earlier situation, no subtle "limits" are
involved here; in fact, the proof of Lemma II.1.1 only required simple algebra. The reader may wish to review the proof of that Lemma before
moving on.

It is now clear how to proceed. We assume that the average velocity
v_0 over a time interval $[c_0, d_0]$ is non-zero, say $v_0 > 0$. Divide the interval
in half. By the observation just made, v_0 cannot exceed the maximum of
the average velocities over each half interval. In other words, the average
velocity v_1 over at least one of these half intervals must be at least as large
as v_0, i.e., $v_1 \geq v_0$. Label that half by $[c_1, d_1]$. Now repeat this process
starting with v_1 and the interval $[c_1, d_1]$, and then repeat over and over.
At the *nth* step one obtains an interval $[c_n, d_n] \subset [c_{n-1}, d_{n-1}] \subset [c_0, d_0]$ of
length $(d_0 - c_0)/2^n$, so that the average velocity v_n over $[c_n, d_n]$ satisfies
$v_n \geq v_{n-1} \geq v_0$. The nested interval theorem (Theorem I.1.3), which is a
consequence of the completeness of $\mathbb{R}$, guarantees that there is at least one
point t^* contained in all these intervals. Then the instantaneous velocity
$v(t^*)$, being the limit of average velocities $v_n \geq v_0$ over shorter and shorter
time intervals shrinking to t^*, must also be greater than or equal to $v_0 > 0$,
and hence $v(t^*) \neq 0$ as needed. (If desired, the last (intuitive) argument
can be made rigorous by invoking the precise limit definition of derivatives
combined with the observation above to pass from average velocities over
$[c_n, d_n] = [c_n, t^*] \cup [t^*, d_n]$ to intervals with one endpoint at t^*.)

What if $v_0 < 0$? Then the preceding argument still gives $v(t^*) \geq v_0$,
although now this does not imply $v(t^*) \neq 0$. Yet surely the whole argument
can be modified to find another $t^{\#}$ with $v(t^{\#}) \leq v_0$. We therefore have
verified the desired conclusion: If the average velocity over a time interval

is not zero, then at some time during that interval the instantaneous velocity has to be non-zero as well.

III.2.2 *The Mean Value Inequality*

Let us recast the conclusion we just obtained in the setting of an arbitrary differentiable function. Given such a function f defined on an interval I, we define the average rate of change of f over $[a, b] \subset I$, where $a < b$, by

$$\Delta(f, [a, b]) = \frac{f(b) - f(a)}{b - a}.$$

The argument we just went through (in the language of velocities) proves the second inequality in the following theorem. The first inequality follows by applying that result to $-f$ in place of f. We refer to this result as the *Mean Value Inequality*.

Theorem 2.1. *(**Mean Value Inequality**) Assume that f is differentiable on I. If $[a, b] \subset I$, then there exist x_{low} and $x_{high} \in [a, b]$ such that*

$$D(f)(x_{low}) \leq \Delta(f, [a, b]) \leq D(f)(x_{high}).$$

As we saw already at the end of the previous section, the result that prompted the whole discussion follows immediately.

Corollary 2.2. *If f is differentiable on the interval I with $f' \equiv 0$ on I, then f is constant.*

Proof. By the Theorem, the hypothesis implies $0 \leq \Delta(f, [a, b]) \leq 0$, i.e., $\Delta(f, [a, b]) = 0$, and hence $f(b) = f(a)$ for all $a, b \in I$. We have thus verified that the only solutions of the differential equation $y' = 0$ are indeed just the constant functions. ∎

Corollary 2.3. *If two differentiable functions f_1 and f_2 satisfy $f_1'(x) = f_2'(x)$ for all x in an interval I, then f_1 and f_2 differ by a constant, i.e., there is $C \in \mathbb{R}$ such that $f_2(x) = f_1(x) + C$ for all $x \in I$.*

Proof. The function $f_2 - f_1$ has derivative $(f_2 - f_1)' = f_2' - f_1' = 0$, and hence is a constant C. ∎

We therefore know *all* solutions to the differential equation $y' = g$ as soon as we know *one* solution.

Examples.

i) Find all functions that satisfy $y' = 2x^4$. By the standard differentiation rules, $(ax^5)' = a5x^4$ for any constant a. Choose a so that $5a = 2$, i.e., $a = 2/5$. Then the function $f(x) = \frac{2}{5}x^5$ satisfies $f'(x) = 2x^4$, so is one solution. All other solutions are therefore of the form $\frac{2}{5}x^5 + C$, where C is a constant.

ii) All solutions of $y' = \sin x$ are of the form $f(x) = -\cos x + C$.

iii) All solutions of $y' = 3^x$ are of the form $g(x) = \frac{1}{\ln 3}3^x + C$.

Another important consequence of Theorem 2.1 is the following estimate.

Corollary 2.4. *Suppose f is differentiable on I and that its derivative $D(f)$ is bounded on the interval $[a, b] \subset I$, i.e., there exists K such that $|D(f)(x)| \leq K$ for $x \in [a, b]$. Then*

$$|f(x_1) - f(x_2)| \leq K|x_1 - x_2|$$

for all $x_1, x_2 \in [a, b]$.

Proof. Since $|D(f)(x_{low})| \leq K$ and $|D(f)(x_{high})| \leq K$, the Mean Value Inequality implies $|\Delta(f, [x_1, x_2])| \leq K$. The desired estimate then follows by multiplying with $|x_1 - x_2|$.   ■

Finally, the Mean Value Inequality easily implies also the following theorem, which is known as the *Mean Value Theorem*.

Corollary 2.5. (Mean Value Theorem) *Suppose f is differentiable on the interval I and that its derivative $D(f)$ is continuous on I. Given $[a, b] \subset I$, there exists a point $c \in [a, b]$ such that*

$$D(f)(c) = \Delta(f, [a, b]) = \frac{f(b) - f(a)}{b - a}.$$

Proof. By the Mean Value Inequality one has

$$D(f)(x_{low}) \leq \Delta(f, [a, b]) \leq D(f)(x_{high}).$$

Since $D(f)$ is assumed to be continuous, the Intermediate Value Theorem II.4.4 gives the existence of the desired solution c of the equation $D(f)(x) = \Delta(f, [a, b])$.   ■

III.2.3 *Antiderivatives*

We introduce a new name and notation to describe a generalization of the equation $D(F)(x) = 0$ for all x in some interval, as follows. A function F

on an interval I is called an *antiderivative of f* if F is differentiable and $F'(x) = f(x)$ for all $x \in I$. An antiderivative of f is denoted by the symbol $\int f(x)dx$. Therefore

$$F = \int f(x)dx \iff F' = f.$$

Sometimes the symbol $\int f(x)dx$, also called *an indefinite integral of f*, is used to denote the collection of *all* antiderivatives of f.

By simply reversing known rules of differentiation one obtains the following formulas for antiderivatives.

i) $\int e^x dx = e^x + C$;

ii) $\int b^x dx = \frac{1}{\ln b} b^x + C$, for any $b > 0$ and $\neq 1$;

iii) $\int \sin x \, dx = -\cos x + C$; $\int \cos x \, dx = \sin x + C$;

iv) $\int x^r dx = \frac{1}{r+1} x^{r+1} + C$ for $r \neq -1$ and $x > 0$. In particular, if n is a positive integer, then $\int x^n dx = \frac{1}{n+1} x^{n+1} + C$ for all $x \in \mathbb{R}$.

v) $\int \frac{1}{x} dx = \ln x + C$ for $x > 0$; $\int \frac{1}{x} dx = \ln(-x) + C$ for $x < 0$.

vi) $\int [af(x) + bg(x)]dx = a \int f(x)dx + b \int g(x)dx$ for any constants a, b.

We shall see in Chapter IV that *every* continuous function f on an interval I has an antiderivative on I.

III.2.4 *Solutions of $y' = ky$*

It is now easy to also describe *all* solutions of the differential equation $y' = ky$. Suppose that $y = f(x)$ is a solution of this equation on an interval I, i.e., f satisfies $f'(x) = kf(x)$ for all $x \in I$. We want to show that $f(x) = Ce^{kx}$ for some constant C. We therefore consider the function h defined by

$$h(x) = f(x)/e^{kx} = f(x)e^{-kx}.$$

Differentiation (use the product rule, the differential equation for f, and the chain rule) gives

$$h'(x) = f'(x)\,e^{-kx} + f(x)[e^{-kx}]'$$
$$= kf(x)\,e^{-kx} + f(x)[e^{-kx}](-k)$$
$$= e^{-kx}[kf(x) - kf(x)] = 0.$$

By Corollary 2.2 above, h is a constant C and we are done. In particular, it follows that every solution of $y' = ky$ is defined on the whole real line.

Remark. Notice that if a solution $f(x) = Ce^{kx}$ of $y' = ky$ takes on the value 0 at some point x_0, then necessarily $C = 0$, since $e^{kx} \neq 0$ for all x.

Therefore $f(x) = 0$ for all x. So the only solution that takes on the value zero at some point is the constant function $f \equiv 0$. All other solutions are never zero.

Let us summarize the main conclusions.

Theorem 2.6. *Let f be a solution of the equation $y' = ky$ on the interval I. Then there exists a constant C, so that $f(x) = Ce^{kx}$ for all $x \in I$. If $f(x_0) = 0$ for some point $x_0 \in I$, then $f(x) = 0$ for all $x \in I$.*

Note that for $k = 0$ this theorem includes the earlier result that a function whose derivative is always zero is necessarily constant.

III.2.5   *Initial Value Problems*

We saw in the preceding section that the solutions of differential equations such as $y' = g(x)$ and $y' = ky$ are determined up to a constant. By prescribing the value y_0 of a solution at one fixed point x_0 one obtains an additional condition that typically will be satisfied by one and only one choice of that constant. In this way one singles out a particular solution. Combining the differential equation with such a choice (x_0, y_0) determines what is called an *initial value problem*.

Example. Solve the initial value problem $y' = 2\cos x$ with $y(\pi/2) = 1$.
Solution. The differential equation has solutions $\int 2\cos x\, dx = 2\sin x + C$. The initial value condition requires

$$2\sin\frac{\pi}{2} + C = 1,$$

or $2 + C = 1$, so that $C = 1 - 2 = -1$. So the desired solution f is given by $f(x) = 2\sin x - 1$.

Geometrically, we see that the graphs of the family of all solutions of $y' = 2\cos x$ are obtained by parallel translation in the vertical direction (i.e., by adding a constant to the y-coordinate) of the graph of a particular solution. Specifying an initial value (x_0, y_0) therefore selects the one graph that goes through that particular point.

Example. Describe, for varying times, the number of bacteria in a culture that is of size 2000 at 1 p.m. and which doubles every 5 hours. Use an exponential growth model.
Solution. Let $P(t)$ denote the size of the culture, where t is measured in hours, so that $t = 0$ corresponds to 1 p.m. Assuming an exponential

growth model, one has $P' = kP$ for a constant k. The solutions are of the form $P(t) = Ce^{kt}$. The initial value condition implies $2000 = P(0) = C$. To determine k we use the information $P(5) = 2P(0)$; this implies $e^{k5} = 2$. Hence $k5 = \ln 2$, so that $k = (\ln 2)/5 = .13862\,944 \approx 0.139$. The desired function that describes the number of bacteria at time t therefore is

$$P(t) = 2000\, e^{0.139\,t}.$$

If we want to describe the populations in terms of the time T given by the clock, note that $T = t + 1$, or $t = T - 1$. Hence $P(T) = 2000e^{0.139\,(T-1)}$.

This example generalizes easily to the following result, which is the prototype of the general existence and uniqueness theorem for initial value problems.

Theorem 2.7. *Given an arbitrary point (x_0, y_0) in the plane, there exists exactly one solution f of the differential equation $y' = ky$ defined on $\mathbb{R}$ that satisfies the initial value condition $f(x_0) = y_0$.*

Proof. We know that any solution f is of the form $f(x) = Ce^{kx}$. The condition $y_0 = f(x_0) = Ce^{kx_0}$ implies that $C = y_0/e^{kx_0} = y_0 e^{-kx_0}$. So

$$f(x) = y_0 e^{-kx_0} e^{kx} = y_0 e^{k(x-x_0)}$$

is the (unique) solution to the initial value problem. ∎

III.2.6 *Exercises*

1. Let $f(t) = \cos(t)$
   a) Determine the average rate of change $R = \Delta(f, [0, \pi])$ of f over the interval between $t = 0$ and $t = \pi$.
   b) Find specific points l and m in $[0, \pi]$ so that $f'(l) \le R \le f'(m)$.
   c) Is there any subinterval $[a, b] \subset [0, \pi]$, with $\Delta(f, [a, b]) > 0$? Explain your answer!

2. Let $p(x) = 4x^3 - 2x^2 + 1$.
   a) Find *all* antiderivatives of $p(x)$.
   b) Find a function F so that $F(1) = 0$ and $F'(x) = p(x)$.

3. Suppose the functions f and g satisfy $f'(x) = 4g'(x)$ for all x. If $f(0) = g(0) = 1$, and $f(10) = 5$, what is $g(10)$?

4. Solve the following initial value problems.
   a) $y' = 2\cos x$, $y(0) = 1$;
   b) $y' = 3^x$, $y(0) = 0$;

c) $y' = 2y$, $y(0) = 10$;

d) $y' = 3y$, $y(0) = 0$;

e) $y' = -y$, $y(2) = 3$.

5. A particle moves along the graph of a function f, so that at each point $(x, f(x))$ on the trajectory the tangent has slope $4x$. Assume the particle goes through the point $(1, 2)$. Determine the function f.

6. Use a graphing calculator to sketch in one window the graphs of the antiderivatives of $g(x) = 3x^2$ that go through the points $(0, -1), (0, 0), (0, 1), (0, 2)$. Describe the geometric relationship of the graphs.

7. Find a function F that satisfies the equation $F' = kF$ for some constant k, and so that $F(0) = 2$ and $F(1) = 5$.

III.3 "Explosive Growth" Models

III.3.1 *Beyond Exponential Growth*

The basic model underlying exponential growth is described by the differential equation $y' = ky$, where $k > 0$ is constant. It implies that both $y(t)$ and the rate of growth $dy/dt = y'$ increase in time. The differential equation requires that dy/dt and y are just a fixed constant multiple of each other. The rate of growth of dy/dt is measured by its derivative $D(dy/dt) = D(ky) = kD(y)$, so the rates of growth $D(dy/dt)$ and $D(y)$ of dy/dt and y are still proportional, with the same factor k. Furthermore, $D^{(2)}(y) = D(ky) = k^2y$. Continuing to differentiate, one sees that for each positive integer n one has

$$D^{(n)}(y) = kD^{(n-1)}(y) = k^2 D^{(n-2)}(y) = \dots = k^n y.$$

Thus there is a *linear* relationship between any two derivatives of a solution. Each derivative of y still satisfies an exponential growth model in relationship to y, although the relevant constant k^n changes according to the number of differentiations involved. This situation expresses a deep regularity of the underlying process. As we know, the solutions are given by exponential functions that are defined for all values of t, and unless external conditions change, the growth process continues indefinitely, leading to $\lim_{t \to \infty} y(t) = \infty$ as soon as $y(t_0) > 0$ at some moment in time.

  The situation changes dramatically if one considers a *non-linear* growth model described, for example, by the differential equation $y' = ky^2$, where

k is again constant. Here y' grows much faster than y. The relationship is not linear but quadratic. To simplify, let us choose $k = 1$. We want to differentiate the equation again. Since the solution $y(t)$ is assumed to be differentiable, it follows that $y'(t) = [y(t)]^2$ is differentiable as well, and therefore $D(y') = D([y(t)]^2) = 2y(t) \cdot D(y)$ by the chain rule. The differential equation then implies that

$$D(y') = 2y \cdot y' = 2y \cdot y^2 = 2y^3.$$

Thus $D(y') = D^2(y)$ grows very much faster in comparison to $D(y) = y'$ as y increases. By differentiating again, and so on, one obtains that

$$D^{(n)}(y) = n \cdot (n-1) \cdot \ldots \cdot [y]^{n+1}$$

for $n = 2, 3, \ldots$. We see that the nth derivative of any solution $y(t)$ must grow like the power y^{n+1}. In contrast to the exponential model $y' = y$, the derivatives in this simple non-linear model grow progressively faster, far exceeding the rate of natural growth of exponential models. This suggests a most unusual behavior of the solutions.

As we will show, there exists a critical point in time T_c that depends on the initial conditions, such that the corresponding solution is defined only for $t < T_c$. Furthermore, the process literally blows up as t approaches the critical point T_c.

III.3.2 *An Explicit Solution of* $y' = y^2$

It turns out that we can quite easily determine a formula for the solution of any initial value problem related to the non-linear differential equation $y' = y^2$. To be specific, suppose that $f(t)$ is a solution defined near the point $t = 0$, and that $f(0) = 1$. By continuity of f it follows that $f(t) > 0$ on a sufficiently small interval I centered at 0. Hence the differential equation $D(f) = f^2$ can be written in the form

$$\frac{D(f)(t)}{f(t)^2} = 1 \text{ for } t \in I.$$

By the reciprocal rule for derivatives we see that $h(t) = -1/f(t)$ is an antiderivative of the left side on the interval I. Since $\int 1 dt = t$ is also an antiderivative of $D(f)/f^2$, it follows that there exists a constant C such that

$$h(t) = t + C \text{ for } t \in I,$$

that is, we have verified that $-1/f(t) = t + C$. The initial value condition implies that $C = -1/f(0) = -1$, and therefore $1/f(t) = 1 - t$. Since $f(t) > 0$ for $t \in I$, we must have $t < 1$ for $t \in I$. It follows that

$$f(t) = \frac{1}{1-t} \qquad (\text{III.1})$$

is the unique solution of $y' = y^2$ with $f(0) = 1$ on the interval I. The formula (III.1) shows that the solution f initially defined on the interval I has a natural extension to the interval $\{t : t < 1\}$ which satisfies the differential equation for all $t < 1$. In fact, the expression on the right side of (III.1) is the *only* extension of the solution f from the interval I to the interval $(-\infty, 1)$ that continues to satisfy the differential equation $y' = y^2$. (See Problem 2 of Exercise III.3.3 for more details.) Furthermore, (III.1) shows that $f(t)$ cannot be extended in any meaningful way to $t = 1$, since $f(t) \to \infty$ as $t \to 1$ from the left side. Thus the solution $f(t)$ "blows up" as t approaches the critical value $T_c = 1$. Figure III.1 shows the "explosive" behavior of the graph of $f(t)$ as $t \to 1^-$. Simple modifications of these techniques show that all solutions of $y' = ky^r$ exhibit similar explosive behavior at corresponding critical points as soon as the exponent r is greater than 1. The (linear) exponential model $y' = ky$ thus is the borderline case beyond which *non-linear* growth exhibits very different properties.

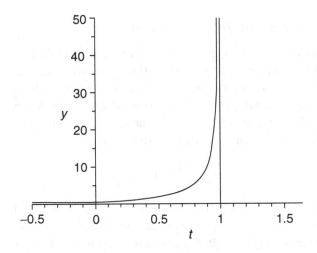

Fig. III.1   The solution $f(t)$ blows up as t approaches 1.

III.3.3 *Exercises*

1. Modify the arguments given in the text for the initial value $f(0) = 1$ to show that if $b \neq 0$, then the initial value problem $y' = y^2$ with $y(t_0) = b$ has a unique solution on some interval I centered at t_0.

2. a) Verify that $g(t) = 1/(1 - t)$ satisfies $g'(t) = g(t)^2$ for $t < 1$ and $g(0) = 1$.

   b) Let f be any solution of the initial value problem in a) on the interval $(-\delta, 1)$ for some $\delta > 0$. Let $\Lambda = \{\lambda \in (-\delta, 1) : f(t) = g(t)$ for all $t \in (-\delta, \lambda)\}$. Show that Λ is not empty. (Hint: Modify the argument in the text.)

   c) Let $\lambda^* = \sup \Lambda$. Suppose that $\lambda^* < 1$. Show that $f(\lambda^*) = g(\lambda^*) \neq 0$ and that any solution $h(t)$ of the initial value problem $y' = y^2$ with $h(\lambda^*) = g(\lambda^*)$ must agree with $g(t)$ for all t in some interval $I = (\lambda^* - \varepsilon, \lambda^* + \varepsilon)$ centered at λ^*. (Hint: Use Problem 1).

   d) Show that c) implies that $f(t) = g(t)$ for all $-\delta < t < \lambda^* + \varepsilon$.

   e) Show that the result in d) implies that $\lambda^* = 1$, so that $f(t) = g(t)$ for all $t < 1$. (Hint: Explain why the conclusion in d) contradicts the assumption $\lambda^* < 1$ made in c).)

3. a) Given $r > 1$, find a solution $f_r(t)$ for the initial value problem $y' = y^r$ with $y(0) = 1$ on some interval centered at 0.

   b) Determine the critical point T_r for f_r, so that $f_r(t)$ is defined for all $t < T_r$ and $\lim_{t \to T_r^-} f_r(t) = \infty$.

4. Follow the steps below to show that if $f(t)$ is a solution of $y' = y^2$ on some interval I centered at 0 which satisfies $f(0) = 0$, then there exists an interval $[-\delta, \delta] \subset I$ such that $f(t) = 0$ for all t with $|t| \leq \delta$. Thus this initial value problem has the *unique* solution $f(t) = 0$ for all t near 0.

   a) Show that there exists $0 < \delta < 1/2$, such that $|f(t)| < 1/2$ for all t with $|t| \leq \delta$.

   b) Show that if for some $k \in \mathbb{N}$ one has $|f(t)| \leq (1/2)^k$ for all t with $|t| \leq \delta$, then $|f(t)| \leq (1/2)^{2k+1}$ for all t with $|t| \leq \delta$. (Hint: Use the differential equation to first estimate $|f'(t)|$, and then apply the Mean Value Inequality to estimate $|f(t)|$ for $|t| \leq \delta < 1/2$.)

   c) Show that a) and b) imply that $|f(t)| = 0$ for $|t| \leq \delta$.

5. Consider the differential equation $y' = y^{1/2}$. Show that there are two *different* solutions of this equation on $\mathbb{R}$ that satisfy the initial value condition $y(0) = 0$. Hence *uniqueness* of solutions fails for this initial value problem. (Hint: Clearly $g(t) = 0$ for all t solves the initial value

problem. Show that the function f defined by $f(t) = (1/4)t^2$ for $t \geq 0$ and $f(t) = 0$ for $t < 0$ is differentiable at $t = 0$ and (trivially) at all other points, and that it also solves the initial value problem.)

III.4　Acceleration and Motion with Constant Acceleration

III.4.1　*Acceleration*

Given a motion described by the position function s which measures the distance from a fixed point in dependence of time t, we had seen that the instantaneous velocity $v(t)$ at time t is defined by the derivative $D(s)(t)$. It is convenient to use the notation $\frac{ds}{dt}(t)$ for this and similar derivatives to remind us that the derivative measures a rate of change.

The rate of change of *velocity* is known as the **"acceleration"** of the motion. Again, one distinguishes between the *average* acceleration

$$\frac{v(t_2) - v(t_1)}{t_2 - t_1}$$

during the time interval $[t_1, t_2]$, where $t_1 \neq t_2$, and the *instantaneous* acceleration $a(t_0)$ at time $t = t_0$ defined by the derivative $D(v)(t_0) = \frac{dv}{dt}(t_0)$ of the velocity function $v(t)$ at t_0. This means that $a(t_0)$ equals the limit of the *average* accelerations, i.e.,

$$a(t_0) = \lim_{t \to t_0} \frac{v(t) - v(t_0)}{t - t_0}.$$

We see that the acceleration at t_0 is approximated very well by the average acceleration over shorter and shorter time intervals ending at t_0. The derivative dv/dt of the velocity function is the *second* order derivative of the position function s, so that

$$a(t) = \frac{d}{dt}\left(\frac{ds}{dt}\right)(t) = s''(t) = D^2(s)(t).$$

Example. A boat floating on the sea bounces up and down with the waves. Suppose the motion is described by the function $h(t) = 12\sin(0.25t)$, where $h(t)$ measures the amount in feet that the boat rises above or drops below a fixed level. At what times does the acceleration have maximal absolute value? What is that maximal acceleration?

Solution. By differentiation one obtains $h'(t) = 12(.25)\cos(0.25t)$ and $h''(t) = -12(.25)^2\sin(0.25t)$. Hence $|h''(t)| = \frac{3}{4}|\sin(0.25t)|$. The maximal value $3/4\,\text{ft/sec}^2$ occurs when $|\sin(.25t)| = 1$. The motion is "roughest" at those moments when the boat is on top of a wave or in the valley between two waves.

III.4.2 *Free Fall*

To illustrate how to apply the general relationship between position, velocity, and acceleration established by the process of differentiation, we shall consider the simplest case when the acceleration is *constant*. In particular, as a consequence of basic physical principles, we shall recover the results of Galileo's investigations about falling objects that we had already discussed in Section 4 of the Prelude. We consider an object (say a rock falling off a cliff) falling towards the ground. We denote by $s(t)$ its height above ground at time t. Neglecting all perturbations due to wind or air resistance, the only force that acts on such an object is the gravitational force F_g exerted by the earth. At heights that are very small compared to the radius of the earth, this force can be assumed to be constant, i.e., independent of the height of the rock. According to Newton's law of motion *force = mass × acceleration*, this force causes an acceleration a (i.e., a change in velocity) on the object, which therefore must also be constant near the surface of the earth. This acceleration a due to earth's gravity is usually denoted by $-g$ with $g > 0$, where the minus sign makes explicit the fact that the force pulls downwards, so that the height is decreasing. Depending on the units chosen, the numerical value of the constant g is 9.81 m/sec^2 or 32 ft/sec^2.

Given information about the acceleration a, we can now determine the position function s by using the tools of calculus. Let $v = ds/dt$ be the velocity of the object. Since $v'(t) = a$, the velocity is an antiderivative of the acceleration. By the results in Section 2.3 one has

$$v(t) = \int a(t)dt = \int -gdt = -gt + C$$

for some constant C, whose value $C = v(0)$ is the "initial velocity" v_0 at time $t = 0$. In particular, it then follows from the above formula that $v(t_2) - v(t_1) = (-g) \cdot (t_2 - t_1)$, that is, the motion of a freely falling body is "uniformly accelerated". This, of course, is the fundamental fact discovered by Galileo early in the 17th century. Next, since $s(t)$ is an antiderivative $\int v(t)\, dt$ of the velocity, it follows that

$$s(t) = \int (-gt + v_0)dt = -g\frac{t^2}{2} + v_0 t + s_0,$$

where $s_0 = s(0)$ is the initial position (at time $t = 0$). Therefore the motion is completely determined by this equation once one knows the values of *initial* position and velocity.

Example. A stone is dropped from the top of a tower, and it hits the ground 4 seconds later. Find the height of the tower.

The model for the motion under constant acceleration applies. Let h denote the (unknown) height of the tower in meters. This is the initial position at time $t = 0$. Since the stone is simply "dropped" at that time, the initial velocity $v_0 = 0$, so that

$$s(t) = -\frac{9.81}{2}t^2 + h \quad \text{meters.}$$

When the stone hits the ground ($t = 4$), the height $s(t)$ is zero, so that $s(4) = -\frac{9.81}{2}4^2 + h = 0$. Solving for h gives

$$h = \frac{9.81}{2} \cdot 16 = 75.48.$$

The tower is approximately 75 meters high.

III.4.3 *Constant Deceleration*

We discuss another situation where constant acceleration occurs.

Example. A car travels along a highway at 50 miles/hour. The driver sees a washed out bridge approximately 100 ft down the road, and immediately applies the brakes with constant pressure. After one second his speed is down to 35 miles/hour. Will the car stop in time?

Solution. Based on the given information, we assume that the car decelerates at a constant rate of a ft/sec^2. Let $t = 0$ correspond to the time when the brakes are first applied, measured in seconds. Conversion of the initial speed of 50 miles/hour to ft/sec gives $50 \cdot 5280/3600 \approx 73.3$ ft/sec. Thus $v(t) = \int -a\,dt = -at + 73.3$. (The minus sign reflects the fact that the braking action slows down the car, which we interpret as a negative acceleration.) Since we are told that

$$v(1) = 35\,\text{m/h} = 35 \cdot 5280/3600 \approx 51.3 \text{ ft/sec},$$

we can determine the deceleration rate a from the equation $51.3 = -a \cdot 1 + 73.3$, resulting in $a = 22$. Next we can determine the time required for the car to stop (assuming that there is no interruption) from the equation $0 = v(t) = -22t + 73.3$. One obtains $t = 73.3/22 \approx 3.3$ seconds. We can now calculate the distance the car would travel until coming to a stop as follows. The position function $s(t)$ satisfies

$$s(t) = \int v(t)\,dt = -22\frac{t^2}{2} + 73.3t + 0.$$

Hence $s(3) = -22 \cdot 9/2 + 73.3 \cdot 3 = 120.9$. Unfortunately, the car will not stop in time before falling over the edge... if only the driver had replaced his worn tires the day before!

III.4.4 *Exercises*

1. The specifications for a shipping box state that the box should withstand an impact against a fixed object up to a speed of 10 ft/sec. What would happen to the box if it is dropped from a balcony 20 ft above the ground?

2. While traveling at 95 miles per hour in your Ferrari you spot a state trooper in the distance and immediately apply the brakes with constant pressure. After 100 ft your speed is down to 70 m.p.h. Continuing with the same deceleration, what will be your speed when you pass the trooper who is an additional 100 ft away? Will you get a ticket? (60 m.p.h. equals 88 ft/sec.)

3. A car accelerates under constant acceleration from 0 to 100 km/h in 6 seconds. Find the acceleration in m/sec^2.

4. A cannon fires its ammunition straight up with an initial speed of 60 m/sec.

   a) How high will the cannon ball reach?
   b) How long will it take for the cannon ball to hit the ground again? (Better move away...).
   c) At what speed does the cannon ball hit the ground?

5. A stone dropped from a tower takes 6 seconds to hit the ground. How high is the tower?

6. On a snowy afternoon Joe Q. travels on the very slippery interstate highway at a speed of 60 ft/sec (roughly 40 m.p.h.), when he suddenly sees a big truck ahead losing control, rolling over and blocking the road. He applies the brakes, but because of the snow his deceleration is just $10\,ft/sec^2$.

   a) Find the speed of Joe's car $t > 0$ seconds after he applies the brakes.
   b) How many seconds would it take for the car to come to a stop?
   c) When Joe applied the brakes, the truck was about 200 ft away. Will Joe be able to stop before hitting the truck? Explain!

III.5 Periodic Motions

III.5.1 *A Model for a Bouncing Spring*

Suppose a steel ball of mass m is attached to a spring that hangs from the ceiling. The weight will stretch the spring by an amount s_0, at which point the ball will be at rest at the equilibrium position. Suppose the ball is now pulled down an additional amount c_0 and released; it will then bounce up and down around the equilibrium point. (See Figure III.2.) We want to find a mathematical model to describe the motion of the ball under the action of the spring. As the motion appears to involve some periodicity, we expect that the model will involve trigonometric functions or perhaps other more complicated periodic functions.

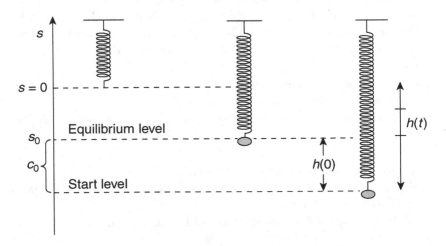

Fig. III.2 The mass stretches the spring by $|s_0|$.

At the equilibrium level $s = s_0$ the ball is at rest. (Recall that according to our choice of orientation, s_0 is a *negative* number.) So the sum $F(s_0)$ of all forces acting on the ball at that position is zero. One force is the gravitational force F_G, which equals $-mg$. The other force F_S comes from the spring. According to Hooke's Law from physics, that force is a multiple $-ks$ of the amount s that the spring has been stretched, where $k > 0$ is the so-called spring constant. In particular, $F_S(s_0) = -ks_0$. Since $s_0 < 0$, one has $F_S(s_0) > 0$, which is consistent with the fact that the spring pulls the ball upwards. The equation $F(s_0) = F_G(s_0) + F_S(s_0) = -mg + (-ks_0) = 0$ then implies that $-mg = ks_0$. Incidentally, this result gives a practical method to determine the spring constant k: If a mass m stretches the spring by the amount $-s_0$, then $k = -mg/s_0 = mg/(-s_0)$. If one uses the standard metric units kilograms for mass, meters for length, and seconds for time, the spring constant k is measured in kg/sec^2. For example, if a mass of 1 kg stretches the spring by 10 cm ($= 0.1$ m), then $k = 1 \cdot 9.81/0.1 = 98.1$ kg/sec^2.[2]

Rather than focusing on the level $s(t)$ of the mass, we shall consider the (signed) distance $h(t) = s(t) - s_0$ of the mass from the equilibrium level given by s_0. Since $h(t)$ and $s(t)$ differ by a constant, one has $h''(t) = s''(t)$. Newton's law of motion and Hooke's law then imply

$$mh''(t) = ms''(t) = F_G + F_S(s(t)) = -mg - ks(t) = ks_0 - ks(t)$$
$$= -k(s(t) - s_0) = -k \cdot h(t)$$

at time t. Therefore the function $h = s - s_0$ satisfies

$$h''(t) + \frac{k}{m}h(t) = 0.$$

If we set $\omega = \sqrt{k/m}$, the position function $h(t)$ relative to the equilibrium point of the bouncing ball must satisfy the differential equation $y'' + \omega^2 y = 0$.

The main result proved in the next section implies that

$$h(t) = A\sin(\omega t) + B\cos(\omega t)$$

for some constants A and B. The constants are determined by the initial position and velocity of the ball at time $t = 0$. In fact, the above equation implies that $h(0) = B$ and $h'(0) = A\omega$. Concretely, suppose the experiment is started by pulling the ball down from the equilibrium level by an amount $c_0 > 0$ and then releasing it at time $t = 0$, so that its velocity right at that moment is zero. This means that $h(0) = -c_0$ and $h'(0) = 0$. Consequently, given these initial conditions, the motion of the ball is described by the function

$$h(t) = -c_0\cos(\omega t) = -c_0\cos(\sqrt{k/m}\,t).$$

Figure III.3 qualitatively visualizes the vertical displacement $h(t)$ of the ball as a function of time t in seconds. The constant c_0 in this equation is

---

[2]The unit kg/sec^2 for the spring constant k may appear quite unnatural. If one introduces a separate unit to measure forces, one obtains a different description for k. More precisely, one defines 1 *Newton* (1 N) to be the size of a force that accelerates a mass of 1 kg by 1 m/sec^2. The unit kg/sec^2 translates to kg·(m/sec^2)/m = N/m. This latter unit for k more directly reflects Hooke's law $F = -k\,s$, which implies $k = -F/s$.

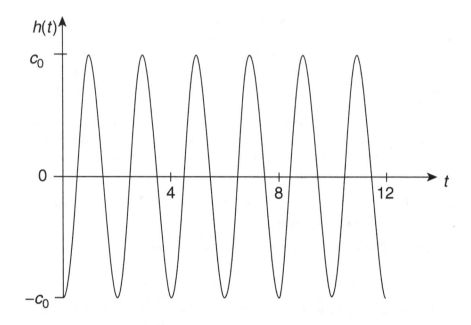

Fig. III.3   The motion of the spring over time t.

called the *amplitude* of the (periodic) motion. It measures the maximum displacement from the equilibrium position. The other quantity that is often used to describe periodic motions is the so-called *frequency* ν, which gives the number of cycles per unit of time. The frequency thus is an aggregate measure of the speed by which the ball bounces up and down, sort of like an average speed.

The reciprocal $T = 1/\nu$ of the frequency is the *period* of the motion, i.e., T is that time it takes the ball to complete a full cycle to return to the initial position at the bottom. In Figure III.3 the period T equals 2 seconds and hence the frequency ν is equal to $1/2$ cycles per second. Since the cosine function has a period 2π, the period T of the bouncing mass must satisfy the equation $\sqrt{k/m}\,T = 2\pi$, that is, $T = 2\pi\sqrt{m/k}$. The result shows, in particular, the effect of the spring constant k and the mass m on the motion of the ball. The period increases if the mass m of the ball is increased. Similarly, if one uses balls of equal weights with two springs of different stiffness, the equation shows that the period for the stiffer spring (i.e., larger k) will be shorter than the one for the softer spring. Stated differently, a stiffer spring produces a much quicker bounce, i.e., higher frequency, than a softer spring.

Remark. A more accurate model of the bouncing spring needs to take into account other factors, such as the resistance encountered by the motion of the ball in the surrounding medium and internal friction of the spring. The main new effect is that energy is lost over time, thereby slowing down the motion until the ball eventually comes to a stop. The resistance gives an additional force F_R acting on the ball. The simplest model assumes that this force is proportional to the velocity y' and acts in the direction opposite to the direction of the motion, i.e., $F_R = -Ry'$ for some constant $R > 0$. The total force $F = my''$ now satisfies $F = -mg + F_R + F_S$. The resulting differential equation for the displacement $y(t)$ from equilibrium then is

$$my'' + Ry' + ky = 0,$$

where all constants are positive. The analysis of the solutions for this equation when $R > 0$ is more involved than the case when $R = 0$. The reader interested in more details should consult a basic text on Differential Equations.

III.5.2 *The Solutions of* $y'' + \omega^2 y = 0$

We saw in the preceding section that the function h that describes the motion of a ball attached to a spring must be a solution of the differential equation $y'' + \omega^2 y = 0$. Recall that $\sin t$ and $\cos t$ satisfy the differential equation $y'' + y = 0$. More generally, we had already remarked in Section II.5.5 that all functions f of the form $f(t) = A \sin(\omega t) + B \cos(\omega t)$, where A and B are constants, satisfy the corresponding equation

$$y'' + \omega^2 y = 0. \tag{III.2}$$

In order to conclude that the position function h of the spring is of this particular form one needs to know that the functions f of the form given above do indeed represent *all* possible solutions of the equation (III.2). The analogous problem of identifying all solutions for the differential equation $y' = ky$ was handled in Section III.2.4 by reducing the problem to the simpler equation $y' = 0$. Similarly, the equation $y'' + \omega^2 y = 0$ can be reduced to that form by a simple procedure. We multiply the equation (III.2) by the derivative y' to obtain

$$y'' y' + \omega^2 y y' = 0.$$

Note that by the chain rule one has $D([y(t)]^2) = 2y(t)D(y(t)) = 2yy'$. Similarly, $D([y'(t)]^2 = 2y'(t)D(y'(t)) = 2y'y''$. Combining these two equations results in

$$D([y'(t)]^2 + \omega^2[y(t)]^2) = 2y'y'' + \omega^2 2yy'$$
$$= 2y'[y'' + \omega^2 y].$$

Therefore, if the function y is a solution of (III.2), then $D([y'(t)]^2 + \omega^2[y(t)]^2) = 0$. This easily implies the following statement.

Lemma 5.1. *Let y be any solution of the differential equation $y'' + \omega^2 y = 0$ on the interval I. Then*

$$[y'(t)]^2 + \omega^2[y(t)]^2 \text{ is constant for all } t \in I.$$

In particular, any such solution y that satisfies $y(t_0) = y'(t_0) = 0$ at some point $t_0 \in I$ must satisfy $y(t) = 0$ for all $t \in I$.

Proof. We saw that for any solution y of (III.2) the function $[y'(t)]^2 + \omega^2[y(t)]^2$ has zero derivative on I. Hence it must be a constant C on I. The conditions $y(t_0) = y'(t_0) = 0$ imply that $C = 0$. Since $[y'(t)]^2 \geq 0$ and $\omega^2[y(t)]^2 \geq 0$, their sum can be zero only if the terms are zero individually. It follows that $\omega^2[y(t)]^2 = 0$, and hence $y(t) = 0$ for all $t \in I$. ∎

We can now readily prove the main result of this section.

Proposition 5.2. *If the function f satisfies the differential equation $y'' + \omega^2 y = 0$ on the interval I, then there exist constants A and B such that*

$$f(t) = A\sin(\omega t) + B\cos(\omega t).$$

Proof. Let us assume first that $0 \in I$. Let

$$h(t) = f(t) - [f'(0)/\omega]\sin(\omega t) - f(0)\cos(\omega t).$$

It follows that h also satisfies $h'' + \omega^2 h = 0$. Furthermore, $h(0) = f(0) - 0 - f(0) = 0$, and since $h' = f' - f'(0)\cos(\omega t) + f(0)\omega\sin(\omega t)$, it also follows that $h'(0) = f'(0) - f'(0) + 0 = 0$. By the Lemma, $h(t) = 0$ for all $t \in I$, and clearly this implies the desired conclusion. An appropriate modification of this argument works in the case $0 \notin I$. See Problem 3 Exercise III.5.4 for details. ∎

III.5.3   *The Motion of a Pendulum*

A pendulum, such as found, for example, in big wall clocks, provides another familiar example of a periodic motion. We shall now investigate the corresponding mathematical model. We consider a pendulum consisting of a weight of mass m attached to the bottom of a rigid rod of length L, whose mass we assume to be negligible compared to m. (See Figure III.4.) The rod swings from a hinge at the top. Neglecting, as usual, factors such as resistance, etc., the only force acting on the pendulum is the gravitational force $F_G = -mg$ that pulls the weight vertically down. The motion of

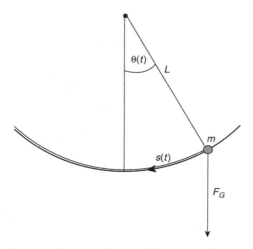

Fig. III.4   Motion of a pendulum of length L.

the pendulum is described by the arc $s(t)$ on the circle of radius L that measures the distance the weight has moved from the central position at the bottom. The orientation is chosen so that $s(t) > 0$ corresponds to a position on the right of the center, while $s(t) < 0$ means that the weight is on the left side. As shown in Figure III.4, the position is also identified by the angle $\theta(t)$ that the rod forms with the vertical line. If $\theta(t)$ is measured in radians, one has $s(t) = L \cdot \theta(t)$.

In order to apply Newton's law of motion we need to identify the force that acts in the direction of the motion of the weight along the circle. According to basic physical principles, at any position $s(t)$, the gravitational force F_G may be decomposed into

$$F_G = F_T + F_N,$$

where F_T is tangential to the circle and F_N is normal, i.e., perpendicular to the circle at the point corresponding to $s(t)$. (See Figure III.5.) Clearly F_N has no effect on the motion of the pendulum: it simply tries to stretch the rod, which is assumed to be rigid. So the only force relevant to the motion of the pendulum is the tangential component F_T. Consequently one has $ms'' = F_T$.

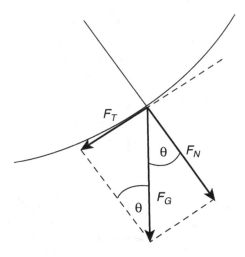

Fig. III.5   Decomposition of the force F_G.

According to Figure III.5, $F_T = F_G \sin \theta(t) = -mg \sin \theta(t)$. Note that the sign of F_T indicates that this force pulls the weight towards the center, regardless of whether $\theta(t)$ (i.e., $\sin \theta(t)$) is positive or negative. Newton's law of motion thus takes the form

$$ms''(t) = F_T = -mg \sin \theta(t).$$

Since $s''(t) = L\theta''(t)$, after rearranging and dividing by mL, one obtains

$$\theta''(t) + \frac{g}{L} \sin \theta(t) = 0.$$

We notice that this equation is more complicated than the corresponding equation for the motion of the bouncing spring studied in Section 5.1, as it involves $\theta(t)$ composed with the sine function. In order to simplify, we rely on the basic principle that differentiability means that in sufficiently small neighborhoods of a point P the graph of a function is very well approximated by the tangent line at P. Applied to the function $\sin \theta$ at $\theta = 0$, whose tangent at that point is given by $l(\theta) = \theta$, this principle implies that $\sin \theta(t) \approx \theta(t)$, and that the approximation improves the smaller θ gets. Similarly, physical principles suggest that if relevant quantities in a process are changed just by small amounts, the corresponding motion will also change by appropriately small amounts. Altogether, if we approximate the tangential component F_T of the gravitational force by

$$F_T = -mg \cdot \sin \theta(t) \approx -mg \cdot \theta(t)$$

the solutions of the resulting differential equation

$$\theta''(t) + \frac{g}{L}\theta(t) = 0$$

will provide an approximation for the actual motion of the pendulum provided the angle θ is sufficiently small.

With $\omega = \sqrt{g/L}$, it then follows from Section 5.2 that

$$\theta(t) = A\sin \omega t + B\cos \omega t.$$

If we assume that the initial position satisfies $\theta(0) = 0$, it follows that $B = 0$, so that

$$\theta(t) = A\sin \omega t = A\sin(\sqrt{g/L}\, t).$$

The *amplitude* A measures the maximal size of the angle in the motion of the pendulum. As indicated above, we need to assume that A is rather small to be assured that this solution gives a good approximation of the motion of the pendulum. Since $s(t) = L\theta(t)$ and $\theta'(t) = A\omega \cos \omega t$, the amplitude is related to the initial velocity $v_0 = s'(0)$, that is, to the velocity of the pendulum right when the weight is at the bottom, by the equation $A = \theta'(0)/\omega = (s'(0)/L)\sqrt{L/g}$, or

$$A = v_0\sqrt{1/Lg}.$$

Just as in the case of the bouncing spring, the period T of the pendulum is determined by $\omega T = 2\pi$, so that

$$T = 2\pi/\omega = 2\pi/\sqrt{g/L} = 2\pi\sqrt{L/g}.$$

What is perhaps surprising is that—in contrast to the bouncing spring—the mass m of the weight attached to the pendulum does not appear in this formula for the period. In other words, changing the weight of a pendulum does not affect its period. On the other hand, the above formula for T clearly shows the effect of the *length* L of the pendulum, that is, of the distance of the weight from the hinge at the top. By increasing that length, the period increases, i.e., the motion of the pendulum is slowed down. This is a phenomenon familiar to anyone who ever attempted to adjust the accuracy of a wall clock.

III.5.4   *Exercises*

1. Consider the spring model discussed in the text. Suppose the displacement c_0 from the equilibrium level is doubled, i.e., c_0 is replaced by $2c_0$. Determine the effect on i) the velocity $v(t)$ of the ball, ii) the *average* velocity of the ball between a low point and the following high point, and iii) the frequency of the motion.

2. A weight of 5 kg is attached to a spring, which causes the spring to stretch by 15 cm. Determine the period of the motion that results after the weight has been given an initial push.

3. This problem completes the proof of Proposition 5.2 in the case $0 \notin I$. Let $f(t)$ be a solution of $y'' + \omega^2 y = 0$ on the interval I, and choose any point $t_0 \in I$.

   a) Show that $h(t) = f(t) - \frac{f'(t_0)}{\omega} \sin \omega(t - t_0) - f(t_0) \cos \omega(t - t_0)$ satisfies the equation $y'' + \omega^2 y = 0$.

   b) Use Lemma 5.1 to show that $h(t) = 0$ for all $t \in I$.

   c) Show that $f(t) = A \sin(\omega t) + B \cos(\omega t)$ for suitable constants A and B. (Hint: Use a) and b), expand $\sin \omega(t - t_0)$ and $\cos \omega(t - t_0)$ by means of trigonometric addition formulas, and rearrange.)

4. Use the result of problem 3 to describe the function f that satisfies the equation $f'' + 9f = 0$ and the initial conditions $f(\pi/6) = 0$ and $f'(\pi/6) = 4$.

5. In order for a large wall clock to give accurate time the frequency of its pendulum needs to be exactly $1/2$ cycles per second. Determine the distance in cm from the hinge at which the weight needs to be placed in order for the clock to be accurate.

III.6   Geometric Properties of Graphs

III.6.1   *Increasing and Decreasing Functions*

A function f whose values are getting larger as the input gets larger is said to be *increasing*. Geometrically, a function is increasing if its graph moves higher as one moves to the right. More precisely, f is *increasing on the interval* I if

$$f(x_1) \leq f(x_2) \text{ for all } x_1, x_2 \in I \text{ with } x_1 < x_2.$$

f is said to be *strictly increasing* on I if

$$f(x_1) < f(x_2) \text{ for all } x_1, x_2 \in I \text{ with } x_1 < x_2.$$

Correspondingly, one has the concepts of *decreasing* and *strictly decreasing* function, as follows. f is decreasing (strictly decreasing) on the interval I if

$$f(x_1) \geq f(x_2) \quad (f(x_1) > f(x_2)) \quad \text{for all } x_1, x_2 \in I \text{ with } x_1 < x_2.$$

Figure III.6 visualizes these concepts.

Fig. III.6   Increasing   Strictly Increasing   Decreasing

Note: Common usage does not distinguish between increasing and *strictly* increasing, so one must be careful with the more precise language used here. For example, a constant function, whose graph is a horizontal line, is both increasing and decreasing according to the definition given here, although it clearly is not strictly increasing or strictly decreasing.[3]

Examples.

i) The functions $f(x) = 2^x$ and $g(x) = x^3$ are strictly increasing on $\mathbb{R}$.

ii) $f(x) = e^{-x}$ is strictly decreasing on $\mathbb{R}$.

iii) $p(x) = x^2$ is strictly decreasing on $(-\infty, 0]$ and strictly increasing on $[0, \infty)$.

iv) $\cos x$ is strictly decreasing on the interval $[0, \pi]$.

All these properties are immediately verified from the familiar graphs of these functions. (See Figure III.7.)

III.6.2   *Relationship with Derivatives*

The derivative of a function allows us to readily characterize the geometric properties illustrated above. From the preceding figures it is clear that an increasing function on I that is differentiable must have derivative ≥ 0.

---

[3] An alternative terminology refers to "increasing", as defined here, as *nondecreasing*, and it uses the term *increasing* for what is called "strictly increasing" here. Correspondingly, one then uses the terms *nonincreasing* and *decreasing* in place of decreasing and strictly decreasing. In this terminology a constant function is both nondecreasing and nonincreasing, which may sound more reasonable than to say that such a function is both increasing and decreasing.

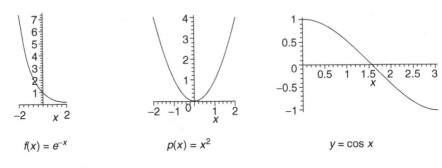

Fig. III.7

The analytic argument is just as simple. Fix a point $a \in I$. Consider the basic factorization formula

$$f(x) - f(a) = q(x)(x - a) \tag{III.3}$$

at $x = a$. If f is increasing on I, then $x - a > 0$ implies that the left side in (III.3) is ≥ 0, which implies that $q(x) \geq 0$ as well. By continuity of q at $x = a$ it follows that $f'(a) = q(a) = \lim_{x \to a^+} q(x) \geq 0$ as expected.

But one should NOT jump to the conclusion that if f is *strictly* increasing, then $f'(a) > 0$. While it is true that in this case (III.3) implies that $q(x) > 0$ for all $x > a$, the limit as $x \to a^+$ may very well turn out to be zero. For example, $f(x) = x^3$ is strictly increasing, yet $f'(x) = 3x^2$ has a zero for $x = 0$.

Similarly, if f is decreasing on I, it follows that $f'(a) \leq 0$ for all $a \in I$. The converse of the above conclusion holds as well.

Lemma 6.1. *If the function f is differentiable on the interval I and satisfies $f'(x) \geq 0$ (or $f'(x) \leq 0$) for all $x \in I$, then f is increasing (resp. decreasing) on I.*

Proof. Pick any two points $x_1, x_2 \in I$ with $x_1 < x_2$. By the Mean Value Inequality (Theorem 2.1) there exists $x_{low} \in [x_1, x_2]$ with

$$f'(x_{low}) \leq \Delta(f, [x_1, x_2]) = \frac{f(x_2) - f(x_1)}{x_2 - x_1}.$$

Since by assumption $f'(x_{low}) \geq 0$, it follows that $f(x_2) - f(x_1) \geq 0$, i.e., $f(x_1) \leq f(x_2)$ as required. The proof for the corresponding result when $f'(x) \leq 0$ uses the existence of x_{high} with $\Delta(f, [x_1, x_2]) \leq f(x_{high}) \leq 0$. ∎

To summarize:

A differentiable function f on an interval I is increasing on I if and only if f'(x) ≥ 0 for all x ∈ I.

By replacing ≥ with > in the Lemma and its proof one also obtains the following result.

If f satisfies f'(x) > 0 on I, then f is strictly increasing on I.

As noted earlier, the converse of this last statement is not correct in general.

By completely analogous arguments one sees that decreasing functions on I are characterized by $f'(x) \leq 0$ on I, and that in the case $f'(x) < 0$ for all $x \in I$, one gets the stronger conclusion that f is *strictly* decreasing on I.

Example. Determine intervals where the function $p(x) = x^3 + \frac{3}{2}x^2 - 18x + 5$ is strictly increasing or decreasing.

Solution. This is easily done by visual inspection of the graph of p obtained with the aid of a graphing calculator. If no graphing calculator is available, we apply the principles we just discussed and consider the derivative $p'(x) = 3x^2 + 3x - 18$. The set of points where $p'(x) \neq 0$ is the complement of the zeroes of p', i.e., of the solutions of

$$p'(x) = 3x^2 + 3x - 18 = 0.$$

These solutions are -3 and 2. (Use the formula for solving quadratic equations.) The real line is thus separated into the intervals $(-\infty, -3)$, $(-3, 2)$, and $(2, \infty)$, on each of which p' has no zero. Furthermore, on each of these intervals p' is either always positive or always negative, since a change of sign would result in an additional zero by the Intermediate Value Theorem (Theorem II.4.4). Note that for large $|x|$ the polynomial $p'(x)$ is positive; also, $p'(0) = -18 < 0$. It follows that p is strictly increasing on the intervals $(-\infty, -3)$ and $(2, \infty)$, and strictly decreasing on $(-3, 2)$. (See Figure III.8.)

III.6.3 *Local Extrema*

Continuing with the last example, note that the points where $p'(x) = 0$ have a special geometric significance. The tangent line at these places is horizontal. As seen from Figure III.8, the function has a high point where $x = -3$ and a low point at $x = 2$. These are examples of *local (or relative) extrema,* which are defined as follows.

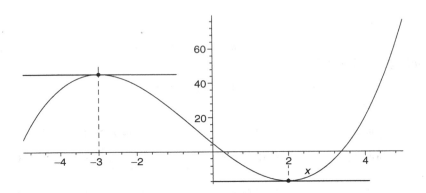

Fig. III.8   Regions where p is increasing or decreasing.

Definition 6.2. *The function $y = f(x)$ has a local maximum (or a local minimum) at the point a if there exists a neighborhood U of a, such that*

$$f(x) \leq f(a) \qquad (\text{or } f(x) \geq f(a)) \qquad \text{for all } x \in U.$$

Geometrically, at a local maximum the graph of the function changes from increasing to decreasing, while at a local minimum the opposite change occurs.

The following result is geometrically evident.

Theorem 6.3. *Suppose f has a local extremum at a. If f is differentiable at a, then $f'(a) = 0$.*

Proof. For completeness' sake we also present the simple analytic proof. It is enough to consider the case that f has a local maximum at $x = a$. The other case follows by an analogous argument. We again consider the factorization

$$f(x) - f(a) = q(x)(x - a).$$

Given that f has a local maximum at a implies that the left side is ≤ 0 for all x near a. If $x < a$ the factorization implies that $q(x) \geq 0$, while for $x > a$ one must have $q(x) \leq 0$. The differentiability of f at a, i.e., the continuity of q at a, implies that $q(a) = \lim_{x \to a} q(x)$. Hence both one-sided limits exist as well and must be equal. Since $\lim_{x \to a^-} q(x) \geq 0$ and $\lim_{x \to a^+} q(x) \leq 0$ by the preceding observations, it follows that $q(a) = \lim_{x \to a} q(x) = 0$. Therefore $f'(a) = q(a) = 0$.   ∎

A function may have a relative extremum at a point a and not be differentiable at a. For example, $g(x) = |x|$ clearly has a local minimum at 0, and g is not differentiable at that point.

One says that a is a *critical point* of the function f if either f fails to be differentiable at a or else $f'(a) = 0$. So the results we just discussed can be summarized by saying that local extrema are found among the critical points. Note, however, that not every critical point is necessarily a point at which there is a local extremum.

Example. $f(x) = x^3$ has derivative $f'(x) = 3x^2$ for all x. So 0 is the (only) critical point of f. But clearly f takes on values that are less than $f(0) = 0$, as well as values that are greater than $f(0)$ in any neighborhood of 0. So f does not have an extremum at 0. For the inverse function $g(x) = x^{1/3}$ of f one sees that g fails to be differentiable at 0, while $g'(x) = \frac{1}{3}x^{-2/3} \neq 0$ for all $x \neq 0$. So 0 is the only critical point of g, yet g has no extremum at 0.

Usually a function of one variable has only finitely many critical points on any given finite interval, and these can be determined readily in many cases. However, in general, to find the solutions of the equation $f'(x) = 0$ may require numerical approximations. In Section 7 we will discuss a numerical method for solving equations that involves another application of derivatives.

III.6.4 *Convexity*

Identification of the critical points of a function f as well as of the intervals on which f is either (strictly) increasing or decreasing gives pretty good information about the graph of f. We shall now examine an additional geometric property of the graphs of functions.

For example, recall the discussion of the polynomial p in the previous section, the results of which are summarized in Figure III.8. Notice that somewhere in the interval $(-3, 2)$ the general shape of the graph must change. As one moves along the graph from left to right, at first the tangents turn clockwise (the slopes are decreasing, corresponding to a right turn), but eventually they must turn counterclockwise, corresponding to increasing slopes, i.e., to a left turn. This property is clearly related to changes in the slope, that is, to the derivative $D(p')$ of p', which is the second derivative p'' of p.

Let us examine that polynomial p with $p'(x) = 3x^2 + 3x - 18$, more in detail. To determine where p' is increasing (left turn) or decreasing (right turn), we carry out the same sort of analysis we did earlier, but now for the function p'. Its derivative

$$D(p')(x) = 6x + 3,$$

i.e., the second derivative $p'' = D^2(p)$ of p, has a zero precisely at $-\frac{1}{2}$. Furthermore, $p''(x) < 0$ for $x < -1/2$, so that p' is strictly decreasing there. Thus the slopes of the tangents get smaller, i.e., the tangents turn to the right, as x moves to the right while staying less than $-1/2$. Also, $p''(x) > 0$ for $x > -1/2$, so that here the slopes increase, i.e., the tangents turn to the left. So we have identified the special point $(-\frac{1}{2}, p(-\frac{1}{2}))$ where the graph changes from a right turn to a left turn. Such a point is called a *point of inflection* of the function. A more accurate graph of p, as shown in Figure III.9, identifies this point of inflection precisely.

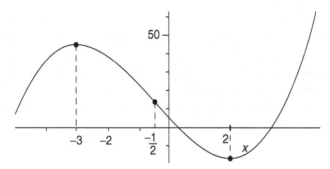

Fig. III.9   Graph of p with inflection point.

The following terminology is used to describe these geometric properties.

Definition 6.4. *The graph of the differentiable function f is said to be concave up (or convex) on the interval I, if its derivative f' is increasing on I. Correspondingly, the graph of f is said to be concave down (or simply concave) if f' is decreasing on I.*

It then follows from the results discussed earlier that in the case where f is two times differentiable on the interval I, the graph of f is convex (resp. concave) if and only if $f''(x) \geq 0$ (resp. $f''(x) \leq 0$) for all $x \in I$.

III.6.5   *Points of Inflection*

As already mentioned while discussing the earlier example, points at which the concavity changes are called *points of inflection*. Assuming that f has a second derivative f'' that is continuous on the interval I, if f has a point of inflection at $x = a$, then it follows from the Intermediate Value Theorem applied to f'' that necessarily $f''(a) = 0$. This condition allows us to identify points of inflection by solving an equation. Note however that NOT every point a at which $f''(a) = 0$ is necessarily a point of inflection. For example, for the function $g(x) = x^4$ one has $g''(x) = 12x^2$, which is always ≥ 0. The graph of g is always convex. (See Figure III.10.) So there is no point of inflection, even though $g''(0) = 0$.

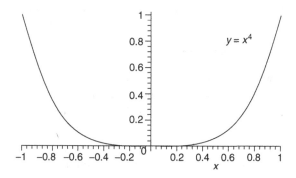

Fig. III.10   Graph of $g(x) = x^4$.

Example. We illustrate these new concepts by analyzing the familiar function $y = \sin x$. Here $y' = \cos x$, which has zeroes at the points $x_k = \frac{\pi}{2} + k\pi$ for any integer k. Since $\sin x$ is periodic, we shall focus on the interval $I = [0, 2\pi]$. The only critical points in I are $x_0 = \frac{\pi}{2}$ and $x_1 = \frac{3}{2}\pi$. Note that $y' = \cos x > 0$ on the intervals $(0, \frac{\pi}{2})$ and $(\frac{3}{2}\pi, 2\pi)$, and that $\cos x < 0$ on the interval $(\frac{\pi}{2}, \frac{3}{2}\pi)$. It follows that $y = \sin x$ has a local maximum at $\frac{\pi}{2}$ and a local minimum at $\frac{3}{2}\pi$.

Next we consider $y'' = -\sin x$, which has zeroes at $0, \pi$ and 2π. Since $-\sin x < 0$ on $(0, \pi)$ and $-\sin x > 0$ on $(\pi, 2\pi)$, it follows that the graph is concave down on $(0, \pi)$ and concave up on $(\pi, 2\pi)$. Therefore there is indeed a point of inflection at $x = \pi$. Using periodicity to add copies of the graph on $[-2\pi, 0]$ and $[2\pi, 4\pi]$, one sees that there are points of inflection also at 0 and 2π. Based on the analysis of the first and second derivative one obtains a pretty accurate picture of the graph of the sine function. The results are visualized in Figure III.11.

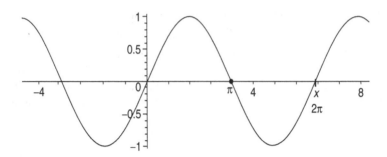

Fig. III.11   Graph of $y = \sin x$ with extrema and inflection points.

III.6.6   *Graphing with Derivatives*

We discuss a few more examples to illustrate how derivatives can be used to identify significant properties of the graph of a function.

Example. Identify relevant geometric features of the graph of $f(x) = x^3 - 3x + 4$.

Solution. We start by looking at the derivative $f'(x) = 3x^2 - 3$, which has zeroes at -1 and 1. Notice that $f'(x) > 0$ (and hence f is increasing) on $(-\infty, -1)$ and $(1, \infty)$, while $f'(x) < 0$ (f decreasing) on $(-1, 1)$. Therefore f has a relative maximum at -1, with value $f(-1) = 6$, and f has a relative minimum at 1, with value $f(1) = 2$. To determine the concavity, we note that $f''(x) = 6x$, which has a zero at 0. Since $f''(x) < 0$ for $x < 0$, f is concave down to the left of 0, and similarly, f is concave up to the right of 0. So f has a point of inflection at 0, with $f(0) = 4$. The information so obtained is shown in Figure III.12.   No other significant features appear

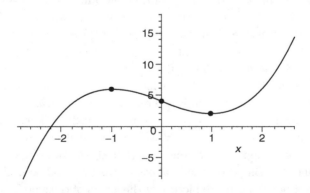

Fig. III.12   Important features of the graph of f.

on the graph of f. On the interval $(1, \infty)$ both f' and f'' are positive, so the graph will remain increasing and concave up for all $x > 1$. Similarly, the graph is everywhere increasing and concave down for $x < -1$.

Example. Identify relevant features of the graph of $g(x) = xe^{-x}$.

Solution. We evaluate

$$g'(x) = 1e^{-x} + xe^{-x}(-1) = e^{-x}(1 - x).$$

Hence $g'(x) = 0$ at 1, and since $e^{-x} > 0$ for all x one sees that $g'(x) > 0$ for $x < 1$ and $g'(x) < 0$ for $x > 1$. It follows that g has a local maximum at the point 1, with value $g(1) = e^{-1} \approx .37$. Next we analyze

$$g''(x) = -e^{-x}(1 - x) + e^{-x}(-1) = e^{-x}(x - 2).$$

Clearly $g''(x) = 0$ precisely for $x = 2$, and $g''(x) < 0$ for $x < 2$ and $g''(x) > 0$ for $x > 2$, so that g has a point of inflection at $x = 2$. Note that $g(2) = 2e^{-2} \approx .27$. We also observe that $g(0) = 0$, and that $g(x) < 0$ for $x < 0$ and $g(x) > 0$ for $x > 0$. All this information leads to the graph of g shown in Figure III.13.

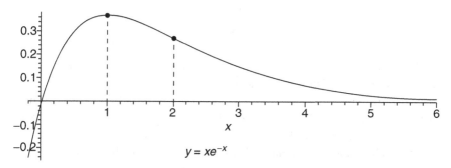

Fig. III.13   The graph of $y = xe^{-x}$.

Example. Identify relevant features of the graph of $q(x) = x^4 - 6x^3$. Here $q'(x) = 4x^3 - 18x^2 = 2x^2(2x - 9)$ and $q''(x) = 12x^2 - 36x = 12x(x - 3)$. We see that q has critical points at 0 and 4.5, and that $q'(x) \leq 0$ for $x < 4.5$ and $q'(x) > 0$ for $x > 4.5$. This shows that there is a local minimum at 4.5. However, even though $q'(0) = 0$, there is no local extremum at 0 since the function is decreasing for all $x < 4.5$. Furthermore, $q''(x) = 0$ for $x = 0$ and $x = 3$, and

$$q''(x) \text{ is } \begin{cases} > 0 \text{ for } x < 0 \\ < 0 \text{ for } 0 < x < 3 \\ > 0 \text{ for } x > 3 \end{cases}.$$

This shows the concavity properties of q. In particular, one sees that the concavity changes at $x = 0$ and $x = 3$, so that q has points of inflection at these points. In order to sketch the special points so identified, we evaluate $q(0) = 0$, $q(3) = -81$, and $q(4.5) \approx -137$. The results of this analysis are shown in Figure III.14. Again, no significant changes in the behavior of the graph occur outside the interval shown.

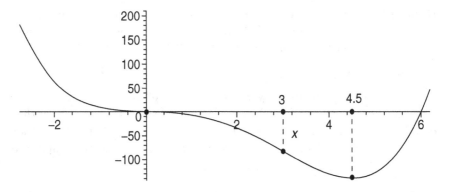

Fig. III.14   The graph of $q(x) = x^4 - 6x^3$.

Of course, all the features that we have identified in the preceding examples are readily visible once the graph of the function is plotted with the help of a graphing calculator. The point of the present discussion is to show how geometric properties of the graphs of functions relate to derivatives, and to understand how these tools allow us to find special points on the graph, such as local extrema and points of inflection. In the days before graphing calculators became widely available—say 30 to 35 years ago—the kind of analysis we discussed in this section was the principal technique used to understand basic features of functions and to plot their graphs. Typically, in order to obtain a reasonably accurate graph of a function other properties needed to be investigated as well, such as its zeroes, its behavior near points where the function fails to be defined, or the behavior for very large or very small values of x. However, these latter properties do not directly relate to the derivatives of a function, so that there is no point in discussing the details in a text that focuses on calculus. Today, the graphs of functions are predominantly investigated by using graphing calculators or more powerful computational technology. Yet the tools we discussed here are still relevant. For example, without any guidance, the viewing window that is displayed may miss important features of the graph of a function. So it is important to use tools of calculus as in the preceding examples, in order to first identify the approximate location of special points on the graph.

III.6.7 *Exercises*

Do not use the graphing functions of your calculator to work on these problems.

1. Suppose g satisfies $g'(x) < 0$ for $1 < x < 3$, $g'(3) = 0$, and $g'(x) > 0$ for $3 < x < 4$.

   a) Does g have a local extremum at 3? If so, is it a local maximum or minimum?

   b) Assume $g(3) = -1$. Sketch a possible graph of a function g that satisfies all the given properties.

2. Verify that $p(x) = x^5$ satisfies $p'(0) = 0$. Does p have a local extremum at 0? If so, is it a local maximum or minimum? Explain!

3. Make a sketch of the graph of a function F that satisfies the following conditions:

   $F(0) = -2$, $F(2) = +2$, $F'(x) > 0$ for $0 < x < 2$, and $F(x) > 0$ and $F'(x) < 0$ for all $x > 2$.

4. Figure III.15 shows the graph of the *derivative* $D(g) = g'$ of a function g. Suppose that $g(0) = 0$. Make a sketch of a possible graph of g that matches the properties of the derivative shown in Figure III.15.

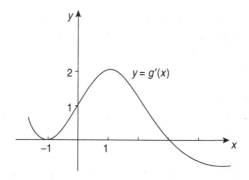

Fig. III.15   Graph of derivative of g.

5. Let $f(x) = x \cdot 4^{-x}$.

   a) Where is f decreasing, where is it increasing?

   b) Does f have any local extrema? If yes, find them and describe their type.

6. Answer the same questions as in problem 5 for the function $f(x) = x^2 \cdot e^x$. (Hint: Factor f' as much as possible, and analyze the sign of the various factors.)

7. a) Where is $G(x) = x - 4\sqrt{x}$ increasing, where is it decreasing?

   b) Explain why $G'(x)$ is close to 1 when x is getting very large.

   c) Use the information in a) and b) to sketch the graph of G.

8. Does $y = x + \sin x$ have any local extrema? Explain.

9. Use the first derivative to determine the intervals where the function $f(x) = 3x^4 - 4x^3 - 12x^2 + 5$ is strictly increasing or strictly decreasing.

10. In the graphs shown in Figure III.16, identify the intervals where the graphs are concave up, and where they are concave down. Identify any points of inflection.

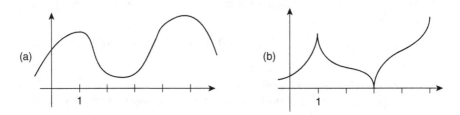

Fig. III.16   Graphs for Problem 10.

11. Find all points of inflection of the function $f(x) = x^4(x-1)^3$. (Hint: Do not expand the products... use the product rule for derivatives!)

12. Sketch a possible graph of a function f that satisfies $f(2) = 1$, $f'(2) = -1$, and $f''(2) > 0$.

13. Consider a function f that is differentiable at all points $x \neq 0$, and that satisfies $f(0) = 1$, $f'(x) > 0$ for $x < 0$, $f'(x) < 0$ for $x > 0$ and $f''(x) > 0$ for all $x \neq 0$.

   a) Could such a function have an inflection point at 0? Explain!

   b) Make a possible sketch of the graph of such a function.

   c) Could a function with the properties stated above also be differentiable at 0? (Hint: What would $f'(0)$ be?)

14. Explain why a function g whose second derivative is continuous near the point 1, and which satisfies $g'(1) = 0$ and $g''(1) = -1$, must have a local maximum at 1.

15. Explain why a polynomial of degree 3 has exactly one inflection point.

16. For the two functions f and g whose second derivatives f'' and g'' are given below, find the intervals where the graphs are concave up and concave down, and find all points of inflection.

   a) $f''(x) = (x - 2)(x + 1)^2 x^3$;
   b) $g''(x) = x \cos^2 x \, \sin x$.

17. a) Plot the function $N(x) = e^{-x^2}$ with a graphing calculator.

   b) How many points of inflection can be identified from the graph? Estimate their x coordinate(s).

   c) Find the precise coordinates of the inflection points (Use the calculator!)

18. Use a graphing calculator to plot the graph of $P(x) = 1000x^3 - 3051x^2 + 3102x$.

   a) Can you see any local extrema? If so, find their coordinates.

   b) Find all points of inflection of $P(x)$.

   c) Graph the function in a small window centered at the point of inflection of P. What do you see now?

III.7 An Algorithm for Solving Equations

III.7.1 *Newton's Method*

Given a function g, it is often necessary to find the solutions of the equation $g(x) = 0$. For example, local extrema of functions are found among the solutions of $f'(x) = 0$. Algebraic techniques allow us to find the solutions in the case where g is a linear function, or a polynomial of degree 2. In the latter case, solutions are obtained by the familiar formula for solving quadratic equations. But as soon as one deals with a polynomial of degree larger than two, there is no analogous elementary formula. Simple equations involving transcendental functions, such as $2^x + x = 0$, present other problems and can typically not be solved exactly. It is therefore of interest to develop numerical techniques that allow us to find at least approximate solutions. Computers handle such problems very efficiently. We shall now describe such a technique that is based on an application of tangent lines, i.e., differentiation, and which usually produces good approximations quite rapidly.

The idea is quite natural. Based on rough geometric information about the graph of g pick any (reasonable) first "guess" x_0 for a solution of the equation $g(x) = 0$. Unless you are very lucky, $g(x_0) \neq 0$. Assuming g is differentiable, replace g by its linearization $L_{x_0}(x)$, i.e., by its tangent line at the point $(x_0, g(x_0))$. If that tangent is not horizontal, it will cross the x–axis somewhere, i.e., the *linear* equation $L_{x_0}(x) = 0$ will have a solution x_1 that can easily be found, and that will (hopefully) be a better approximate solution to the original equation. Figure III.17 illustrates the matter in a typical situation. Then repeat the process with x_1 in place of x_0, to obtain an even better approximation x_2, and so on.

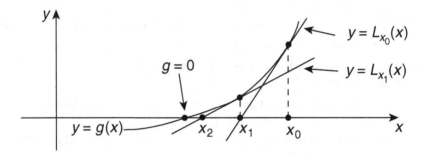

Fig. III.17   Two successive approximations for the solution of $g(x) = 0$.

Let us carry out the details. Recall that the equation for the tangent line is given by $L_{x_0}(x) = g(x_0) + g'(x_0)(x - x_0)$. Therefore, if $g'(x_0) \neq 0$, one readily sees that the equation

$$L_{x_0}(x) = g(x_0) + g'(x_0)(x - x_0) = 0$$

has the solution

$$x_1 = x_0 - \frac{g(x_0)}{g'(x_0)}.$$

After repeating this step, the next approximation is then given by

$$x_2 = x_1 - \frac{g(x_1)}{g'(x_1)},$$

and in general, if we have found x_n, then

$$x_{n+1} = x_n - \frac{g(x_n)}{g'(x_n)}.$$

The values $x_1, x_2, ..., x_n$ can be calculated quickly with a calculator, and if one recognizes that $x_n \to L$, then L will be a solution of $g(x) = 0$, i.e., $g(L) = 0$. This procedure is widely known as **Newton's Approximation Method**.

III.7.2 *Numerical Examples*

Let us now check the effectiveness of the method with some numerical examples. We start with the simple equation $g(x) = x^2 - 2 = 0$, whose solutions are $\sqrt{2}$ and $-\sqrt{2}$. In order to find the positive solution by the process we just described, let us begin with $x_0 = 1$. Since $g'(x) = 2x$, we obtain

$$x_1 = 1 - \frac{1-2}{2 \cdot 1} = 1 + \frac{1}{2} = 1.5.$$

Repeat the process to obtain

$$x_2 = 1.5 - \frac{g(1.5)}{g'(1.5)} = 1.5 - \frac{.25}{3} = 1.41666666...$$

Note that

$$g(x_2) = (x_2)^2 - 2 = 0.006944444...,$$

so already the second step takes us quite close to a solution. In fact, since the exact solution $L = \sqrt{2} = 1.414213562...$, the error is $x_2 - L = 0.00245... < 3 \cdot 10^{-3}$. With the help of a programmable calculator one readily continues the process to obtain the numbers shown in Table III.1.

$$x_3 = 1.414215686,$$
$$x_4 = 1.414213562,$$
$$x_5 = 1.414213562,$$
$$x_6 = 1.414213562,$$
$$x_7 = 1.414213562.$$

Table III.1. Approximations to $\sqrt{2}$.

We see that already the 4th iteration matches all first 10 digits of the exact solution.

By applying this procedure to the function $g(x) = x^2 - A$ one obtains an efficient algorithm for calculating the square root of any positive number

A as follows. Start with $x_0 = 1$, or, to speed up the process, take x_0 any integer whose square x_0^2 is reasonably close to A. For example, if $A = 30$, $x_0 = 5$ would be a reasonable start. Then evaluate the sequence $x_1, x_2, x_3, \ldots$ by successively applying the rule

$$x_{n+1} = x_n - \frac{x_n^2 - A}{2x_n} \text{ for } n = 0, 1, 2, 3, \ldots .$$

Usually it will then follow that

$$x_n \to \sqrt{A}$$

as n gets larger and larger. Let us check one more example. Take $A = 100$, and start with $x_0 = 9$. With the help of a calculator one obtains the values

$$x_1 = 9 - \frac{9^2 - 100}{2 \cdot 9} = 10.05555556,$$

$$x_2 = 10.05555556 - \frac{10.05555556^2 - 100}{2 \cdot 10.05555556} = 10.00015347,$$

$$x_3 = 10.00015347 - \frac{10.00015347^2 - 100}{2 \cdot 10.00015347} = 10.00000000.$$

So already x_3 gives the exact result up to 8 digits past the decimal point!

Next, let us check $A = 110$, starting again with $x_0 = 9$. The process gives the numbers shown in Table III.2.

$$
\begin{aligned}
x_1 &= 10.61111111, \\
x_2 &= 10.48880163, \\
x_3 &= 10.48808851, \\
x_4 &= 10.48808848, \\
x_5 &= 10.48808848, \\
x_6 &= 10.48808848.
\end{aligned}
$$

Table III.2. Approximations to $\sqrt{110}$.

This time the iterations remain stable after x_4, so we conclude that $\sqrt{110} = x_4 \pm 10^{-8}$. In fact, checking with a calculator gives $\sqrt{110} = 10.48808848\ldots$, confirming the accuracy of the algorithm.

Finally, we apply the algorithm to solve the transcendental equation $\cos x = x$. Such a solution x is also called a *fixed point* of the cosine function. We try to find the solution of the equation $f(x) = \cos x - x = 0$, with $f'(x) = -\sin x - 1$, and we use $x_0 = 0$ as our starting point.

The successive approximations are defined recursively by the formula $x_n = x_{n-1} - f(x_{n-1})/f'(x_{n-1})$. One obtains

$$x_1 = 0 - \frac{\cos 0 - 0}{-\sin 0 - 1} = 1, \text{ and}$$

$$x_2 = 0.7503638679,$$
$$x_3 = 0.7391028909,$$
$$x_4 = 0.7390851334,$$
$$x_5 = 0.7390851332,$$
$$x_6 = 0.7390851332.$$

Therefore the desired solution is $x_5 = 0.7390851332...$.

III.7.3 *"Chaotic" Behavior*

There are certain unusual situations that may occur. For example, the equation $g(x) = x^2 - 2 = 0$ we considered earlier has two solutions, and the initial "guess" has to be chosen reasonably close to the particular solution one wants to evaluate more precisely. Starting with $x_0 = 1$, we had approximated the solution $+\sqrt{2}$. In order to approximate the other solution $-\sqrt{2}$, one needs to start with a negative "guess", such as $x_0 = -1$. Note that $x_0 = 0$ is not admissible, since $g'(0) = 0$, and a horizontal tangent will not intersect the x-axis. On the other hand, $x_0 = 0$ separates the regions where the approximations lead to $\sqrt{2}$, respectively $-\sqrt{2}$. Inspection of the graph of g confirms that any very small *negative* initial value will produce a sequence that eventually approximates the *negative* solution, while starting with any small *positive* initial value will lead to an approximation of the other solution. This very simple construction thus exhibits the phenomenon that *arbitrarily small* changes from the initial starting point 0 lead to *large* consequences. Corresponding phenomena occur in general in the application of Newton's method, whenever there are more than one solution. In fact, the situation can become extremely complicated, leading to what has been called "chaotic" behavior. Even more startling phenomena occur when one considers analogous processes allowing *complex* numbers as inputs.[4]

Another problem that may arise is that the sequence x_n that is obtained by the process we described fails to converge to a fixed value L. For example,

---

[4]The reader may consult H. O. Peitgen and P. H. Richter, *The Beauty of Fractals*, Springer Verlag, Berlin 1986, for more details and stunning pictures.

let us consider $g(x) = x^{1/3}$; here the only solution of $g(x) = 0$ is $L = 0$. Let us examine the process graphically, starting with $x_0 = 1$, as illustrated in Figure III.18.

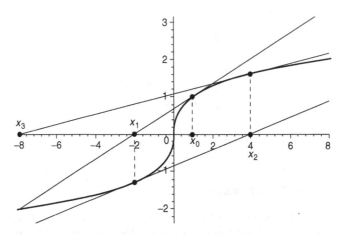

Fig. III.18   An example where the method does not converge.

It is clearly visible that in this case the process does not converge at all, since the sequence of approximations $x_1 = -2$, $x_2 = 4$, $x_3 = -8, \ldots$ steadily moves *away* from zero. Let us analyze the process numerically. Recall that by the power rule one has $g'(x) = \frac{1}{3}x^{-2/3}$ for $x \neq 0$. Hence

$$x_{n+1} = x_n - \frac{g(x_n)}{g'(x_n)}$$

$$= x_n - \frac{x_n^{1/3}}{1/3\,x_n^{-2/3}} = x_n - \frac{x_n^{1/3}\,x_n^{2/3}}{1/3}$$

$$= x_n - 3x_n = -2x_n.$$

This formula shows that no matter which initial value $x_0 \neq 0$ is chosen, the resulting sequence will never converge. In fact, $|x_n| = 2^n\,|x_0| \to \infty$ as $n \to \infty$. Hence Newton's method fails completely in this case.

III.7.4   *Exercises*

You should use a calculator to work out these problems. However, DO NOT use the *solve* command or any similar command or procedure built into your calculator.

1. Use Newton's method as in the text to approximate $\sqrt{200}$ within an error less than 10^{-8}.

2. Set up Newton's method to approximate the fifth root $\sqrt[5]{200}$ within an error less than 10^{-8}.

3. Find the solution of $x + 2^x = 0$ within an error less than 10^{-6}.

4. a) Use the intermediate value theorem to show that the equation $g(x) = x^7 + x - 1 = 0$ has one solution in the interval $[0, 1]$.

   b) Show that $g'(x) > 0$ for all x, and explain why this implies that $g(x) = 0$ does not have any other solutions in $\mathbb{R}$.

   c) Use Newton's method to approximate the solution in a) within an error less than 10^{-8}.

5. Find an approximation to the *non-zero* solution of $\cos x - e^x = 0$ within an error less than 10^{-5}.

6. a) Explain why e is the only solution of the equation $\ln x = 1$.

   b) Use Newton's method in the equation in a) to determine the first 10 digits of e.

III.8   Applications to Optimization

III.8.1   *Basic Principles*

We discuss a few examples to illustrate how the tools of calculus can be used in practical problems to identify (local) extrema at which relevant quantities either have a maximum or a minimum value.

In order to understand the mathematical process, let us consider a simple function, such as $f(x) = -3x^2 + 12x - 4$. We are familiar with the general shape of its graph. It is a parabola that opens to the bottom. The maximal value of f clearly occurs at the vertex. This is the only place where f has a horizontal tangent. To find this point, we need to solve the equation $f'(x) = 0$, i.e., $f'(x) = -6x + 12 = 0$. The solution is $x = 2$. The value of f at this point is $f(2) = -12 + 24 - 4 = 8$.

What about a minimal value of f? From the graph (see Figure III.19) we see that there is no minimum. The problem is different when one restricts the function f to a closed bounded interval. For example, suppose that we want to find the minimal value of $f(x)$ for $x \in [-4, 4]$. Since f has no other critical points except $x = 2$, the only place where the minimum can occur is at the boundary of the interval. So we evaluate $f(-4) = -100$ and $f(4) = -4$. Clearly the minimal value is taken at the point $x = -4$. Of course, the graph shown in Figure III.19 readily confirms this.

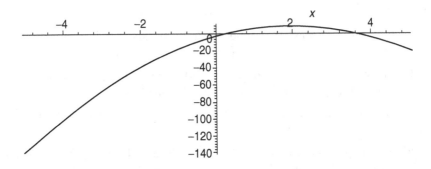

Fig. III.19   Graph of $f(x) = -3x^2 + 12x - 4$.

Let us summarize the concepts and results that are relevant to this type of question. First of all, let us recall the basic existence Theorem II.4.6 we proved in Chapter II.

Theorem 8.1. *Let f be a continuous function on a closed and bounded interval $[a, b]$. Then f takes on a maximal and a minimal value on $[a, b]$, i.e., there are points $x_{\min}, x_{\max} \in [a, b]$, such that*

$$f(x_{\min}) \leq f(x) \leq f(x_{\max}) \text{ for all } x \in [a, b].$$

We note that this theoretical result is really very important. It gives specific criteria that make it meaningful to search for maximal and minimal values of a function. The hypotheses listed in the theorem are all necessary, that is, if one of them fails, then there may not be any maximum or minimum, and hence it would be futile to start searching for such values. For example, the function $f(x) = 1/x$ is continuous on $(0, 1)$ (no endpoints here!), yet there obviously is no maximal value, since $\lim_{x \to 0+} f(x) = \infty$. In fact there also is no *minimal* value on $(0, 1)$! This fact is more subtle, since one may jump to the conclusion that 1 is such a minimal value. While it is true that $f(x) \geq 1$ for all $x \in (0, 1)$, and that $\lim_{x \to 1-} f(x) = 1$, there is no point $x_{\min}$ in $(0, 1)$ for which $f(x_{\min}) = 1$; $x = 1$ is the only number on the real axis for which $f(x) = 1$, but $1 \notin (0, 1)$. This example clearly shows the importance of including the boundary points for the theorem to be correct. Of course, it is true in this case that f has a minimal value on the half open interval $(0, 1]$. Similar examples show that one cannot drop any of the other hypotheses in the theorem.

The preceding discussion shows that the precise verification of this result is not an obvious matter. In fact, the proof given in Chapter II makes essential use of the *completeness* of the real numbers.

How do we find maximal and minimal values? Recall from Section 6.3. that if there is a (local) maximum or minimum for the function f at the *interior* point a in the domain of f, then a must be a *critical* point of f, i.e., either f fails to be differentiable at a, or else $f'(a) = 0$. The tools of calculus can then be used to identify such critical points. The only other points left for a (possible) extremal value are the boundary points of the domain. One can then check systematically the values of the function at its (interior) critical points and at the boundary points, and determine maximal and minimal values by direct comparison.

Example. Determine the maximal and minimal values of $f(x) = x^4 - 4x^3$ on the interval $[-1, 5]$.

Solution. Since $f'(x) = 4x^3 - 12x^2 = 4x^2(x - 3)$ for all x, the only critical points are $x = 0$ and $x = 3$. We then evaluate $f(0) = 0$ and $f(3) = -27$. At the boundary points one has $f(-1) = 5$ and $f(5) = 125$. We can now readily conclude that the minimal value of f on $[-1, 5]$ is $f(3) = -27$, while the maximal value $f(5) = 125$ occurs on the boundary.

III.8.2 *Some Applications*

Optimization problems occur in numerous applications. In order to apply the techniques of calculus we just discussed one must first translate the particular question into a precise mathematical statement. The next step then involves finding maximal and/or minimal values of a function according to the techniques discussed above. We discuss two such examples.

Example. A farmer has bought 750 feet of fencing. He wants to enclose a rectangular area and divide it up into 4 pens with fencing parallel to one side of the rectangle. What is the largest area that he can enclose, and what are the dimensions?

Solution. It is helpful to sketch a graph of the situation. (See Figure III.20.) We denote the length (in feet) of the sides of the rectangle by x

and w. Then the area is given by $A = x \cdot w$. Clearly no fencing material

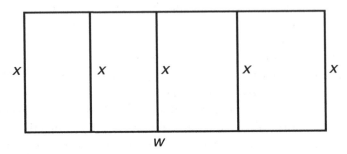

Fig. III.20   Dividing the rectangular garden into 4 pieces.

should be left over. Therefore one also has the equation $750 = 5x + 2w$. Solving for w gives $w = (750 - 5x)/2 = 375 - \frac{5}{2}x$. It follows that for a given x, $0 \le x \le 750$, the area A is given by $A(x) = x(375 - \frac{5}{2}x) = 375x - \frac{5}{2}x^2$. The (only) critical point of $A(x)$ satisfies

$$A'(x) = 375 - 5x = 0,$$

i.e., $x = 375/5 = 75$. Since $A(x)$ has no minimum, this critical point therefore is the value where the maximum occurs. The other side $w = w(x)$ is then given by $w = 375 - \frac{5 \cdot 75}{2} = 187.5$. The maximal area enclosed then is $A = 75 \cdot 187.5 = 14,062.5$ sqft.

Example. A textile mill determines that the total cost in dollars of producing x yards of a particular fabric is closely modeled by the function

$$C(x) = 1200 + 12x - 0.1 \cdot x^2 + 0.0005 \cdot x^3.$$

Marketing analysis shows that the price $p(x)$ per yard in dollars at which x yards of that fabric can be sold per month is given by $p(x) = 29 - 0.00021 \cdot x$. What should be the monthly production level in order to maximize the profits?

Solution. Profit equals revenue minus cost. Selling x yards of fabric per month results in the revenue $R(x) = x \cdot p(x) = 29x - 0.00021\, x^2$. Consequently, the profit when x yards are produced and sold in a month is given by

$$P(x) = R(x) - C(x) = -1200 + 17x + 0.09979\, x^2 - 0.0005\, x^3.$$

We need to find the maximal value $P(x)$ on the interval $(0, \infty)$. Since $\lim_{x \to \infty} P(x) = -\infty$ and $P(1000)$ is negative, it clearly is enough to find the maximum on the closed and bounded interval $[0, 1000]$. Since $P(0) < 0$, the (positive) maximum will occur at a critical point in the interior of $[0, 1000]$. We calculate $P'(x) = 17 + 0.19958\, x - 0.0015\, x^2$. The solutions of $P'(x) = 0$ are given by

$$\frac{0.0015 \pm \sqrt{(0.19958)^2 + 4 \cdot 0.0015 \cdot 17}}{-2 \cdot 0.0015} \approx -59 \text{ or } 192$$

Since $x = 192$ is the only critical point of $P(x)$ in $[0, 1000]$, the value $P(192) \approx 2203$ is the maximal profit that can be achieved in a month. The monthly production level should be set at or near 192 yards per month.

III.8.3 *Exercises*

1. Find the maximum and minimum values of the function $f(x) = x^3 - 9x^2 + 24x$ on the interval $[0, 4]$.
2. A home owner wants to build a rectangular flower bed with one side along his garage, and he wants to border it on the remaining 3 sides with a low brick wall. He has enough bricks to build a wall at most 50 feet long. How should he choose the dimensions in order to maximize the area of the flower bed?
3. Show that among all rectangles with the same perimeter the square is the one which has the largest area. (Hint: Let L be the fixed perimeter. If x is the length of one side, what is the length u of the side perpendicular to it?)
4. In an effort to lower its costs after the latest increase in aluminum price, a brewery wants to redesign its standard 12 ounce beer can. In order to maintain minimal strength, the thickness of the aluminum sheet cannot be reduced. On the other hand, by changing the radius and height of the can, different amounts of aluminum are needed. How should the dimensions be chosen in order to minimize the amount of aluminum required to make one 12 ounce can? (1 US fluid ounce is about 1.8 cubic inches. The volume of a can equals the product of the area of the bottom multiplied with the height.)
5. Work out a more general version of Problem 4, as follows. Determine the relationship between radius r and height h that minimizes the surface area among all cylindrical cans with the same volume $V = \pi r^2 h$.
6. Cylindrical mailing tubes for posters come in different sizes, depending on the length l and the circumference c. The sum $g = l + c$ is called the girth. The post office does not accept tubes for mailing whose girth

exceeds 108 inches. Find the dimensions of the tube of largest volume that can be mailed through the postal system.

7. Based on data collected over several years, the manager of a movie theater found that the number of tickets $N(x)$ sold for the showing of popular movies if the price per ticket is x dollars, is approximated by $N(x) = x^2 - 40x + 300$ for $0 \leq x \leq 10$. How should he price the tickets, so as to maximize the total revenue $R(x) = x\,N(x)$? (Note that in this example the cost is essentially fixed, that is, it does not depend on the number of people seated. Thus profits are maximized at the price level that maximizes revenue.)

III.9   Higher Order Approximations and Taylor Polynomials

III.9.1   *Quadratic Approximations*

Recall from Section II.3.4 the geometric interpretation of the statement that the function f is differentiable at the point $x = a$: it means that near the point $(a, f(a))$ the graph of f is approximated very closely by a line, i.e., by its tangent line, whose equation is given by

$$L_a(x) = f(a) + f'(a)(x - a).$$

Clearly the "bending" of the graph of f affects the precision of the approximation. (See Figure III.21.) As we saw in Section 6.4, this bending (or concavity) of the graph is measured by the second derivative f'' of f. In fact, we shall see later that the difference $f(x) - L_a(x)$ between the function and the tangent line can be expressed precisely in terms of f''.

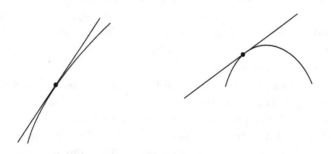

Fig. III.21   Less bending gives better linear approximation.

The bending of the graph of f suggests that a better approximation might be obtained by replacing the tangent line by a parabola, i.e., by the graph of a quadratic function that is carefully chosen to match the geometric properties of f at the point $(a, f(a))$. Suppose the quadratic function is described by

$$P_2(x) = b_0 + b_1 x + b_2 x^2.$$

What properties would determine the unknown coefficients? Clearly we would like P_2 and f to have the same tangent line at $x = a$, so we need $P_2(a) = f(a)$ and $P_2'(a) = f'(a)$. The final condition we impose is that the "bending" of the graph of P_2 at $x = a$ agrees with the "bending" of the graph of f. The simplest way to enforce this is to require that $P_2''(a) = f''(a)$. As we shall see shortly, these three conditions uniquely determine the coefficients of the quadratic function P_2.

In order to simplify the algebra, it is convenient to rearrange the terms of P_2 so as to highlight the point a, that is, we write

$$P_2(x) = c_0 + c_1(x - a) + c_2(x - a)^2.$$

We need to determine the coefficients c_0, c_1, and c_2, so that the conditions we had identified are satisfied. In detail, since $P_2(a) = c_0$, the first condition implies that we must choose $c_0 = f(a)$. Next, we calculate (use the chain rule)

$$P_2'(x) = c_1 + 2c_2(x - a).$$

This shows that $P_2'(a) = c_1$, and therefore we must choose $c_1 = f'(a)$. Lastly, we note that $P_2''(x) = 2c_2$, so that the third condition implies that $2c_2 = f''(a)$, i.e., we must choose

$$c_2 = \frac{1}{2}f''(a).$$

We have thus determined that the unique quadratic function P_2 that matches the requirements is given by

$$P_2(x) = f(a) + f'(a)(x - a) + \frac{1}{2}f''(a)(x - a)^2.$$

Notice that the linear part of P_2 is precisely the linear part (i.e., the tangent line) of f. Also, for the quadratic function P_2 the difference

$$P_2(x) - [f(a) + f'(a)(x - a)] = \frac{1}{2}f''(a)(x - a)^2$$

between P_2 and its linear approximation is given by a term that involves the second derivative $P_2''(a) = f''(a)$ of P_2.

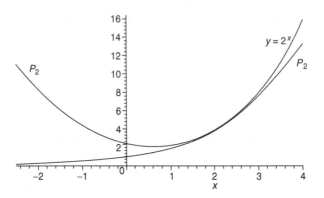

Fig. III.22   Quadratic approximation of $y = 2^x$ near $(2, 4)$.

Figure III.22 shows the graph of $y = 2^x$ and its quadratic approximation P_2 at the point $(2, 4)$. A simple calculation shows that

$$P_2(x) = 4 + \ln 2 \cdot 4 \, (x - 2) + \frac{1}{2} (\ln 2)^2 4 (x - 2)^2,$$

that is, $P_2(x) = 4 + 2.772 \, (x - 2) + .961(x - 2)^2$.

Figure III.23 shows the graph of $y = \sin x$ together with its quadratic approximation at the point $(\pi/2, 1)$. Here

$$P_2(x) = \sin(\pi/2) + \cos(\pi/2)(x - \pi/2) + \frac{1}{2}(- \sin(\pi/2))(x - \pi/2)^2,$$

that is,

$$P_2(x) = 1 - 1/2(x - \pi/2)^2.$$

III.9.2   *Higher Order Taylor Polynomials*

Once one has moved beyond the approximation by the tangent line, there is really no reason to stop with approximations by *quadratic* polynomials. The process clearly can be extended to higher order. Specifically, given the function f, a point a in its domain, and a positive integer n, one looks for a polynomial P_n of degree n that satisfies the conditions

$$P^{(j)}(a) = f^{(j)}(a) \text{ for } j = 0, 1, 2, ..., n.$$

By arguments completely analogous to those that we have used in the case $n = 2$, one verifies that there is exactly one such polynomial P_n that

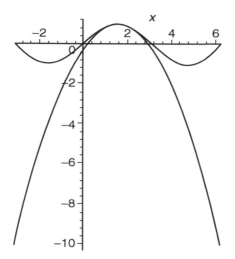

Fig. III.23   Quadratic approximation of $y = \sin x$ near $x = \pi/2$.

satisfies these conditions, and that it is given by the formula

$$P_n(x) = f(a) + f'(a)(x - a) + \frac{1}{2}f''(a)(x - a)^2$$

$$+ \frac{1}{2 \cdot 3}f'''(a)(x - a)^3 + \ldots + \frac{1}{2 \cdot 3 \cdot \ldots \cdot n}f^{(n)}(a)(x - a)^n,$$

or, more briefly (recall that $0! = 1$), by

$$P_n(x) = \sum_{j=0}^{n} \frac{1}{j!}f^{(j)}(a)(x - a)^j.$$

The polynomial P_n defined above is called the *Taylor polynomial of degree* n for the function f centered at the point a. If one wants to explicitly indicate the point a where the approximation is taken, one may write $P_{n,a}$.

It should not come as a surprise that as n increases, for most functions f the Taylor polynomials give increasingly better approximations. This approximation technique is particularly useful for transcendental functions such as exponential functions and trigonometric functions, for which there is no direct and effective procedure for calculating values at specific points. In contrast, note that the Taylor polynomials can easily be evaluated at any point x by simple algebraic operations (at least for special choices of the point a).

III.9.3   *Taylor Approximations for the sine Function*

To illustrate the effectiveness of the approximation property, we consider the Taylor polynomials of $f(x) = \sin x$. Here the simplest choice for the center is $a = 0$, since all derivatives of $\sin x$ can easily be evaluated at 0, as follows. Note that

$$
\begin{aligned}
f(x) &= \sin x, \text{ so that} & f(0) &= 0, \\
f'(x) &= \cos x, \text{ so that} & f'(0) &= 1, \\
f''(x) &= -\sin x, \text{ so that} & f''(0) &= 0, \\
f'''(x) &= -\cos x, \text{ so that} & f'''(0) &= -1, \\
f^{(4)}(x) &= \sin x, \text{ so that} & f^{(4)}(0) &= 0.
\end{aligned}
$$

Since $f^{(4)}(x) = \sin x = f(x)$, the pattern repeats, and no further calculations are needed. We notice that all terms in the Taylor polynomials with even power of x will have coefficient *zero*, so only odd powers occur. Consequently, if $n = 2m - 1$ is odd, there is no difference between P_{2m} and P_{2m-1}. Let us now evaluate a few of the Taylor polynomials explicitly.

$$
\begin{aligned}
P_1(x) &= x, \\
P_2(x) &= x + \frac{1}{2} \cdot 0 \cdot x^2 = x = P_1(x), \\
P_3(x) &= P_4(x) = x - \frac{1}{3!}x^3, \\
P_5(x) &= P_6(x) = x - \frac{1}{3!}x^3 + \frac{1}{5!}x^5, \\
P_7(x) &= P_8(x) = x - \frac{1}{3!}x^3 + \frac{1}{5!}x^5 - \frac{1}{7!}x^7, \\
P_9(x) &= P_{10}(x) = x - \frac{1}{3!}x^3 + \frac{1}{5!}x^5 - \frac{1}{7!}x^7 + \frac{1}{9!}x^9.
\end{aligned}
$$

Figure III.24 shows the graph of $\sin x$ together with several of the Taylor polynomials. It is remarkable how the approximation improves as the degree of the Taylor polynomial increases.

For comparison, in Figure III.25 we show the graphs of $\sin x$ and $P_{29}(x)$.

The graphical evidence suggests very strongly that for any value $x \in \mathbb{R}$ one has

$$
\sin x = \lim_{n \to \infty} P_{2n+1}(x) = \lim_{n \to \infty} \sum_{j=0}^{n} (-1)^{j-1} \frac{x^{2j+1}}{(2j+1)!}.
$$

We will see in Section IV.9 that this result is indeed correct.

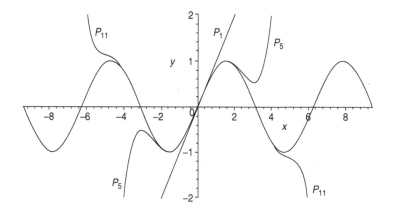

Fig. III.24 Graph of $y = \sin x$ and some Taylor polynomials.

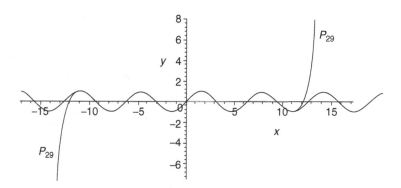

Fig. III.25 Graph of $y = \sin x$ with Taylor polynomial of degree 29.

III.9.4   *The Natural Exponential Function*

Next we consider the function $E(x) = e^x$. Because of the complicated numerical value of the base e, there is no direct way to determine even values as simple as e^2 or e^3, not to mention $e^{1/2} = \sqrt{e}$, and so on. So the approximation by Taylor polynomials is an important theoretical and practical tool. Differentiation is as easy as it gets, since $E^{(k)}(x) = E(x)$ for all $k = 1, 2, 3, \ldots$. Choosing the center for the approximation again as 0, one obtains $E^{(k)}(0) = 1$ for all k. Consequently,

$$P_1(x) = 1 + x,$$

$$P_2(x) = 1 + x + \frac{1}{2}x^2,$$

$$P_3(x) = 1 + x + \frac{1}{2}x^2 + \frac{1}{3!}x^3,$$

$$\vdots$$

$$P_n(x) = 1 + x + \frac{1}{2}x^2 + \frac{1}{3!}x^3 + \ldots + \frac{1}{(n-1)!}x^{n-1} + \frac{1}{n!}x^n,$$

or, by using the summation notation,

$$P_n(x) = \sum_{j=0}^{n} \frac{1}{j!}x^j.$$

Figure III.26 shows the graphs of $E(x) = e^x$ and of several of its Taylor polynomials. Again, the approximation is striking. Over the interval $[-4, 4]$ the graphs of e^x and of the Taylor polynomial P_{10} of degree 10 are indistinguishable. Figure III.26 provides strong evidence for the statement

$$E(x) = e^x = \lim_{n \to \infty} P_n(x) = \lim_{n \to \infty} \sum_{j=0}^{n} \frac{1}{j!}x^j \text{ for all } x \in \mathbb{R}.$$

We shall see in Section IV.9 that this result is indeed correct.

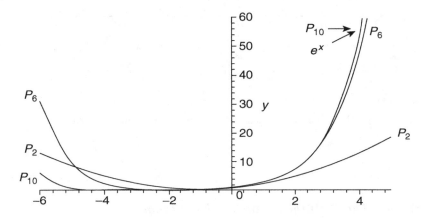

Fig. III.26   Graph of $y = e^x$ with some Taylor polynomials at $x = 0$.

In particular, for $x = 1$ one obtains the following remarkable expression for e

$$e = e^1 = \lim_{n \to \infty} P_n(1) = \lim_{n \to \infty} \sum_{j=0}^{n} \frac{1}{j!}.$$

Let us check this numerically. In Table III.3 we list the values of $P_n(1)$ (obtained with a computer) for $n = 2, 3, ..., 11$.

$$P_2(1) = 2.500000000$$
$$P_3(1) = 2.666666667$$
$$P_4(1) = 2.708333334$$
$$P_4(1) = 2.716666667$$
$$P_6(1) = 2.718055556$$
$$P_7(1) = 2.718253969$$
$$P_8(1) = 2.718278771$$
$$P_9(1) = 2.718281527$$
$$P_{10}(1) = 2.718281803$$
$$P_{11}(1) = 2.718281828.$$

Table III.3. Approximations of e by Taylor polynomials.

For comparison, we note that the first 10 digits of e are given by

$$e = 2.718281828...,$$

so that the approximation by Taylor polynomials is really quite efficient. The numerical evidence thus confirms the existence of $\lim_{n\to\infty} P_n(1)$. Such a converging "infinite sum" is an example of what mathematicians call an "infinite series"; the shorthand notation is

$$e = \sum_{j=0}^{\infty} \frac{1}{j!}.$$

In concluding, let us collect the formulas for e that we have obtained so far into one equation

$$e = 2^{1/\ln 2} = \lim_{n\to\infty} (1 + \frac{1}{n})^n = \sum_{j=0}^{\infty} \frac{1}{j!}.$$

Furthermore, recall that e is also identified as that number that satisfies

$$\lim_{h\to 0} \frac{e^h - 1}{h} = 1.$$

III.9.5 *Exercises*

1. Let $f(x) = x^3 - 2x^2 + 3x - 1$.

   a) Determine the Taylor polynomial $P_{2,1}$ of degree 2 for f centered at 1.

   b) Determine the Taylor polynomial $P_{3,1}$ for f .

   c) Note that f itself is a polynomial of degree 3, which surely gives the best possible approximation of f. How do $P_{3,1}$ and f differ? Explain!

2. a) Sketch the graph of the function $q(x) = \frac{1}{1-x}$ for $-2 \le x \le 2$, $x \ne 1$. (Use a graphing calculator.)

   b) Find the derivatives $q^{(n)}(x)$ of order $n = 1, 2, 3, 4$. Can you see the pattern? What is $q^{(n)}(x)$ for arbitrary positive integer n? (Hint: It is easier if you write $q(x) = (1 - x)^{-1}$.)

   c) Use b) to find the Taylor polynomials $P_{3,0}$ and $P_{4,0}$ centered at 0 for the function $q(x)$. What is the corresponding expression for $P_{n,0}$ for n arbitrary?

   d) Use a graphing calculator to plot q, $P_{3,0}$, and $P_{4,0}$ in one window. Over what interval does $P_{4,0}$ give a fairly good approximation to $q(x)$?

   e) By plotting Taylor polynomials $P_{n,0}$ for larger n, say $n = 10$ and 11, try to recognize for which x the approximation of $q(x)$ by $P_{n,0}(x)$ will be hopeless, no matter how large n is chosen.

3. a) Determine the Taylor polynomials $P_n(x)$ centered at 0 for the function $f(x) = \cos x$. (Hint. Modify the procedure discussed in the text in the case of $\sin x$.)

   b) Explain why $P_{n+1} = P_n$ in the case where $n = 2m$ is an even number.

   c) Use a graphing calculator to plot $\cos x$, P_4, and P_8 in one window.

Chapter IV

The Definite Integral

IV.1   The Inverse Problem: Construction of Antiderivatives

IV.1.1   *Antiderivatives and New Functions*

In Section III.2 we had already considered the problem of finding all antiderivatives $\int f(x)\,dx$ of a given function f. We had seen that the main difficulty is to find *at least one* function F that satisfies $F' = f$, since any other antiderivative is then of the form $F + c$ for some constant $c \in \mathbb{R}$.

In many cases an antiderivative is readily found by just reversing a particular differentiation formula. For example, the formula $(\sin x)' = \cos x$ can be read to say that $\sin x$ is an antiderivative of $\cos x$, or the formula $(x^n)' = nx^{n-1}$ translates into the statement that x^n/n is an antiderivative of x^{n-1} as long as $n \neq 0$. Sums of functions are easily handled by working with each summand individually, since if F and G are antiderivatives of f and g, respectively, the differentiation rule $(F + G)' = F' + G' = f + g$ readily translates into the formula

$$\int (f + g)dx = F + G = \int f\,dx + \int g\,dx$$

for antiderivatives. Similarly one sees that $\int af(x)dx = a \int f(x)dx$ if a is a constant. Unfortunately, the chain rule, the inverse function rule, and the product/quotient rules for differentiation, which are fundamental tools for finding derivatives of more complicated functions, do not have simple reverse versions that allow us to find antiderivatives of compositions or products if one knows the antiderivatives of the individual ingredients. Only in selected cases is it possible to use these rules in reverse, and it takes much practice and experience to recognize the many special situations that may arise. For example, as we will see later, by reversing the chain rule one obtains $\int 2xe^{x^2}dx = e^{x^2} + c$, a formula that can readily be verified by differentiation. However, no one has been able to find a formula for $\int e^{x^2} dx$ that only involves familiar elementary functions. In fact, it has been proven that such an explicit formula does not exist. Similarly, careful application of the product rule in reverse allows us to find antiderivatives of $x \sin x$ or $2^x \sin x$, but such methods fail, for example, for $\ln x \cdot \sin x$.

Does that mean that some functions have antiderivatives, while others that may look even simpler, do not? Note that just because we cannot readily *find* an antiderivative by using familiar functions and techniques, that does not mean that an antiderivative *does not exist*. For example, if all one knew were power functions, then

$$\int x^n dx = \frac{1}{n+1} x^{n+1}$$

is easily verified, provided $n \neq -1$. But this leaves open the case $n = -1$, i.e., find $\int \frac{1}{x} dx$. As we have seen, the answer here involves another type of function, namely the natural logarithm. This phenomenon is quite typical: the search for antiderivatives will often involve the discovery of completely new functions.

As we shall discuss in this chapter, attempts to construct an antiderivative of a given function lead to a new kind of limit process, the so-called *definite integral*. It turns out that this new limit arises in many other contexts as well, independent of the search for antiderivatives. In fact, definite integrals have numerous important applications of their own, involving concepts such as areas and volumes, probability distribution functions, the work done by a force, or the value of an income stream, to name just a few. The underlying connection between definite integrals and derivatives (or rather antiderivatives) is so central to the ideas of calculus that it is referred to as the *Fundamental Theorem of Calculus*. We shall formulate explicit versions of this result after we have thoroughly discussed the new limit process that appears in the construction of antiderivatives.

IV.1.2   *Finding Distance from Velocity*

Let us begin by considering the problem in the context of motion, that is, we assume that the given function is the (instantaneous) *velocity* v of a vehicle moving along a road over a time interval $[0, M]$. To be specific, let us assume that $v(t) \geq 0$, and that time t is measured in minutes. Fix an initial position $s_0 = s(0)$ indicated by a highway marker along the road. If $s(t)$ denotes the position of the vehicle at time t, measured in kilometers, then $s(t) - s(0)$ measures the distance the vehicle has traveled in t minutes. We know that the velocity v is the derivative of the function s, in other words, the distance function s is an antiderivative of the velocity function v, or $s(t) = \int v(t)\, dt$. Our task thus is to determine the values $s(t)$ from the known (varying) instantaneous velocity $v(t)$, that is, to develop a *mathematical odometer* that determines the *position* $s(t)$ from the information shown by the speedometer.

Note that in the case where the velocity is *constant*, say $v(t) = v_0$ for all $t \in [0, M]$, then the average velocity $\frac{s(t) - s(0)}{t}$ over the interval $[0, t]$ is equal to this constant v_0 for all t. This implies that $s(t) - s(0) = v_0 t$, or $s(t) = s(0) + v_0 t$, and the problem is solved. The general case is based on the principle that for t very close to a point t_0, the (instantaneous) velocity $v(t_0)$ is very close to the *average* velocity over the small interval $[t_0, t]$. This means that

$$\frac{s(t) - s(t_0)}{t - t_0} \approx v(t_0) \text{ for } t \text{ sufficiently close to } t_0,$$

or

$$s(t) - s(t_0) \approx v(t_0)\,(t - t_0) \text{ for } t \text{ sufficiently close to } t_0.$$

In order to apply this *local* approximation, we fix $T \in [0, M]$ and decompose the interval $[0, T]$ into a succession of very short time intervals, to each of which one applies the corresponding approximation. Precisely, for a given positive integer n, we decompose $[0, T]$ into n time intervals $[t_{j-1}, t_j]$, $j = 1, ..., n$, of equal length T/n, where $t_0 = 0$ and $t_n = T$. Then $s(t_j) - s(t_{j-1})$ measures the distance traveled during the time interval $[t_{j-1}, t_j]$, so that the total distance $s(T) - s(0)$ traveled in the time interval $[0, T]$ surely equals the sum

$$[s(t_1) - s(t_0)] + [s(t_2) - s(t_1)] + ... + [s(t_{j+1}) - s(t_j)] + ... + [s(t_n) - s(t_{n-1})],$$

that is,

$$s(T) - s(0) = \sum_{j=0}^{n-1} [s(t_{j+1}) - s(t_j)].$$

If the time intervals $[t_j, t_{j+1}]$ are very short—as will be the case if n is chosen quite large—then one has $s(t_{j+1}) - s(t_j) \approx v(t_j)(t_{j+1} - t_j)$, so it appears reasonable that

$$s(T) - s(0) \approx \sum_{j=0}^{n-1} v(t_j)(t_{j+1} - t_j),$$

with all approximations improving as $T/n \to 0$, i.e., as n is chosen larger and larger. The situation is distilled into the (symbolic) limit statement

$$s(T) - s(0) = \lim_{n \to \infty} \sum_{j=0}^{n-1} v(t_j)(t_{j+1} - t_j) . \qquad \text{(IV.1)}$$

This limit process solves our problem. We have found a way to recover the distance $s(T) - s(0)$ traveled in T minutes from the velocity $v(t)$ at each moment t in the time interval $[0, T]$. This procedure could be programmed into a computer chip which is then built into the odometer.

More significantly, we see that the search for a process to reverse differentiation naturally leads us to the approximating sums in formula (IV.1) and to the corresponding limit. Hence it becomes important to investigate these concepts carefully and thoroughly. Looking ahead, let us introduce the notation

$$\int_0^T v(t)dt = \lim_{n \to \infty} \sum_{j=0}^{n-1} v(t_j)(t_{j+1} - t_j),$$

which captures the essential structure of the approximating sums, with the finite sum symbol $\sum$ on the right replaced by the symbol $\int$ to indicate that a limit process has occurred. This new quantity $\int_0^T v(t)dt$ is called the *definite integral of the function* v *from* 0 *to* T. Since the velocity v is the derivative $D(s)$ of the distance function, the result obtained is then summarized by the formula

$$s(T) - s(0) = \int_0^T D(s)(t)dt.$$

IV.1.3 *Control of the Approximation*

While it indeed appears quite plausible that the limit process introduced in the preceding section is meaningful and leads to a correct result, it does require a careful justification. In fact, the matter is quite subtle, because the n "small" errors made in each of the *local* approximations could add up to something that possibly may no longer be small at all as $n \to \infty$. That they do not, follows from the fact that each of the errors made in the local approximations is "much smaller" than the length $|t_{j+1} - t_j| = T/n$ of the corresponding short time intervals.

In order to understand this more precisely, recall the basic principle that differentiability means that locally the function is well approximated by the linear function that describes the tangent line. The critical local approximation $s(t_{j+1}) - s(t_j) \approx v(t_j)(t_{j+1} - t_j)$ is based on this principle. As we saw in Section II.3.4, the error

$$\mathcal{E}_{t_j}(t) = s(t) - [s(t_j) + v(t_j)(t - t_j)]$$

between the function and its tangent is of the form $\mathcal{E}_{t_j}(t) = g_j(t)(t - t_j)$, where $\lim_{t \to t_j} g_j(t) = 0$.

Figure IV.1 illustrates the relevant estimation of

$$\left| s(T) - s(0) - \sum_{j=0}^{n-1} v(t_j)(t_{j+1} - t_j) \right|$$

$$= \left| \sum_{j=0}^{n-1} \{[s(t_{j+1}) - s(t_j)] - v(t_j)(t_{j+1} - t_j)\} \right| \le \sum_{j=0}^{n-1} \left| \mathcal{E}_{t_j}(t_{j+1}) \right|.$$

very clearly in the case $n = 6$.

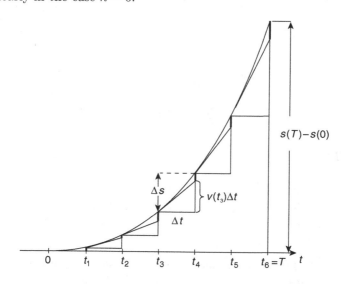

Fig. IV.1 Approximation of total distance $s(T) - s(0)$.

The total change $s(T) - s(0)$ is approximated by the sum of the increases $v(t_j)\Delta t$ along the tangent lines, with the total error estimated by the sum of the lengths of the heavy short line segments. The reader should firmly record this picture in her mind.

As $t_{j+1} \to t_j$ when $n \to \infty$, $g_j(t_{j+1})$ will indeed be quite small when n is large. Since in the limit process not only do the points t_j change, but they also increase in number, in the end one has to deal with infinitely many different error terms $\mathcal{E}_{t_j}(t)$. It is therefore critically important to measure the size of $\mathcal{E}_{t_j}(t)$, and in particular of $|g_j(t_{j+1})|$, in a manner that is *uniform* for all the small intervals and points t_j, and that is also independent of the number n of these intervals. As we shall verify in detail below, for suitable functions s one has such an estimate in the precise form

$$\left|\mathcal{E}_{t_j}(t_{j+1})\right| \leq K\,|t_{j+1} - t_j|^2 \tag{IV.2}$$

with a constant K that does not depend on n and on the particular points $t_j, j = 0, ..., n$. Recalling that $|t_{j+1} - t_j| = T/n$, it then follows that

$$\left|\sum_{j=0}^{n-1}[s(t_{j+1}) - s(t_j)] - \sum_{j=0}^{n-1} v(t_j)(t_{j+1} - t_j)\right| \leq \sum_{j=0}^{n-1}\left|\mathcal{E}_{t_j}(t_{j+1})\right|$$

$$\leq K\sum_{j=0}^{n-1}|t_{j+1} - t_j|^2$$

$$= nK(\frac{T}{n})^2 = (KT^2)\frac{1}{n}.$$

One therefore obtains the estimate

$$\left|[s(T) - s(0)] - \sum_{j=0}^{n-1} v(t_j)(t_{j+1} - t_j)\right| \leq KT^2/n.$$

Since $KT^2/n \to 0$ as n gets larger and larger, this estimate surely proves that

$$\sum_{j=0}^{n-1} v(t_j)(t_{j+1} - t_j) \to s(T) - s(0) \text{ as } n \to \infty$$

as expected.

In order to prove the estimate (IV.2) we shall consider first the case that the distance function $s = s(t)$ is an algebraic function in the class $\mathcal{A}$. We will then recognize what particular hypotheses need to be made so that the arguments work for more general functions as well. As the relevant arguments are somewhat technical, the reader should feel free to skip these details on first reading.

Recall the factorization $s(t) - s(t_j) = q_j(t)(t - t_j)$, where the factor q_j is also in the class $\mathcal{A}$, and $q_j(t_j) = D(s)(t_j) = v(t_j)$. One therefore obtains

$$\mathcal{E}_{t_j}(t) = s(t) - s(t_j) - v(t_j)(t - t_j)] = [q_j(t) - q_j(t_j)](t - t_j). \tag{IV.3}$$

The factor g_j in the error term $\mathcal{E}_{t_j}(t) = g_j(t)(t - t_0)$ is thus given by $g_j(t) = q_j(t) - q_j(t_j)$. Let us fix $t > t_j$. Since $q_j(t)$ equals the average velocity over the interval $[t_j, t]$, the Mean Value Theorem (Corollary III.2.5) produces a number $c_j \in [t_j, t]$ such that

$$v(c_j) = q_j(t).$$

Since the derivative $D(v)$ of $v \in \mathcal{A}$ is again in the class $\mathcal{A}$, by Theorem I.6.5 there exists a constant K such that $|D(v)(t)| \leq K$ for all t in $[0, M]$. Therefore Corollary III.2.4 gives the estimate

$$|q_j(t) - q_j(t_j)| = |v(c_j) - v(t_j)| \leq K |c_j - t_j| \leq K |t - t_j|. \qquad \text{(IV.4)}$$

By introducing this estimate into (IV.3), we obtain the desired *uniform* approximation

$$s(t_{j+1}) - s(t_j) \approx v(t_j)(t_{j+1} - t_j)$$

in the precise form

$$
\begin{aligned}
|[s(t_{j+1}) - s(t_j)] - v(t_j)(t_{j+1} - t_j)| &= \left| \mathcal{E}_{t_j}(t_{j+1}) \right| \\
&\leq |q_j(t_{j+1}) - q_j(t_j)| \, |t_{j+1} - t_j| \\
&\leq K |t_{j+1} - t_j|^2,
\end{aligned}
$$

i.e., we have proved the estimate (IV.2).

Finally, in the case where s is a more general non-algebraic function, the arguments in the preceding discussion remain correct provided one assumes that the velocity $v = D(s)$ has a derivative $D(v) = D^2(s)$ that is bounded over the interval $[0, M]$. The preceding arguments therefore provide a complete proof of the following important preliminary version of the so-called Fundamental Theorem of Calculus.

Theorem 1.1. *Suppose F is a two times differentiable function on the interval I, and that its second derivative $D(D(F))$ is bounded over the interval $[a, b] \subset I$. Then*

$$\int_a^b D(F)(t)dt = F(b) - F(a).$$

IV.1.4 *A Geometric Construction of Antiderivatives*

The preceding discussion resulted in a process to recover the distance function s (i.e., an antiderivative of the velocity) from the known velocity function v. This is very useful indeed, but the proof explicitly made use of the concrete and familiar distance function, i.e., it used knowledge of the antiderivative to begin with.   Therefore it is not readily apparent how this process would help to actually *construct* an antiderivative in general, especially when such an antiderivative is unknown at the very beginning. The situation may be loosely compared to finding square roots of numbers. This is quite simple if the given number is already of the form r^2 for some rational number. But when presented with a number which is not known to be of that form, say the number 2 for example, finding $\sqrt{2}$ is considerably more difficult. Not only does one need some approximation scheme, but more significantly, one needs to go beyond the rational numbers.

We shall now introduce a slight variation of the process that we considered for velocities. It is a geometric version of the former process that—most significantly— does *not require a priori* knowledge of an antiderivative. We will see that the proof of the existence of the corresponding limit is—as expected—somewhat more complicated than when the value of the limit is known explicitly from the beginning in terms of an antiderivative—such as the distance function in the case where one starts with velocity. However, this version of the process has the advantage that it ultimately will allow us to "construct" an antiderivative for any reasonable function, and thereby prove, in particular, that any such function does indeed have an antiderivative.

Given a function f on an interval I, as we look for an antiderivative F of f, the essential (and only) information available to us is that the *slope of the graph of F* at a point $(c, F(c))$ for some $c \in I$ is precisely $f(c)$. By the general geometric interpretation of differentiability, in a small neighborhood of that point the graph of F essentially coincides with its tangent line which has slope $f(c)$. Given $x \in I$ we shall use this information to build successive approximations for the value $F(x)$ starting with some initial point $(a, F(a))$.

More in detail, in order to find an antiderivative F of f, we shall begin by constructing "approximate antiderivatives" F_n for f by piecing together n short line segments whose slopes are determined by the values of f at successive points, i.e., by the known derivative of the function we are trying to find. Figures IV.2 and IV.3 illustrate this construction in the case of 5 short pieces of the relevant tangents.

We choose the approximations to all satisfy the same initial value condition $F_n(a) = 0$. The goal then is to produce the value $F(x)$ of an exact antiderivative F at the fixed point x by taking the limit of these approximations, that is, by setting $F(x) = \lim_{n\to\infty} F_n(x)$.

We now carry out the procedure in a systematic way. We fix a point

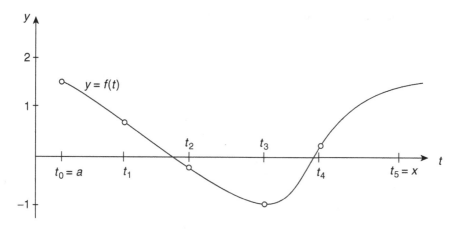

Fig. IV.2 The given function $y = f(t)$.

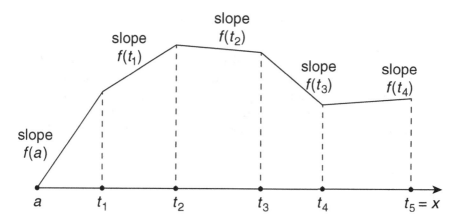

Fig. IV.3 Approximation of an antiderivative of f (Figure IV.2) by lines with slopes $f(t_j)$.

$x \in I$, $x > a$, and for each $n = 1, 2, 3, \ldots$ we construct an approximation $F_n(x)$ to the (still unknown) value $F(x)$ by using n line segments. The first approximation to the graph of F beginning at the point $(a, 0)$ is just given by its tangent line at that point; since it is required that $F'(a) = f(a)$, that tangent line is the graph of the function

$$F_1(t) = f(a)(t - a).$$

Its value $F_1(x) = f(a)(x - a)$ at x is the desired *first* approximation of the unknown value $F(x)$. (See Figure IV.4.) Of course, $F_1(x)$ will typically be a poor approximation—after all, the tangent provides a good approximation to the graph of F only close to the point a.

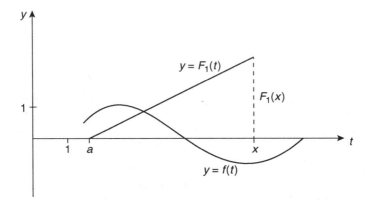

Fig. IV.4 $y = f(t)$ and its first approximation $F_1(t)$.

For the second approximation with $n = 2$, we take the midpoint $t_1 = a + (x - a)/2$ of $[a, x]$ to divide the interval $[a, x]$ into two equal pieces of length $\Delta_2 t = (x - a)/2$, and we use the line with slope $f(a)$ (≈ 0.5) on the first half $[a, t_1]$, while for the second half $[t_1, x]$ we use the line at $(t_1, F_1(t_1))$ with slope $f(t_1)$ (≈ 0.3). (See Figure IV.5.)   The formula for F_2 on the

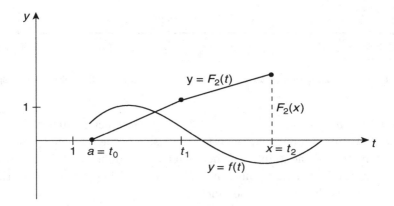

Fig. IV.5   Graph of the approximation $F_2(t)$.

interval $[a, x]$ is given by

$$F_2(t) = \begin{cases} f(a)(t-a) & \text{for } a \leq t \leq t_1, \\ f(a)(t_1 - a) + f(t_1)(t - t_1) & \text{for } t_1 \leq t \leq x. \end{cases}$$

It follows that

$$F_2(x) = f(a)(t_1 - a) + f(t_1)(x - t_1)$$
$$= f(a)\Delta_2 t + f(t_1)\Delta_2 t.$$

It is now clear how to proceed. We set $t_0 = a$, and for any $n \geq 2$ we divide the interval $[a, x] = [t_0, x]$ into n equal pieces $[t_0, t_1]$, $[t_1, t_2]$, ..., $[t_{n-1}, t_n]$ of equal length $t_{j+1} - t_j = \Delta_n t = (x - a)/n$. One has

$$t_0 = a, \ t_1 = a + \Delta_n t, \ t_2 = a + 2\Delta_n t, \ ..., \ t_j = a + j\Delta_n t, \ ..., \ t_n = a + n\Delta_n t = x.$$

Starting at $(a, 0)$, we move along the polygon consisting of n line segments with successive slopes $f(t_0), f(t_1), ..., f(t_{n-1})$, ending up at the point $(x, F_n(x))$, where

$$F_n(x) = f(t_0)\Delta_n t + f(t_1)\Delta_n t + ... + f(t_{n-1})\Delta_n t$$
$$= \sum_{j=0}^{n-1} f(t_j)\Delta_n t.$$

By construction, each F_n also satisfies $F_n(a) = 0$ for all n. Figure IV.6 shows the graph of f and the approximation $F_n(x)$ for the value $F(x)$ of the antiderivative for $n = 5$.

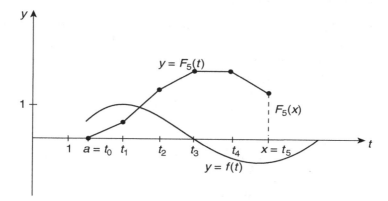

Fig. IV.6   The approximation $F_5(t)$ and the value $F_5(x)$.

It seems reasonable that the values $F_n(x)$ should approximate the corresponding value $F(x)$ for the (still unknown) antiderivative F of the original function f. After all, near each point $(t_j, F(t_j))$ the graph of the presumed antiderivative would be very closely approximated by a short piece of the tangent line at that point—remember, *locally* the graph of a differentiable function is nearly indistinguishable from its tangent line. At $(t_j, F(t_j))$ this tangent has slope $F'(t_j) = f(t_j)$. By construction, the approximating polygon F_n has exactly that same slope $f(t_j)$ at the point $(t_j, F_n(t_j))$. So this polygon closely matches the tangents of the graph of the antiderivative F that we try to determine.

Note that the expression

$$F_n(x) = \sum_{j=0}^{n-1} f(t_j)\Delta_n t$$

that approximates $F(x)$ is, except for a change in notation, identical to the expression

$$\sum_{j=0}^{n-1} v(t_j)(t_{j+1} - t_j)$$

that appeared earlier in the approximation of the total distance $s(T) - s(0)$. This structural match reinforces the importance of these sums and the need to thoroughly study their limits. Clearly these limits are central to the process of constructing antiderivatives.

IV.1.5 *A Simple Example*

Let us explicitly work out this approximation process for the simple function $f(x) = x^2$. We fix a value $x > 0$ and carry out the details of the procedure just discussed on the interval $[0, x]$. Thus $\Delta_n t = x/n$, and $t_j = j \cdot \Delta_n t = j \cdot x/n$ for $j = 0, 1, 2, ..., n$, so that

$$F_n(x) = \sum_{j=0}^{n-1} f(t_j)\Delta_n t = \sum_{j=0}^{n-1} (j \cdot x/n)^2 (x/n)$$

$$= \sum_{j=0}^{n-1} x^2 \frac{j^2}{n^2} \frac{x}{n} = x^3 \frac{\sum_{j=0}^{n-1} j^2}{n^3}$$

$$= x^3 R(n),$$

where we have denoted the expression $\sum_{j=0}^{n-1} j^2/n^3$ by $R(n)$. It is remarkable that the same factor $R(n)$ works independently of the choice of x. This reduces the problem to just determining $\lim_{n\to\infty} R(n)$. With the help of a calculator one obtains

$$R(10^2) = .3283500000$$
$$R(10^3) = .3328335000$$
$$R(10^4) = .3332833350$$
$$R(10^5) = .3333283333$$
$$R(10^6) = .3333328333$$
$$R(10^7) = .3333332833$$
$$R(10^8) = .3333333283$$

This numerical evidence strongly suggests that

$$R(n) \to 0.33333... \ = \frac{1}{3} \quad as \ n \to \infty.$$

It is possible to confirm the correctness of this statement by a formal argument that does not rely on numerical evidence. (See Problems 5 and 6 of Exercise IV.1.7.) We therefore are justified to conclude that

$$F(x) = \lim_{n\to\infty} F_n(x) = x^3 \lim_{n\to\infty} R(n) = x^3 \frac{1}{3}.$$

As we observed, this process works for any *positive* $x \in \mathbb{R}$, without any new computations. Of course $F(0) = \lim F_n(0) = 0$. If $x < 0$, the same formulas apply, with the only difference that $\Delta_n t = x/n$ is now negative, and consequently the polygons that begin at $a = 0$ are now built up by moving towards the left side. Once x^3 has been factored out from the approximating sums, the same formula for $R(n)$ remains. We thus have *constructed* the function $F(x) = \lim_{n\to\infty} F_n(x) = x^3/3$ for every real number x. We notice that F is indeed an antiderivative of $f(x) = x^2$, as expected. In fact, since antiderivatives with a specific initial value are determined uniquely, F is the *unique* antiderivative of $f(x) = x^2$ that satisfies the initial condition $F(0) = 0$.

IV.1.6 *The Definite Integral*

What we have done for this specific example can—in principle—be carried out for more general functions. However, no "exact" evaluation of the limits, as was done in the example above, will be possible in general. More complicated arguments based on the completeness of $\mathbb{R}$ and involving careful estimations are required to prove the main result, as follows.

Theorem 1.2. *Suppose the function f is continuous on the interval I, and fix a point $a \in I$. Then for each $x \in I$ the expressions*

$$F_n(x) = \sum_{j=0}^{n-1} f(t_j)\Delta_n t,$$

where $\Delta_n t = (x-a)/n$, and $t_j = a + j\Delta_n t$, have a limit $F(x)$ as $n \to \infty$.

Remark. The result is valid also for $x < a$. In that case the quantity $\Delta_n t$ will be negative.

Consistent with the terminology introduced in Section 1.2 in the case of the velocity function, we shall call this limit the *definite integral of f from a to x* and write

$$\int_a^x f(t)dt = \lim_{n\to\infty} \sum_{j=0}^{n-1} f(t_j)\Delta_n t.$$

The proof of the theorem in the case of an arbitrary *continuous* function requires technical concepts and arguments that go beyond this introductory discussion, and hence will be omitted.[1] However, if one assumes that f is *differentiable* on I and that its derivative $D(f)$ is bounded over any closed bounded interval $[a, b] \in I$, the proof is much more elementary than in the general case and will be discussed later on in Section 6. Note that all algebraic functions, the elementary transcendental functions, most other "natural" functions, and standard combinations of these, and so on, enjoy this additional good property, so that for all practical purposes this restriction is not at all significant. The general case of continuous functions is mainly of interest for abstract theoretical investigations.

Of course, if the function f has a bounded derivative $D(f)$, and if it is also known that f *does have an antiderivative* G, i.e., $D(G) = f$, then the discussion in the preceding section already gives a simple proof of the existence of the limit that defines the definite integral $\int_a^x f(t)dt$. In fact, it even gives a precise formula for its value, namely

$$\int_a^x D(G)(t)dt = G(x) - G(a). \tag{IV.5}$$

The key novelty therefore is that the present theorem does not require any such *assumption about the existence* of an antiderivative. Instead, it is the principal ingredient to *prove the existence* of an antiderivative.

In order to better understand the meaning of formula (IV.5), let us denote by $I_a(f)$ the function defined by $I_a(f)(x) = \int_a^x f(t) \, dt$ for $x \in I$. Both I_a and D are operations that may be applied to a suitable function to produce new functions. Formula (IV.5) then takes the form

$$I_a(D(G)) = G \quad \text{for all reasonable functions } G \text{ with } G(a) = 0.$$

---

[1] In particular, the proof requires the fact that continuous functions on closed bounded intervals have an additional special property known as *uniform* continuity.

This formula summarizes that taking integrals reverses the process of differentiation. The main question left therefore is whether the order of the operations can be interchanged, that is, whether differentiation reverses the process of integration. Alternatively, is $D(I_a(f)) = f$ valid for all reasonable functions f? In particular, this would confirm that the limit process that defines definite integrals indeed produces an antiderivative for f.

It turns out that given Theorem 1.2 above and a few other natural and useful properties of definite integrals, it will be quite easy to answer this question, as follows.

Theorem 1.3. *Suppose f is continuous on the interval I. Then the function $F = I_a(f)$ defined on the interval I by $F(x) = \int_a^x f(t)\, dt$ is differentiable and satisfies $D(F)(x) = f(x)$ for all $x \in I$.*

Corollary 1.4. *Every continuous function on an interval I has an antiderivative on I.*

This "inverse relationship" between differentiation and definite integrals—which are defined as limits of certain sums—is the central result that is known as the *Fundamental Theorem of Calculus*. We shall postpone the proofs of these results until we have gained more familiarity with definite integrals and have explored some of their important applications that reach well beyond the construction of antiderivatives. In particular, in the next section we will discuss an extremely useful geometric interpretation of the approximating sums F_n that involves the concept of area. This will help us to understand the limit process better, and it will allow us to identify a concrete geometric interpretation of the relevant limit, that is, of definite integrals.

IV.1.7 *Exercises*

1. Find explicit formulas for the following antiderivatives:
   a) $\int e^t dt$; b) $\int 2^t dt$; c) $\int (\sin t + 4t^2)\, dt$;
   d) $\int \frac{1}{x^2}\, dx$; e) $\int x^{100}\, dx$.

2. a) Show that for $p(x) = x^3$ the corresponding approximating functions F_n satisfy
$$F_n(x) = x^4\, R_4(n),$$
   where $R_4(n)$ is an expression that depends on n, but does not involve x. (Hint. Carefully modify the procedure discussed in the

text for the function $f(x) = x^2$ so that it applies to the function $p(x) = x^3$.)

b) Use a calculator (a programmable one would be very helpful) to estimate $R_4(n)$ numerically as n gets larger. Can you recognize $\lim_{n\to\infty} R_4(n)$? What do you think is the correct value for this limit?

3. a) Make a reasonably accurate graph of $y = \cos x$ on the interval $[0, \pi]$. (Use a graphing calculator.)

b) Starting at the point $(0,0)$, insert into the graph in a) a sketch of the polygon whose endpoint gives the value $F_6(\pi)$, by dividing the interval $[0, \pi]$ into 6 equal pieces, and using the graph of $y = \cos x$ to estimate the slopes $\cos t_j$ of the individual line segments for $j = 0, 1, 2, ..., 5$. (See also Figure IV.6.)

c) Estimate the value of $F_6(\pi)$ based on your construction in b). Use a calculator to evaluate $F_6(\pi)$ precisely. What do you think should be the value of $\lim_{n\to\infty} F_n(\pi)$. Explain!

4. Let $S^{(2)}(n) = 1^2 + 2^2 + ... + n^2 = \sum_{j=1}^{n} j^2$.

a) Show that $S^{(2)}(n) = 1 + \sum_{j=1}^{n}(j+1)^2 - (n+1)^2$.
b) Show that $\sum_{j=1}^{n}(j+1)^2 = S^{(2)}(n) + 2\sum_{j=1}^{n} j + n$.
c) Use a) and b) to conclude that $2\sum_{j=1}^{n} j = (n+1)^2 - n - 1$.
d) Use c) to show that $S(n) = \sum_{j=1}^{n} j = \frac{n(n+1)}{2}$.

(The process outlined above can be adapted to prove other identities of this type. See Problem 5.)

5. Show that $S^{(2)}(n) = \sum_{j=1}^{n} j^2 = \frac{n(n+1)(2n+1)}{6}$ by suitably modifying the procedure in Problem 4, as follows.

a) Verify that $\sum_{j=1}^{n} j^3 = 1 + \sum_{j=1}^{n}(j+1)^3 - (n+1)^3$.
b) Verify that $\sum_{j=1}^{n}(j+1)^3 = \sum_{j=1}^{n} j^3 + 3S^{(2)}(n) + 3\sum_{j=1}^{n} j + n$. (Hint: Use $(j+1)^3 = j^3 + 3j^2 + 3j + 1$.)
c) Substitute b) into a) on the right side, simplify, and solve for $S^{(2)}(n)$, making use of the formula $S(n) = \sum_{j=1}^{n} j = n(n+1)/2$ obtained in Problem 4 d).

6. Use the result obtained in Problem 5 to show that

$$\lim_{n\to\infty} \frac{\sum_{j=1}^{n} j^2}{n^3} = \frac{1}{3}.$$

IV.2 The Area Problem

IV.2.1 *Approximation by Sums of Areas of Rectangles*

We shall now discuss a geometric interpretation of the approximating sums $F_n(x) = \sum_{j=0}^{n-1} f(t_j)\Delta_n$ for an antiderivative of the function f that we constructed in the preceding section. This will provide a "visual justification" for the convergence of $F_n(x)$ as $n \to \infty$, as well as a corresponding interpretation of the limit. For simplicity, we assume that $f(t) \geq 0$ for all t in the interval I. For $b > a$, we can visualize $F_n(b)$ as follows. Let us plot the graph of f as usual in a coordinate system and divide the interval $[a, b]$ into n equal pieces determined by the points $a = t_0 < t_1 < ... < t_n = b$. Then $f(t_j)$ represents the length of the vertical line segment from the point t_j on the t-axis to the curve described by f. Since $\Delta_n t$ is the length $(b - a)/n$ of the interval $[t_{j-1}, t_j]$ along the t-axis, the product $f(t_j) \cdot \Delta_n t$ represents the area of the thin rectangle with base $[t_{j-1}, t_j]$ of width $\Delta_n t$ and of height $f(t_j)$. It follows that $F_n(b)$ is the sum of the areas of these rectangles for $j = 0, 1, ..., n - 1$. (See Figure IV.7.)

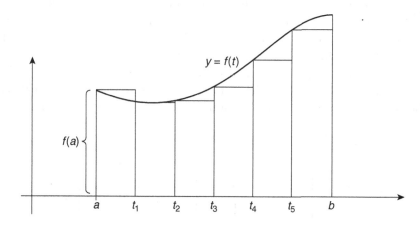

Fig. IV.7   Approximating sum represents the sum of areas of rectangles.

These rectangles approximate the region above the t-axis that lies below the graph of f between the points a and b. We thus interpret the sum $F_n(b)$ as an approximation of the area of that region. It is now apparent—given suitable hypothesis on f—that as n gets larger, the difference between the rectangles and the region under the graph gets smaller, i.e., the errors made in the approximation get smaller. This fact suggests very strongly that the limit

$$F(b) = \int_a^b f(t)\,dt = \lim_{n \to \infty} \sum_{j=0}^{n-1} f(t_j)\Delta_n$$

indeed exists and that it provides a measure for the area under the graph of f identified above.

Note that the area of regions in the plane whose sides are line segments, such as triangles, rectangles, parallelograms, etc., is an elementary concept for which simple formulas exist. On the other hand, to measure areas of regions with curved boundary is much more complicated. Recall, for example, that the area of a disc involves the mysterious transcendental number π. The procedure we just developed provides a systematic technique (or algorithm) to find areas of a very large number of regions described by graphs of functions, and it reveals that the area problem is closely connected with the problem of finding antiderivatives.

IV.2.2 *Rectangles and Triangles*

In order to feel comfortable with this new procedure to calculate areas, we shall examine a couple of familiar regions for which the area is well known, and verify that the (more complicated) limit process we just introduced leads indeed to the familiar correct answers.

Let us first consider a rectangle. In order to implement the limit process, we place the rectangle R as shown in Figure IV.8, so that the rectangle is the area under the graph of the constant function $f(x) = h$ over the interval $[a, b]$.

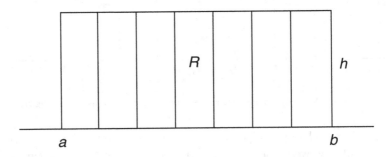

Fig. IV.8   Area of rectangle decomposed into smaller rectangles.

The area of R is then equal to $(b - a) \cdot h$. Let us now evaluate this area as the definite integral $\int_a^b f(t)dt = \int_a^b h\,dt$. According to the procedure we developed, we fix a positive integer n, set $\Delta_n t = (b-a)/n$, and consider

$$A(n) = \sum_{j=0}^{n-1} f(t_j)\Delta_n t = \sum_{j=0}^{n-1} h \, \Delta_n t$$

$$= h \cdot \sum_{j=0}^{n-1} \frac{b-a}{n} = h \cdot n \cdot \frac{b-a}{n}$$

$$= h \cdot (b-a).$$

So $A(n)$ does not depend on n, and clearly

$$\int_a^b f(t)dt = \lim_{n\to\infty} A(n) = h \cdot (b-a),$$

which is exactly the expected result.

Next, we consider the triangle shown in Figure IV.9 that is the area under the graph of $f(x) = \frac{h}{b}x$ over the interval $[0, b]$.

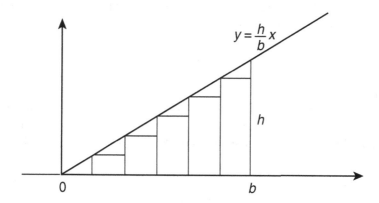

Fig. IV.9   Area of triangle approximated by rectangles.

By the standard formula $Area = base \times height/2$, the area of this triangle equals $bh/2$. According to the procedure based on definite integrals, this area should be given by

$$\int_0^b f(t)dt = \lim_{n\to\infty} \sum_{j=0}^{n-1} f(t_j)\Delta_n t \,,$$

where for each $n = 1, 2, ...$ one has $\Delta_n t = b/n$ and $t_j = j \cdot b/n$ for $j = 0, 1, ..., n$. If we represent $\sum_{j=0}^{n-1} f(t_j)\Delta_n t$ as the sum of areas of rectangles (see Figure IV.9 in the case $n = 6$), it is obvious that—in contrast to the preceding case of a rectangle—this sum is NOT equal to the area of the triangle. On the other hand, it does look reasonable that the sum of all the thin rectangles will cover the whole triangle more and more closely as n gets larger and larger.

In order to examine this precisely, we note that in the present situation

$$\sum_{j=0}^{n-1} f(t_j)\Delta_n t = \sum_{j=0}^{n-1} (\frac{h}{b} t_j)\frac{b}{n}$$

$$= \sum_{j=0}^{n-1} (\frac{h}{b} \cdot j\frac{b}{n})\frac{b}{n} = bh\frac{\sum_0^{n-1} j}{n^2},$$

so that

$$\int_0^b f(t)dt = bh \lim_{n\to\infty} \frac{\sum_0^{n-1} j}{n^2}.$$

The value of $\lim_{n\to\infty}(\sum_0^{n-1} j)/n^2$, while not immediately obvious, is still simple enough that we can work it out precisely. The answer turns out to be $1/2$ (see below), and therefore

$$\int_0^b f(t)dt = \frac{bh}{2},$$

as expected.

In order to verify the limit statement, we shall use a clever argument to find a useful formula for $S(n) = \sum_0^n j = 1 + 2 + ... + n.$[2] Note that by reversing the order we of course also have $S(n) = n + (n-1) + ... + 3 + 2 + 1$. We evaluate

$$
\begin{array}{rccccccc}
S(n) + S(n) = & 1 & + & 2 & + & ... & + ... + (n-1) + & n \\
& +n & + (n-1) & + (n-2) & + ... + & 2 & + & 1 \\
= & (n+1) & + (n+1) & + & ... & + ... + (n+1) & + (n+1) \\
= & (n+1)n. & & & & & &
\end{array}
$$

From $2S(n) = (n+1)n$ one obtains

$$S(n) = 1 + 2 + ... + n = \frac{1}{2}(n+1)\,n.$$

(A different proof for this formula was outlined in Problem 4 of Exercise IV.1.7. The proof there is somewhat more complicated, but it can be

---

[2]It is reported that C. F. Gauss (1777 - 1855), the most eminent mathematician of the 19th century, came up with this argument while in school at age 10. His teacher was so impressed that he took immediate steps to ensure that his remarkable pupil would get the best mathematical training available.

adapted to handle more general cases.) By replacing n with $n - 1$, it follows that

$$\frac{\sum_0^{n-1} j}{n^2} \doteq \frac{S(n-1)}{n^2} = \frac{1}{2} \frac{(n-1)\, n}{n^2}$$

$$= \frac{1}{2} \frac{(n-1)}{n} = \frac{1}{2} - \frac{1}{2n}.$$

Since evidently $\frac{1}{2n} \to 0$ as $n \to \infty$, we have indeed verified that

$$\lim_{n \to \infty} \frac{\sum_0^{n-1} j}{n^2} = \frac{1}{2}.$$

IV.2.3 *Area under a Parabola*

A somewhat more complicated example involves the area under a parabola. Here the answer is no longer given by elementary geometric techniques. However, it was already determined by Archimedes over 2000 years ago, by a procedure remarkably close to the one we developed here, that the parabola splits off one third of the area of the rectangle that encloses it, as shown in Figure IV.10.

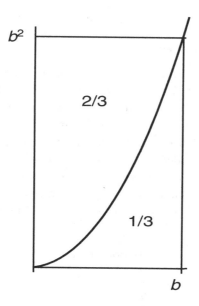

Fig. IV.10   The area under the parabola is 1/3 of the area of the rectangle.

In order to obtain this result by means of definite integrals, we describe the parabola as the graph of $f(x) = x^2$ between 0 and b. (See Figure IV.10.) The enclosing rectangle is also shown; it has area $b \cdot b^2 = b^3$. Recall that in Section 1.5 we had already examined the relevant limit process for this function f. According to the procedure developed there, the area under the parabola between 0 and b equals

$$\int_0^b f(t)\, dt = \lim_{n \to \infty} \sum_{j=0}^{n-1} f(t_j) \Delta_n t$$

$$= b^3 \lim_{n \to \infty} \frac{\sum_{j=0}^{n-1} j^2}{n^3}.$$

Based on numerical evidence, we had determined in Section 1.5 that the value of the latter limit equals $1/3$. This conclusion can indeed be confirmed by a precise theoretical argument, just as we were able to evaluate the corresponding limit in the earlier example concerning the area of a triangle. (See Problems 5 and 6 of Exercise IV.1.7 for an outline.) Thus the area under the parabola is indeed $1/3$ of the area b^3 of the rectangle.

IV.2.4   *Area of a Disc*

Finally we consider the intriguing case of the area of a disc of radius $r > 0$. Here, too, the answer *Area* $= A(r) = \pi r^2$ has been known for over 2000 years. The appearance of the irrational number π makes it evident that this formula is far from elementary, and that it must be the result of a limit process. Since the full circle—the boundary of a disc—is not the graph of a function, we consider the area of the part of the disc that lies in the first quadrant, which is exactly one fourth of the full disc. (See Figure IV.11.) The boundary circle satisfies the equation $x^2 + y^2 = r^2$; the portion in the first quadrant is the graph of $y = \sqrt{r^2 - x^2}$ for $0 \le x \le r$. Hence

$$A(r) = 4 \times area \text{ of part in first quadrant}$$

$$= 4 \times \int_0^r \sqrt{r^2 - t^2}\, dt.$$

For $n = 1, 2, \ldots$ we set $\Delta_n t = r/n$ and then $t_j = j \cdot r/n$ for $j = 1, \ldots, n$. The corresponding approximating sum S_n that represents the sum of the areas

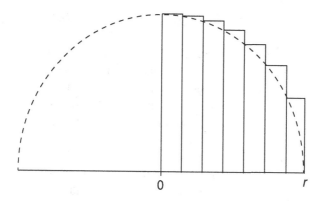

Fig. IV.11   Approximation to the area of a disc.

of the rectangles is given by

$$S_n = \sum_{j=0}^{n-1} \sqrt{r^2 - (j\frac{r}{n})^2} \cdot \frac{r}{n}$$

$$= \sum_{j=0}^{n-1} \sqrt{\frac{r^2}{n^2}(n^2 - j^2)} \cdot \frac{r}{n}$$

$$= r^2 \sum_{j=0}^{n-1} \sqrt{n^2 - j^2} \cdot \frac{1}{n^2}.$$

Thus

$$\int_0^r \sqrt{r^2 - t^2}\, dt = r^2 \lim_{n \to \infty} \sum_{j=0}^{n-1} \sqrt{n^2 - j^2} \cdot \frac{1}{n^2}.$$

The critical expression

$$\sum(n) = \frac{\sum_{j=0}^{n-1} \sqrt{n^2 - j^2}}{n^2},$$

whose limit we must find, is even more complicated than in previous cases—this should not be a surprise—and there is no elementary technique to determine the limit exactly. We therefore resort to numerical approximations. With the help of a programmable calculator, one readily obtains the values shown in Table IV.1.

n	$\sum(n)$	$4 \times \sum(n)$
$n = 10$	.8261295815	3.304518326
$n = 20$	.8071162199	3.228464880
$n = 30$	.8002774553	3.201109821
$n = 40$	.7967369335	3.186947734
$n = 50$	.7945671277	3.178268511
$n = 60$	.7930992417	3.172396967
$n = 70$	.7920392518	3.168157007
$n = 80$	.7912374493	3.164949797
$n = 90$	.7906095043	3.162438017
$n = 100$	.7901042581	3.160417032

Table IV.1. Riemann sum approximations for area of disc.

We expect that $4 \times \lim_{n\to\infty} \sum(n) = \pi = 3.14159\,27...$; notice that the evidence points in that direction, although we are still off by about 2/100. Let us consider a few larger values for n; the results are shown in Table IV.2.

n	$\sum(n)$	$4 \times \sum(n)$
$n = 1000$	.7858888662	3.143555465
$n = 10000$	.7854478701	3.141791480

Table IV.2. More approximations for area of disc.

We are getting closer, but clearly this process is not particularly efficient for approximating π. Still, we recognize that the approximation of the relevant definite integral indeed seems to converge to the expected answer

$$A(r) = 4 \cdot \int_0^r \sqrt{r^2 - t^2}\, dt = r^2\, 4 \cdot \lim_{n\to\infty} \sum(n) = r^2\pi.$$

Taking $r = 1$ and turning matters around, one obtains the following limit statement for π : ·

$$\pi = 4 \cdot \int_0^1 \sqrt{1 - t^2}\, dt = 4 \cdot \lim_{n\to\infty} \frac{\sum_{j=0}^{n-1} \sqrt{n^2 - j^2}}{n^2}.$$

We see from these examples that the very general procedure developed here indeed gives results that are consistent with familiar formulas for areas. This confirms that definite integrals are a useful tool for determining areas of very general classes of regions in the plane. Usually, evaluation has to be handled by numerical methods. However, based on Theorem 1.1, in the case where the function f is the derivative $D(F) = f$ of a known function F, the value of the definite integral can be found exactly by means of the antiderivative F.

Remark. Recall that according to Theorem 1.3 definite integrals produce antiderivatives of functions. In particular, using $r = 1$ in the formula above, that result implies that the function

$$F(x) = \int_0^x \sqrt{1 - t^2} \, dt$$

is the unique antiderivative of $\sqrt{1 - x^2}$ that satisfies $F(0) = 0$; furthermore, based on the geometric interpretation we observed that it satisfies $F(1) = \frac{\pi}{4}$. Unfortunately, F cannot be found explicitly in terms of algebraic functions alone, so this information still does not give any simple means for evaluating π exactly. We will see in Section 7.3 that the explicit formula for this antiderivative F involves the *inverse* of the sine function on the interval $[0, \frac{\pi}{2}]$.

IV.2.5 *Exercises*

1. a) Use the procedure discussed in the text with $n = 6$ to estimate the area under the graph of $y = \sin x$ between $x = 0$ and $x = \pi$. (Hint: Add up the areas of the appropriate rectangles.)
   b) Identify the definite integral whose value is exactly the area considered in a).

2. a) Make a sketch of the area whose value is given by $\int_1^3 \frac{1}{t} \, dt$.
   b) Approximate the area in a) by using approximating sums of rectangles. Use $n = 10$.

3. Explain by geometric arguments why

   a) $\int_0^2 2^t \, dt < \int_0^2 3^t \, dt$,   and

   b) $\int_0^1 2^t \, dt + \int_1^3 2^t \, dt = \int_0^3 2^t \, dt$.

IV.3 More Applications of Definite Integrals

IV.3.1 *Riemann Sums*

According to the discussion in Section 1.6, the definite integral $\int_a^b f(t)\,dt$ is a well defined quantity that arises in the construction of antiderivatives of a continuous function f. We then recognized that in the case $f \geq 0$ the definite integral can be interpreted as a certain area. The important fact to remember is that the definite integral arises as a limit of certain approximating sums that are built up according to a rather simple precise process from the given function f. Similar sums arise in numerous other situations, and consequently these sums and their limits, that is, definite integrals, have many significant applications.

Before looking at some more examples, let us first review the structure of the approximating sums that one considers. In fact, it is convenient to allow a bit more flexibility in these sums than what we had used earlier. Instead of partitions of the interval $[a, b]$ into *equal* pieces, one may choose an arbitrary partition $\mathcal{P}_n$ given by $a = t_0 < t_1 < ... < t_{n-1} < t_n = b$, where the points $t_0, t_1, ..., t_n$ are not necessarily evenly spaced. In order to control the differing sizes of the intervals $[t_{j-1}, t_j]$, one introduces the "norm" $\|\mathcal{P}_n\| = \max\{|t_{j+1} - t_j|, j = 0, ..., n-1\}$ of a partition $\mathcal{P}_n$. The approximation will then involve a sequence $\{\mathcal{P}_n, n = 1, 2, 3, ...\}$ of partitions with $\|\mathcal{P}_n\| \to 0$ as $n \to \infty$. Next, one chooses "sampling" points $t_j^* \in [t_j, t_{j+1}]$ for $j = 0, ..., n-1$. For a function $f : [a, b] \to \mathbb{R}$ one can then consider the sum

$$S_n = S_n(f, \mathcal{P}_n, \{t_j^*\}) = \sum_{j=0}^{n-1} f(t_j^*)(t_{j+1} - t_j).$$

Such a sum is called a *Riemann sum* for the function f over the interval $[a, b]$. Note that the approximating functions F_n that we considered in Section 1 involve special sums of this type, where all the intervals $[t_j, t_{j+1}]$ have equal length $\Delta_n t = (b - a)/n$, and the sampling points t_j^* were chosen to be the left endpoint t_j of the interval. In the case $f \geq 0$ one can again interpret a Riemann sum as the sum of the areas of (thin) rectangles whose union approximates the area under the graph of f; in contrast to the earlier case, the rectangles may have varying widths, and their heights do not necessarily match the value of the function at the left endpoint of the corresponding interval. (See Figure IV.12.)

Based on this geometric interpretation it thus appears plausible that these more general Riemann sums will still approach the definite integral

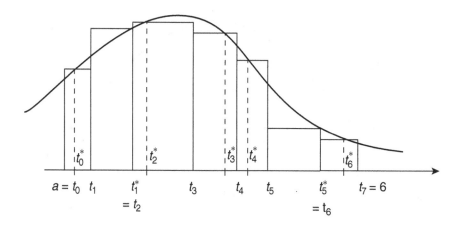

Fig. IV.12 Geometric interpretation of a Riemann sum.

of f over the interval $[a, b]$ as $\|\mathcal{P}_n\| \to 0$, i.e., as the maximal width of all the thin rectangles goes to zero. In the next section we shall formulate the relevant precise limit theorem and discuss the formal properties of definite integrals. In the remainder of this section we will discuss several examples to illustrate how these approximating sums and definite integrals arise in a variety of situations.

IV.3.2 *Areas Bounded by Graphs*

We begin by considering some geometric problems.

Example. Determine the area enclosed by the $x-$ axis and the graph of $\sin x$ between $x = 0$ and π. (See Figure IV.13.)

The relevant area is given by $\int_0^\pi \sin t \, dt$. Since the sine function has the antiderivative $F(t) = -\cos t$, and since its derivative $D(\sin)(t) = \cos t$ is bounded, Theorem 1.1 applies. Hence

$$\int_0^\pi \sin t \, dt = -\cos(\pi) - [-\cos(0)] = 1 + 1 = 2.$$

Example. Determine the area of the shaded region in Figure IV.14.

The area of interest is given by the area of the rectangle with vertices at $(-2, 0), (2, 0), (2, 4), (-2, 4)$ minus the area bounded by the $x-$axis and the graph of $y = x^2$ between $x = -2$ and $x = 2$. Hence

$$Area\,(R) = 4 \cdot 4 - \int_{-2}^2 x^2 \, dx.$$

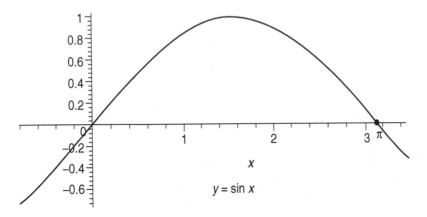

Fig. IV.13   Area under the graph of $y = \sin x$.

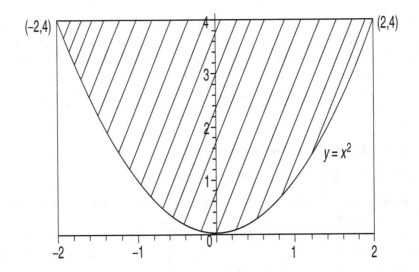

Fig. IV.14   Area inside a parabola.

By the result in Section 2.3 the area under the parabola, i.e., $\int_0^2 x^2\,dx$, equals $\frac{1}{3} \times$ *Area rectangle* $= \frac{1}{3}(2 \cdot 4) = \frac{8}{3}$. By symmetry, $\int_{-2}^0 x^2\,dx = \frac{8}{3}$ as well, so that

$$\int_{-2}^2 x^2\,dx = \int_{-2}^0 x^2\,dx + \int_0^2 x^2\,dx = \frac{8}{3} + \frac{8}{3} = \frac{16}{3}.$$

It follows that *Area* $(R) = 16 - \frac{16}{3} = \frac{2}{3} \times 16 = \frac{32}{3}$.

IV.3.3 *Volume of a Sphere*

Consider a solid ball of radius R, say $R = 10$ cm, and cut it in half. Next, cut one of the halves into 100 very thin slices each 1 mm thick, with all cuts being parallel to the first cut. Place the stack of slices flat on the table. (See Figure IV.15.)

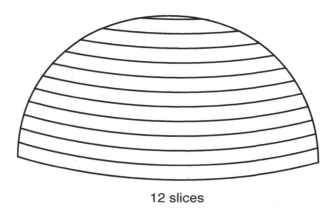

12 slices

Fig. IV.15   Half sphere approximated by a stack of slices.

Each slice looks like a thin disc of a certain radius r and thickness $1/10$ cm, so its volume is approximately $0.1\pi r^2$ cm^3. The volume of half the ball is approximately equal to the sum of the volumes of these 100 slices. In order to keep track of the varying radius, we number the slices, starting with 0 from the bottom. The *jth* slice is at height $t_j = j \times 10/100 = j/10$, so its radius r_j satisfies $t_j^2 + r_j^2 = 10^2$, i.e., $r_j^2 = 100 - t_j^2$, for $j = 0, 1, 2, ..., 99$. Adding up these volumes, we obtain an estimate for the volume $V_{1/2}$ of half the ball, that is,

$$V_{1/2} \approx \sum_{j=0}^{99} \pi(100 - t_j^2)\frac{1}{10} = \sum_{j=0}^{99} \pi(100 - t_j^2)\Delta_n t.$$

We recognize that this sum is a Riemann sum for the integral

$$\int_0^{10} \pi(100 - t^2)dt.$$

In order to determine its value, note that the function $f(t) = \pi(100 - t^2)$ is a polynomial with antiderivative $F(t) = \pi(100t - t^3/3)$. We therefore can apply Theorem 1.1 to obtain

$$V_{1/2} = \int_0^{10} \pi(100 - t^2)dt = F(10) - F(0) = \pi\frac{2}{3}10^3.$$

The volume V of the whole ball therefore is $2 \times V_{1/2} = \frac{4}{3}\pi 10^3$. All this works just as well in the case of arbitrary radius R, in which case the result is

$$V = \frac{4}{3}\pi R^3.$$

Note that the sphere is obtained by rotating a semi circle, i.e., the graph of $y = \sqrt{R^2 - x^2}$, around its diameter. A completely analogous process by "slicing" can be applied to the "solid of revolution" obtained by revolving the graph of any function $y = f(x) \geq 0$, as follows. (See Figure IV.16.)

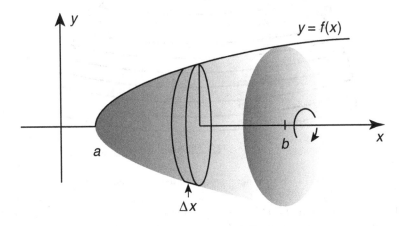

Fig. IV.16   A solid of revolution with one thin slice.

The volume of a thin slice at the point x_j is approximately $\pi[f(x_j)]^2 \cdot \Delta x$. Adding up all the volumes of the thin slices gives a Riemann sum

$$S(\pi[f(x)]^2, \mathcal{P}) = \sum_j \pi[f(x_j)]^2 \, \Delta x$$

which approximates the definite integral

$$V = \int_a^b \pi(f(x))^2 \, dx.$$

This formula thus gives the volume of such a solid of revolution.

IV.3.4   *Work of a Spring*

Next we discuss an application that arises in the physical sciences. The concept of "work" has a precise technical meaning in physics. In the simplest setting, suppose you want to lift a box weighing p lbs to a height of

d feet above ground. This requires some "work", since you must overcome the force of gravity. The weight in pounds is a measure of that force. In this case the total work done in the lifting is measured by

$$W = p \times d = force \times distance \text{ foot-pounds.}$$

This formula assumes that the force is constant along the path of motion. Near the surface of the earth the gravitational force can be assumed constant, but in the case where one considers the motion of a space ship, one needs to take into consideration that the force changes in dependence of the distance from the earth.

A variable force arises with a spring. According to Hooke's law (see also Section III.5.1), the force $F(s)$ generated by a spring that has been compressed (or stretched) by an amount s is proportional to s, at least for small values of s. So $F(s) = -ks$, where the constant $k > 0$ depends on the particular spring. (k is called the *spring constant*.) Small values of k suggest a soft spring, while large values of k correspond to stiffer springs. The minus sign reflects the fact that the force generated by the spring points in the direction that is opposite to the direction of the displacement of the spring.

In order to determine the work done by the force generated by the spring as it is compressed by an amount d, we notice that for very short displacements Δs the force is *close to constant*. So we divide the interval d into a large number n of small segments $[0, s_1], [s_1, s_2], ..., [s_{n-1}, s_n]$, each of length $\Delta_n s = d/n$. For $j = 1, ..., n$, along the *jth* segment $[s_{j-1}, s_j]$ the force is approximately $F(s_j)$, so the work done on that segment is approximately

$$W_j \approx F(s_j)(s_j - s_{j-1}).$$

The total work is then

$$W = W_1 + W_2 + ... + W_n \approx \sum_{j=1}^{j=n} F(s_j)\Delta_n s.$$

Again, the last sum is recognized as a Riemann sum (with sampling points given by the right endpoints of each small interval) which approximates a particular definite integral. We thus define the work done by the variable force $F(s)$ from $s = 0$ to $s = d$ by

$$W = \int_0^d F(s)ds.$$

In the simple case of a spring one has $F(s) = -ks$, so one obtains

$$W = \int_0^d -k\,s\,ds = -k\frac{d^2}{2},$$

where in the last equation we again have used Theorem 1.1.

IV.3.5   *Length of a Curve*

Suppose a curve is described as the graph of the differentiable function f defined on the interval $[a, b]$. In order to measure its length, we observe that locally the graph is well approximated by short line segments, for example pieces of the tangents. We use this fact to estimate the length of the curve by considering the length of suitable approximating polygons that result from a partition $\mathcal{P}_n$ of $[a, b]$ into n equal pieces $[t_0, t_1], [t_1, t_2], ..., [t_{n-1}, t_n]$, each of length $\Delta_n t = (b - a)/n$, as shown in Figure IV.17.

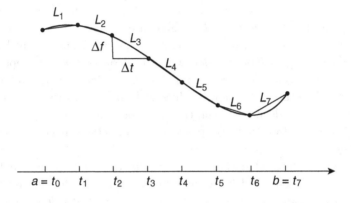

Fig. IV.17   Approximation of a curve by a polygon.

The length of the polygon shown in Figure IV.17 is given by the sum of the lengths $L_0, ..., L_{n-1}$ of the segments of the polygon. By Pythagoras Theorem, it follows from Figure IV.17 that

$$L_j^2 = (\Delta t)^2 + (\Delta f)^2 = (t_{j+1} - t_j)^2 + (f(t_{j+1}) - f(t_j))^2.$$

We now introduce the factorizations $f(t_{j+1}) - f(t_j) = q_j(t_{j+1})(t_{j+1} - t_j)$, so that

$$L_j^2 = [1 + (q_j(t_{j+1}))^2](\Delta_n t)^2.$$

Since f is differentiable, q_j is continuous at t_j, so that $q_j(t_{j+1}) \approx q_j(t_j) = D(f)(t_j)$, for $j = 0, 1, 2, ..., n - 1$, with the approximations improving as $n \to \infty$, since then $\Delta_n t = (b - a)/n \to 0$. It therefore seems reasonable that

$$L_j^2 \approx [1 + (f'(t_j))^2](\Delta_n t)^2.$$

In fact, assuming that both $|D(f)|$ and the second derivative $|D^2(f)|$ of f are bounded by a constant K over the interval $[a, b]$, variations of the arguments used in Section 1 and estimates for the linear approximation for $\sqrt{u}$ can be used to prove that

$$\left| \sum_{j=0}^{n-1} L_j - \sum_{j=0}^{n-1} \sqrt{1 + (f'(t_j))^2} \Delta_n t \right| \le K^* \|\mathcal{P}_n\|$$

for some other constant K^*. This shows that as n gets larger (i.e., $\mathcal{P}_n \to 0$), the difference between the lengths $\sum_{j=0}^{n-1} L_j$ of the approximating polygons and the corresponding Riemann sums for the function $\sqrt{1 + D(f)^2}$ goes to zero. Since the polygons approximate the curve more and more closely, we are led to define

$$L = Length\ of\ curve = \lim_{n \to \infty} \sum_{j=0}^{n-1} L_j$$

$$= \lim_{n \to \infty} \sum_{j=0}^{n-1} \sqrt{1 + (f'(t_j))^2} \Delta_n t = \int_a^b \sqrt{1 + [f'(t)]^2}\, dt.$$

In simple cases, for example if the curve is a line segment (i.e. the graph of a linear function), one can check that this formula indeed leads to the familiar answer.

IV.3.6 *Income Streams*

The lottery agency announces that the jackpot is $100 million. Are you really going to receive that amount if you hold the only winning ticket? Probably not. For one, you will have to pay taxes, but let us ignore that aspect. Reading the fine print reveals that the amount is going to be paid out in 20 equal payments of $5 million each over the next 20 years. Experience suggests that money today is more valuable than money a year from now, so the value *today* of this sequence of payments is going to be less than the full $100 million today. We shall now express the "present value" of this income stream as a certain Riemann sum that in turn approximates a corresponding definite integral.

Let us assume that interest is calculated by continuous compounding at an annual rate r. In order to receive $5 millions k years from now, one would need to invest today an initial amount of $A(k)$ millions that satisfies the relationship $A(k)e^{rk} = 5$. We can solve for

$$A(k) = 5e^{-rk}.$$

This formula gives today's value $A(k)$ (in millions) of \$5 million k years from now. In general, the value today of A dollars at some future time t years from now is

$$A(t) = Ae^{-rt}.$$

So today's value of the lottery winning (assuming that the first \$5 million is paid out at the end of the first year), is given by

$$V_{20} = A(1) + A(2) + ... + A(20) = \sum_{k=1}^{20} 5e^{-rk}.$$

The lottery agency may pay out the winnings in monthly installments of 5/12 millions per month. Now there are $12 \times 20 = 240$ months, and the monthly interest rate is $r/12$. Hence today's value of \$5/12 millions k months from now is given by $\$5/12\, e^{-\frac{r}{12}k}$. Today's total value of the *monthly* income stream therefore equals

$$V_{240} = \sum_{k=1}^{240} \frac{5}{12} e^{-\frac{r}{12}k}.$$

This looks similar to an approximating Riemann sum. To identify the corresponding definite integral, note that the interval $[0, 20]$ has been divided into 240 intervals of equal length $\Delta t = 1/12$. The right endpoint of the kth subinterval is given by the point $t_k = k\frac{1}{12}$. Hence we can rewrite the above sum as

$$V_{240} = \sum_{k=1}^{240} 5e^{-r \cdot t_k} \Delta t.$$

We now recognize that this expression is as a Riemann sum for the function $f(t) = 5\, e^{-rt}$ over the interval $[0, 20]$. By shortening the payout period even more, say to one day, one would get another Riemann sum, and so on. In the limit, one gets a "continuous" income stream, in which money is paid out each moment. Today's value V_c of the latter income stream is given by the definite integral

$$V_c = \int_0^{20} 5e^{-rt} dt.$$

With the help of a calculator one can explicitly evaluate the Riemann sums to obtain approximations for V_c. On the other hand, Theorem 1.1 again provides the exact answer more easily. In fact, the function $f(t) = 5e^{-rt}$

has the antiderivative $F(t) = 5e^{-rt}/(-r)$, whose second derivative $F''(t) = (-r)5e^{-rt}$ is bounded over any bounded interval. Consequently

$$V_c = \int_0^{20} 5e^{-rt}dt = F(20) - F(0)$$

$$= 5\frac{1}{-r}[e^{-r20} - 1] = \frac{5}{r}(1 - e^{-r20}).$$

In order to get a numerical answer, let us assume that the interest rate is 4%, i.e., $r = 0.04$. One then obtains $V_c = 5(1-e^{-0.8})/0.04 \approx 68.8$. By using this value for V_c one can estimate today's value V_{20} of the annual payout stream to be approximately $69 millions, assuming a constant interest rate of 4% over the next 20 years. So the advertised jackpot of $100 million really is worth—today—only about 2/3 of that amount.

IV.3.7 *Probability Distributions*

In probability theory one studies how a particular numerical quantity (e.g., the height of individuals in centimeters) is distributed over a given population. A common graphical presentation of the data is given by a so-called "histogram", as shown in Figure IV.18.

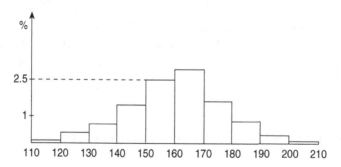

Fig. IV.18   Distribution of height among a population.

The area of each vertical box indicates the percentage of the total population whose height in cm falls within the interval given by the base of the box. For example, the vertical box over the interval $[150, 160]$ has height 2.5, so its area is $2.5 \cdot 10 = 25$. This means that 25% of the population under consideration has height between 150 cm and 160 cm. On the other hand, we see that only 9% of the population has height between 180 cm and 190 cm. The length of the intervals may be shortened to refine the details about the distribution of height.

The upper boundary of the histogram approximates a curve which is the graph of a function p that is known as the *probability density function* for the distribution. (See Figure IV.19.)

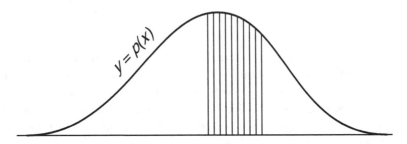

Fig. IV.19   The histogram approximates the probability density function.

The density function p is defined by the property that the area under its graph between two points a and b measures the percentage of the population whose height h lies between a and b. Since that area is given by a definite integral, one obtains that

$$\text{percentage of population with } a \leq h \leq b \; = \int_a^b p(x)\, dx.$$

We note that a histogram is just a graphical representation of a particular Riemann sum for the integral $\int_0^M p(x)\, dx$, where M is the maximal height observed in the population. Since the whole population has height between 0 and M, one clearly must have

$$\int_0^M p(x)\, dx = 100\% \; (= 1 \text{ as numerical value}).$$

An important probability density function is the *normal density* function

$$N(x) = \frac{1}{\sqrt{2\pi}} e^{-x^2/2},$$

whose graph is the "bell shaped" curve shown below in Figure IV.20 that describes the *standard normal distribution* with mean 0 and standard deviation 1. One can show that for M large (say $M = 5$) the integral

$$\int_{-M}^M N(x)\, dx$$

is very close to 1. In fact, the limit as $M \to \infty$ equals 1 exactly. This means that essentially for the whole population the characteristic measured by x falls within the interval $[-M, M]$ when M is large.

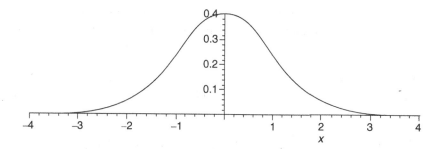

Fig. IV.20   Graph of the standard normal distribution function.

Those familiar with normal distributions may recall the rule of thumb that about 68% of the population falls within one standard deviation of the mean, and that about 95% falls within two standard deviations. Since the standard normal distribution has standard deviation 1 and mean 0, these statements are equivalent to

$$\int_{-1}^{1} \frac{1}{\sqrt{2\pi}} e^{-x^2/2} \, dx \approx 0.68 \quad \text{and} \quad \int_{-2}^{2} \frac{1}{\sqrt{2\pi}} e^{-x^2/2} \, dx \approx 0.95 \,.$$

IV.3.8   *Exercises*

1. For each part below, make a sketch of the area that is described and identify a definite integral whose value equals that area. Do not attempt to determine the value.

   a) The area bounded by the $x-$axis, the graph of $f(x) = \sqrt{x}$, and the line $x = 4$.

   b) The area under the graph of $y = \ln x$ between $x = 1$ and $x = 5$.

   c) The area that is enclosed by the graph of $g(x) = x^2 - x - 2$ and the x-axis.

   d) The area bounded by the line $x = -3$, the coordinate axes, and the graph of $f(x) = 2^{-x}$.

2. By rotating the segment of the line $y = \frac{1}{2}x$ between $x = 0$ and $x = 4$ around the x-axis one generates a circular cone C.

   a) Make a sketch that illustrates the cone that is generated.

   b) Determine a definite integral whose value equals the volume of the cone.

   c) Evaluate the definite integral in b) (Hint: Use the results obtained in Section 2.3. concerning the area under a parabola.)

d) Generalize the procedure in parts a) - c) to determine the formula for the volume of a circular cone of height h and whose base is a circle of radius r.

3. A weight of 10 lbs is attached to a spring suspended from a hook. As a consequence, the spring stretches by 9 inches. Assume that Hooke's law holds.

   a) Determine the value of the spring constant k in lbs/ft.
   b) Determine a definite integral whose value equals the work (in ft-lbs) that has been done by the weight.
   c) Evaluate the definite integral in b).

4. a) Use the formula given in the text for the length of the graph of a function in order to find the length of the line segment from the point $(0, 1)$ to the point $(5, 2)$. (Hint: What is the function whose graph contains the line segment?)
   b) Identify an explicit definite integral whose value gives the length of the parabola $y = x^2$ between 0 and s. (Do NOT try to evaluate!)

5. Identify a definite integral whose value equals the length of the upper half of the circle of radius 10 centered at the origin. (Hint: Find a function whose graph is the given curve.)

6. Consider the lottery income stream discussed in the text. Assume an annual interest rate of 6%. Use a calculator to evaluate today's value of the income stream if $5 million is paid out at the end of each of the next 20 years by adding up the appropriate sum. Compare your answer with the estimate given via definite integrals according to the process discussed in the text.

IV.4   Properties of Definite Integrals

IV.4.1   *Riemann Integrable Functions*

In the preceding section we considered several applications where definite integrals arise naturally. Let us now discuss the relevant concept more precisely. We consider a bounded function f on an interval $[a, b]$. As was mentioned in the preceding section, a partition $\mathcal{P}_n$, described by $a = t_0 < t_1 < ... < t_n = b$, and a choice of sampling points $t_j^* \in [t_{j-1}, t_j]$, determine a Riemann sum

$$S(f, \mathcal{P}_n, \{t_j^*\}) = \sum_{j=0}^{n-1} f(t_j^*)(t_{j+1} - t_j)$$

for the function f, which is an approximation for a definite integral. The type of limit that needs to be considered in this more general setting is somewhat more complicated than in the case when all small intervals in the partition have equal length. We must ensure that even when the lengths differ among the intervals, in the limit the intervals have to shrink down to a point in a "uniform" way. This is made precise by using the *norm* $\|\mathcal{P}_n\|$ of the partition $\mathcal{P}_n$, defined as the *maximum* of the length of the intervals $[t_j, t_{j+1}]$ in the partition, i.e.,

$$\|\mathcal{P}_n\| = \max\{|t_{j+1} - t_j|, j = 0, 1, ..., n - 1\}.$$

Definition 4.1. *The function f is said to be Riemann integrable[3] over the interval $[a, b]$ if*

$$\lim_{\|\mathcal{P}\| \to 0} S(f, \mathcal{P}, \{t_j^*\})$$

exists, independently of the particular choice of partitions and sampling points.

Note that as $\|\mathcal{P}\| \to 0$, the number n of intervals in the partition must necessarily get larger and larger, i.e., $n \to \infty$. More precisely, the above limit statement means that there exists a number L, so that all Riemann sums will be within an arbitrarily small prescribed distance from L for all partitions $\mathcal{P}$ whose norm $\|\mathcal{P}\|$ is sufficiently small, independently of the choice of sampling points. We write symbolically that $S(f, \mathcal{P}, \{t_j^*\}) \to L$ as $\|\mathcal{P}\| \to 0$. As in other situations involving limits, the limit L is determined uniquely in this case as well. (See Problem 1 of Exercise IV.4.4.)

As in Section 1, whenever the limit exists, its value L is called the definite integral of f over $[a, b]$, and it is denoted by

$$\int_a^b f(t)dt = \lim_{\|\mathcal{P}\| \to 0} S(f, \mathcal{P}, \{t_j^*\}).$$

---

[3]The name *Riemann* (B. Riemann, 1826 - 1866) distinguishes the particular notion of *integrable* considered here from other more general notions of "integrable" used in advanced mathematics.

Once it is known that a function f is integrable over $[a, b]$, then its definite integral $\int_a^b f(t)dt$ may be evaluated by choosing special sequences of convenient Riemann sums, for example, by the procedure discussed in Section 1 involving partitions into pieces of equal length, with the left endpoints chosen as sampling points.

The following result identifies a large class of functions that are Riemann integrable.

Theorem 4.2. *Every function f that is monotonic (either increasing or decreasing) over the interval $[a, b]$ is Riemann integrable over that interval.*

We shall prove this theorem in Section 6. Note that this result covers most functions that we have encountered, as long as one restricts them to appropriate intervals on which they are monotonic. For example, $f(x) = x^n$ is monotonic on $[0, \infty)$, and consequently, by the theorem, f is Riemann integrable over each interval $[a, b]$ if $0 \le a < b$. f is also monotonic on $(-\infty, 0]$, so that f is integrable on $[c, d]$ if $c < d \le 0$. By general properties discussed in the following section, it then follows that f is integrable over each interval $[c, b]$; furthermore, by taking linear combinations it then follows also that every polynomial is integrable over any closed bounded interval.

More generally, one has the following theorem that extends Theorem 1.2 from the special partitions considered in that section to the more general ones introduced here.

Theorem 4.3. *If f is continuous on the closed and bounded interval $[a, b]$, then f is Riemann integrable over $[a, b]$.*

We shall prove this theorem in Section 6 under the additional hypothesis that f is differentiable and its derivative $D(f)$ is bounded on $[a, b]$.

Remark. It is not necessary for a function to be continuous at all points of the interval in order for it to be integrable. For example, an increasing function (integrable by Theorem 4.2) may have jumps (i.e., simple discontinuities) at many points. Also, the function f defined by $f(x) = \sin(1/x)$ for $x \ne 0$ and $f(0) = 0$, is Riemann integrable over $[0, 1]$, even though its graph is pretty wild near $x = 0$ (take a look at it with a graphing calculator!). More generally, based on Theorem 4.3, it is quite easy to verify that a bounded function g that is continuous on $[a, b]$ except for finitely many points, is integrable over $[a, b]$. In advanced mathematics one introduces the technical concept of (Lebesgue) *measure* that generalizes the concept of length of intervals to more general sets. For example, the measure of an interval $[a, b] \subset \mathbb{R}$ is equal to its length $b - a$, while the (infinite) set of all rational numbers in $\mathbb{R}$ has measure zero! (Just another way of reminding us that the rational numbers, although everywhere dense, are a negligibly small subset of the real numbers.) One can then prove the following complete characterization. *A bounded function f is Riemann integrable over $[a, b]$ if and only if the set of points $E \subset [a, b]$ where f is NOT continuous is a set of "measure 0".*

IV.4.2 *Basic Rules for Integrals*

We now list some general rules that are very useful when working with definite integrals. The relevant statements formalize several quite natural properties, and they can be verified by routine, though somewhat tedious and uninspiring, arguments. Typically, one verifies such a statement for Riemann sums, and then one passes to the limit. In all statements we assume that the given functions are integrable over the relevant intervals. More precisely, in Rule i) we assume that the functions on the right side are integrable over the relevant interval, and it then follows that the integral on the left exists as well.

Rule 0) **Constant function integral**: *If $k \in \mathbb{R}$, then $\int_a^b k \, dt = k(b-a)$.*

Rule i) **Linearity:** $\int_a^b (cf(t)+dg(t))dt = c \int_a^b f(t) \, dt + d \int_a^b g(t) \, dt$, *where $c, d \in \mathbb{R}$ are constants.*

Rule ii) **Monotonicity:** *If $f(x) \leq g(x)$ for all $x \in [a, b]$, then*

$$\int_a^b f(t)dt \leq \int_a^b g(t)dt.$$

Rule iii) **Additivity:** *If $a < c < b$, then $\int_a^b f \, dt = \int_a^c f \, dt + \int_c^b f \, dt$.* More precisely, if the integral(s) on one side exist, so do the integral(s) on the other side, and the equality holds.

Remarks. 1) It is convenient to define the definite integral $\int_a^b f(t) \, dt$ also in the case $b < a$ in such a way that Rule iii) continues to hold for any three numbers a, b, c, regardless of their order. In particular, if one wants Rule iii) to hold in all cases, taking $a = b$ in Rule iii) leads to $0 = \int_a^a f \, dt = \int_a^c f(t)dt + \int_c^a f(t) \, dt$. This suggests that one should define

$$\int_a^c f(t) \, dt = - \int_c^a f(t) \, dt \text{ whenever } c < a.$$

One then checks that with this definition the Rule iii) holds for any three numbers a, b, c. For example

$$\int_4^1 x^3 \, dx = \int_4^0 x^3 \, dx + \int_0^1 x^3 \, dx.$$

2) Monotonicity implies the following "standard" estimates:
i) If $f \geq 0$ on $[a, b]$, then $\int_a^b f \, dt \geq 0$;
ii)

$$-\int_a^b |f| \, dt \leq \int_a^b f \, dt \leq \int_a^b |f| \, dt, \text{ and hence } \left| \int_a^b f \, dt \right| \leq \int_a^b |f| \, dt \, ;$$

iii) If $|f(x)| \leq M$ on $[a, b]$, then

$$\left| \int_a^b f \, dt \right| \leq M \cdot (\text{length of } [a, b]).$$

IV.4.3 *Examples*

We consider a few examples.

a) Estimate $\int_1^2 e^{t^2} \, dt$. Since $e = e^1 \leq e^{t^2} \leq e^4$ for $1 \leq t \leq 2$ one obtains

$$\int_1^2 e^{t^2} \, dt \leq e^4(2 - 1) = e^4.$$

Similarly, one also obtains $e \leq \int_1^2 e^{t^2} \, dt$.

b) $0 \leq \int_0^{\pi/2} \sin t \, dt \leq 1 \cdot \frac{\pi}{2}$.

c) $\int_{-\pi/2}^{\pi/2} \sin t \, dt = \int_{-\pi/2}^0 \sin t \, dt + \int_0^{\pi/2} \sin t \, dt$

$$= -\int_{-\pi/2}^0 \sin(-t) \, dt + \int_0^{\pi/2} \sin t \, dt,$$

since $\sin t$ is an *odd* function, i.e., $\sin(-t) = -\sin(t)$. By writing down approximating Riemann sums and rearranging the order of the summands, or by looking at the areas (see Figure IV.21) one readily sees that

$$\int_{-\pi/2}^0 \sin(-t) \, dt = \int_0^{\pi/2} \sin t \, dt.$$

Consequently

$$\int_{-\pi/2}^{\pi/2} \sin t \, dt = 0.^4$$

---

[4] Of course this follows more directly by applying Theorem 1.1.

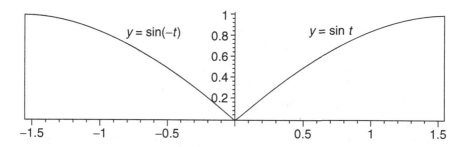

Fig. IV.21 The area on the left obtained by reflection equals the area on the right.

d) Determine $\int_0^5 4f(t)\, dt$ if $\int_0^6 f(t)\, dt = 5$ and $\int_5^6 f(t)\, dt = 2$. By additivity,

$$5 = \int_0^6 f(t)\, dt = \int_0^5 f(t)\, dt + \int_5^6 f(t)\, dt = \int_0^5 f(t)\, dt + 2;$$

hence $\int_0^5 f(t)\, dt = 5 - 2 = 3$, and by linearity it then follows that

$$\int_0^5 4f(t)\, dt = 4 \int_0^5 f(t)\, dt = 4 \cdot 3 = 12.$$

IV.4.4   *Exercises*

1. Show that if a function f is Riemann integrable over $[a, b]$, then the limit L that appears in the definition is determined uniquely. Complete the following steps.

   a) Suppose L_1 and L_2 are two limits of $S(f, \mathcal{P}, t^*)$ as $\|\mathcal{P}\| \to 0$. Show that for each $n \in \mathbb{N}$, there is $\mathcal{P}_n$ with sufficiently small norm, so that $|S(f, \mathcal{P}_n, t^*) - L_j| < 1/2n$ for $j = 1$ and 2.
   b) Use a) and the triangle inequality to conclude that $|L_1 - L_2| < 1/n$ for each n.
   c) Explain why b) implies that $L_1 = L_2$.

2. a) Find lower and upper bounds for the definite integrals $\int_1^2 \frac{1}{t}\, dt$ and $\int_2^3 \frac{1}{t}\, dt$.
   b) Use the estimates in a) and the additivity of the integral to verify that

$$\frac{1}{2} + \frac{1}{3} \leq \int_1^3 \frac{1}{t}\, dt \leq 1 + \frac{1}{2}.$$

  c) Use the process used in a) and b) to obtain lower and upper bounds for $\int_1^n \frac{1}{t} dt$, where n is a positive integer.

3. Suppose $\int_1^6 f(t)dt = 4$ and $\int_3^6 f(t)dt = 6$. What is $\int_1^3 f(t)dt$?

4. a) Which of the integrals $\int_0^1 (1-t)dt$ and $\int_0^1 e^{-t}dt$ is the larger one?

  b) Compare $\int_{-1}^0 (1-t)dt$ and $\int_{-1}^0 e^{-t}dt$. Which one is larger?

  c) Make a sketch that shows appropriate areas in order to illustrate the answers in a) and b).

5. Explain why $\int_0^{2\pi} \sin t\ dt = 0$ by using properties of definite integrals. (Hint: Use $\sin(\pi + t) = -\sin t$ for $t \in [0, \pi]$.)

6. Explain why $\int_0^1 x^3 dx > \int_0^1 x^4 dx$.

7. a) Estimate $\int_1^3 \ln t\ dt$ by using the standard estimate for integrals.

  b) Improve your estimate in a) by using $\int_1^3 \ln t\ dt = \int_1^2 \ln t\ dt + \int_2^3 \ln t\ dt$.

  c) Repeat the process used in b) to obtain lower and upper bounds for $\int_1^n \ln t\ dt$, where n is a positive integer.

IV.5   The Fundamental Theorem of Calculus

The essential features of the inverse relationship between differentiation and integration were already described in Section 1. Here we summarize these results and provide relevant proofs.

IV.5.1   *The Derivative of an Integral*

Recall from Section 1 that definite integrals came up naturally in the construction of antiderivatives. More precisely, the procedure developed in that section suggested that—under suitable hypothesis—the function F defined by $F(x) = \int_a^x f(t)\, dt$ should be an antiderivative of f. Assuming the existence of the definite integral, we shall now see how the basic rules for definite integrals listed in the previous section allow us to justify quite easily this most important result that provided the motivation for introducing the approximating sums that eventually led to the concept of definite integral.

Theorem 5.1. (*Fundamental Theorem of Calculus, Part I*) *If f is continuous on the interval I and $a \in I$, then the function F defined by*

$$F(x) = \int_a^x f(t)\, dt$$

is an antiderivative of f on the interval I, i.e., F is differentiable and $D(F)(x) = f(x)$ for $x \in I$.

Proof. In order to prove this result, recall from Section II.3.3 that F is differentiable at the point c if there exists a factorization

$$F(x) - F(c) = q_c(x)(x - c), \qquad (IV.6)$$

where q_c is continuous at c, or, equivalently, if $\lim_{x \to c} q_c(x)$ exists. One then has $D(F)(c) = q_c(c) = \lim_{x \to c} q_c(x)$.

From the definition of F and the additivity of the integral it follows that

$$F(x) - F(c) = \int_a^x f(t)\,dt - \int_a^c f(t)\,dt$$
$$= \int_c^x f(t)\,dt.$$

If we assume $f(t) \geq 0$ and $x > c$, and set $h = x - c > 0$, the last integral can be visualized by the darker shaded area shown in Figure IV.22. This

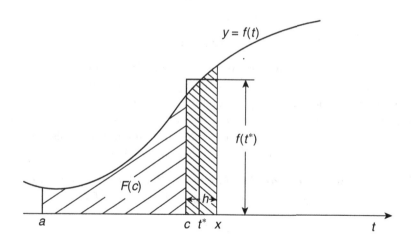

Fig. IV.22   Geometric representation of $F(x) - F(c)$ and $f(t^*)(x - c)$.

suggests that this shaded area is equal to the area $f(t^*)(x-c)$ of a rectangle with base $(x-c)$ and height $f(t^*)$ for an appropriate value t^* in the interval $[c, x]$. So the factor $q_c(x)$ is given by $f(t^*)$, and since f is continuous at c, it follows that $\lim_{x \to c} q_c(x) = \lim_{t^* \to c} f(t^*) = f(c)$. This simple geometric argument captures the main idea of the proof of the theorem.

For completeness, let us turn this intuitive geometric argument into a precise proof. We introduce $f(t) = f(c) + [f(t) - f(c)]$ into the integral and use linearity to obtain

$$\int_c^x f(t)\, dt = \int_c^x f(c)dt + \int_c^x [f(t) - f(c)]\, dt$$
$$= f(c)(x - c) + g(x)(x - c)$$
$$= [f(c) + g(x)](x - c),$$

where for $x \neq c$ the function g is given by $g(x) = \int_c^x [f(t) - f(c)]\, dt/(x - c)$. Therefore the factor q_c in equation (IV.6) is given by $q_c(x) = f(c) + g(x)$. By the basic estimate,

$$\left| \int_c^x [f(t) - f(c)]dt \right| \leq \sup\{|f(t) - f(c)| : |t - c| \leq |x - c|\} \cdot |x - c|.$$

After dividing by $(x - c) \neq 0$ one obtains the estimate

$$|g(x)| \leq \sup\{|f(t) - f(c)| : |t - c| \leq |x - c|\}.$$

Since f is continuous at c, this supremum goes to 0 as $x \to c$. It follows that $\lim_{x \to c} q_c(x) = f(c)$, and we are done.   ∎

IV.5.2   *The Integral of a Derivative*

The Theorem we just proved solves the problem of constructing antiderivatives in the "abstract", although it does not provide a practical tool for finding explicit antiderivatives in concrete situations. On the other hand, whenever one can find an explicit antiderivative of f by some other means, say by inspection or by reversing the rules of differentiation, this information can be used to evaluate definite integrals, as follows.

Theorem 5.2. *(**Fundamental Theorem of Calculus, Part II**) Let f be continuous on $[a, b]$, and let G be any antiderivative of f. Then*

$$\int_a^b f(t)\, dt = G(b) - G(a).$$

Since $f = D(G)$, this theorem says that $\int_a^b D(G)(t)\, dt = G(b) - G(a)$. We had already proved this version in Sections 1.2 and 1.3 under some additional restriction on $D(G)$.

Proof. Since $F(x) = \int_a^x f(t)\, dt$ and G are both antiderivatives of the same function f, they differ by a constant C, i.e., $F(x) = G(x) + C$ for all $x \in [a,b]$. (Recall Corollary III.2.3.) Since $F(a) = 0$, it follows that $G(a) + C = 0$, i.e., $C = -G(a)$. Therefore

$$\int_a^b f(t)\, dt = F(b) = G(b) + C = G(b) - G(a).$$

∎

IV.5.3 *Some Examples*

In Section 3 we had already used a version of Theorem 5.2 to evaluate definite integrals in special cases. For example, the area enclosed by the $x-axis$ and the graph of $y = \sin x$ between 0 and π is given by $\int_0^\pi \sin x\, dx$. Since $-\cos x$ is an antiderivative of $\sin x$, the answer is given by

$$\int_0^\pi \sin x\, dx = -\cos(\pi) - (-\cos 0) = 1 + 1 = 2.$$

Remark on notation. If G is a function defined on the interval $[a,b]$, we introduce the notation

$$[G]_a^b = G\,|_a^b = G(b) - G(a).$$

Evaluation of definite integrals via antiderivatives is then written compactly as in the following examples.

$$\int_0^{\pi/2} \cos t\, dt = \sin t \,|_0^{\pi/2} = \sin\frac{\pi}{2} - \sin 0 = 1,\ \text{and}$$

$$\int_1^2 (4x^3 - 3x)dx = [x^4 - \frac{3}{2}x^2]_1^2 = (16 - \frac{3}{2}4) - (1 - \frac{3}{2}) = 10\frac{1}{2}.$$

Warning. Here, as well as in most analogous evaluations of definite integrals, one must be very careful with the signs. It is safest to use brackets to enclose all terms that need to be subtracted.

By using this notation, the evaluation of the volume of a sphere of radius R (see Section 3.3) is obtained in compact form as follows:

$$V = 2 \cdot \int_0^R \pi(R^2 - t^2)\, dt$$

$$= 2\pi[R^2 t - t^3/3]_0^R = 2\pi[(R^3 - R^3/3) - 0]$$

$$= 2\pi(\frac{2}{3}R^3) = \frac{4}{3}\pi R^3.$$

Example. We had also seen that the area of a disc of radius R is given by

$$A = 4 \int_0^R \sqrt{R^2 - t^2} dt.$$

In this case we encounter a major difficulty, since there is no simple way to identify an antiderivative of $\sqrt{R^2 - t^2}$. In particular, it is not at all evident from this formula how the number π makes its appearance in the (known) answer $A = \pi R^2$. We will determine an appropriate antiderivative in Section 7.3.

Further examples are found in the exercises.

IV.5.4 *Exercises*

1. Find the derivative of the following functions.
   a) $F(x) = \int_1^x \cos(t^2 + 1) dt$;   b) $G(x) = \int_1^{x^3} \cos(t^2 + 1) dt$;
   c) $L(s) = \int_1^s \frac{dt}{t}$. (Hint: In b), write G as the composition of two functions.)

2. a) Find the derivative of $q(x) = \int_x^0 \sqrt{t^4 + 1} dt$;
   b) Find the derivative of $G(x) = \int_{-x}^0 f(t) dt$;
   c) Find the derivative of $S(x) = \int_{-x}^x f(t) dt$.(Hint: Begin by writing the integral as a sum of two simpler integrals.)
   (Assume f continuous in b) and c).)

3. Evaluate the definite integrals
   a) $\int_1^3 2^t dt$;   b) $\int_{-1}^1 x^6 dx$;   c) $\int_{-\pi/2}^{\pi/2} \cos t \, dt$;   d) $\int_0^4 \sqrt{x} dx$.

4. Find the area bounded by the graph of $y = 1/x$ and the x-axis between $x = 1$ and $x = e$.

5. The graph of $f(x) = x^2$ between 0 and 2 is rotated around the x-axis.
   a) Make a sketch of the solid that is enclosed by this process.
   b) Find the volume of the solid.

6. Consider the same graph as in Problem 5, but rotate it around the y-axis. The solid obtained is called a *paraboloid*. Find its volume. (Hint: Replace the variables x and y to transform the problem into the setting of solids of revolution considered in Section 3.3.

7. The continuity of f is necessary in Part I of the Fundamental Theorem of Calculus. Consider the function defined by
   $$f(t) = -1 \text{ for } t < 0, \quad f(0) = 0, \quad f(t) = +1 \text{ for } t > 0.$$

a) Sketch the graph of f. Where is f continuous? Where not?

b) Define $H(x) = \int_0^x f(t)\, dt$. Explain why $H(x) = |x|$ for all x.

c) Note that H is not differentiable at 0. (Why?) Show that $H'(x) = f(x)$ for all $x \neq 0$.

8. Find a simple formula for $\int_0^x \frac{df}{dt}\, dt$ in terms of the function f. (Assume f' is continuous.)

IV.6 Existence of Definite Integrals

In this section we shall prove the basic existence theorems for definite integrals that we had announced in Section 4. The proofs are a little bit technical, and the reader who is eager to proceed may very well skip this section.

IV.6.1 *Monotonic Functions*

We first prove Theorem 4.2, which states that monotonic functions (either increasing or decreasing) are Riemann integrable. The main point of the proof is surprisingly simple. Let us assume that f is increasing over $[a, b]$. (The case when f is decreasing is handled by a simple modification of the following arguments.) Suppose $\mathcal{P} = (t_0, t_1, ..., t_n)$, where $t_0 = a$ and $t_n = b$, is an arbitrary partition of $[a, b]$. By using the right endpoints as sampling points, we obtain the *upper sum* $\overline{S}(f, \mathcal{P}) = \sum_{j=0}^{n-1} f(t_{j+1})(t_{j+1} - t_j)$. Similarly, we introduce the corresponding *lower sum* $\underline{S}(f, \mathcal{P}) = \sum_{j=0}^{n-1} f(t_j)(t_{j+1} - t_j)$ by using the left endpoints. Given an arbitrary choice $\{t_j^*\}$ of sampling points, since f is increasing, one clearly has $f(t_j) \leq f(t_j^*) \leq f(t_{j+1})$, and consequently

$$\underline{S}(f, \mathcal{P}) \leq S(f, \mathcal{P}, \{t_j^*\}) \leq \overline{S}(f, \mathcal{P}). \tag{IV.7}$$

We then observe that $\overline{S}(f, \mathcal{P}) - \underline{S}(f, \mathcal{P}) = \sum_{j=0}^{n-1} [f(t_{j+1}) - f(t_j)](t_{j+1} - t_j)$. Since $f(t_{j+1}) - f(t_j) \geq 0$ for each j (f is increasing!), it follows that

$$0 \leq \overline{S}(f, \mathcal{P}) - \underline{S}(f, \mathcal{P}) \leq \left(\sum_{j=0}^{n-1} [f(t_{j+1}) - f(t_j)] \right) \|\mathcal{P}\|$$

where we have used that $|t_{j+1} - t_j| \leq \|\mathcal{P}\|$. Finally notice that

$$\sum_{j=0}^{n-1} [f(t_{j+1}) - f(t_j)] = f(t_n) - f(t_0) = f(b) - f(a),$$

since all other terms cancel. Therefore

$$0 \leq \overline{S}(f, \mathcal{P}) - \underline{S}(f, \mathcal{P}) \leq [f(b) - f(a)] \, \|\mathcal{P}\| \, . \tag{IV.8}$$

The estimates (IV.7) and (IV.8) establish the critical ingredients that follow from the monotonicity of f. To complete the proof we need to find a number L that is contained in each interval $[\underline{S}(f, \mathcal{P}), \, \overline{S}(f, \mathcal{P})]$

As we just saw, $\underline{S}(f, \mathcal{P}) \leq \overline{S}(f, \mathcal{P})$ for any partition $\mathcal{P}$. We will show below that, more generally, $\underline{S}(f, \mathcal{P}) \leq \overline{S}(f, \mathcal{Q})$ for any two partitions $\mathcal{P}$ and $\mathcal{Q}$ of $[a, b]$, that is, every lower sum is less than or equal to any upper sum. It then follows that

$$\sup\{\underline{S}(f, \mathcal{P}) : \mathcal{P}\} \leq \inf\{\overline{S}(f, \mathcal{Q}) : \mathcal{Q}\}.$$

By chosing partitions $\mathcal{P}$ with $\|\mathcal{P}\| \to 0$, the estimate (IV.8) shows that we must have equality in the preceding inequality. In fact, given an arbitrary positive integer n, choose a partition $\mathcal{P}_n$ with $\|\mathcal{P}_n\| < \frac{1}{n[f(b)-f(a)]}$. Estimate (IV.8) then implies that $0 \leq \overline{S}(f, \mathcal{P}_n) - \underline{S}(f, \mathcal{P}_n) < 1/n$, and hence

$$0 \leq \inf\{\overline{S}(f, \mathcal{Q}) : \mathcal{Q}\} - \sup\{\underline{S}(f, \mathcal{P}) : \mathcal{P}\} \leq \overline{S}(f, \mathcal{P}_n) - \underline{S}(f, \mathcal{P}_n) < 1/n.$$

Since n is arbitrary, the desired equality follows.

Let L denote this common value

$$L = \sup\{\underline{S}(f, \mathcal{P}) : \mathcal{P}\} = \inf\{\overline{S}(f, \mathcal{Q}) : \mathcal{Q}\}.$$

Then $\underline{S}(f, \mathcal{P}) \leq L \leq \overline{S}(f, \mathcal{P})$ for any partition $\mathcal{P}$. By using estimates (IV.7) and (IV.8) one then obtains

$$\begin{aligned} \left| S(f, \mathcal{P}, \{t_j^*\}) - L \right| &\leq \overline{S}(f, \mathcal{P}) - \underline{S}(f, \mathcal{P}) \\ &\leq [f(b) - f(a)] \, \|\mathcal{P}\| \, . \end{aligned}$$

Since the term on the right clearly goes to 0 as $\|\mathcal{P}\| \to 0$ we have established that $\lim_{\|\mathcal{P}\| \to 0} S(f, \mathcal{P}, \{t_j^*\}) = L$. So f is indeed integrable, with $\int_a^b f(t)dt = L$.

In order to prove the estimate $\underline{S}(f, \mathcal{P}) \leq \overline{S}(f, \mathcal{Q})$ for any two partitions $\mathcal{P}$ and $\mathcal{Q}$, we first prove the following lemma.

Lemma 6.1. *Suppose* $\mathcal{P}^{\#}$ *is obtained from the partition* $\mathcal{P}$ *by adding a single point* $t^{\#}$ *to* $\mathcal{P}$. *Then*

$$\underline{S}(f, \mathcal{P}) \leq \underline{S}(f, \mathcal{P}^{\#}) \leq \overline{S}(f, \mathcal{P}^{\#}) \leq \overline{S}(f, \mathcal{P}).$$

Proof. We shall prove the inequality on the right. The one on the left follows by an analogous argument. If $t^\# = t_j$ for some j, then $\mathcal{P}^\# = \mathcal{P}$ and there is nothing to prove. So we may assume that there is an index j with $t_j < t^\# < t_{j+1}$, so that the interval $[t_j, t_{j+1}]$ decomposes into the intervals $[t_j, t^\#]$ and $[t^\#, t_{j+1}]$ in $\mathcal{P}^\#$. We estimate the corresponding sum of the two terms in the upper sum $\overline{S}(f, \mathcal{P}^\#)$ as follows. Since f is increasing,

$$f(t^\#)(t^\# - t_j) + f(t_{j+1})(t_{j+1} - t^\#)$$
$$\leq f(t_{j+1})(t^\# - t_j) + f(t_{j+1})(t_{j+1} - t^\#) = f(t_{j+1})[(t^\# - t_j) + (t_{j+1} - t^\#)]$$
$$= f(t_{j+1})(t_{j+1} - t_j).$$

Since all other terms in $\overline{S}(f, \mathcal{P}^\#)$ are identical to those in $\overline{S}(f, \mathcal{P})$, this estimate implies the desired inequality in the Lemma.

Now suppose $\mathcal{P}$ and $\mathcal{Q}$ are any two partitions of $[a, b]$. We denote by $\mathcal{P} \cup \mathcal{Q}$ the partition obtained by combining the points in $\mathcal{P}$ and $\mathcal{Q}$. Note that $\mathcal{P} \cup \mathcal{Q}$ is obtained from either $\mathcal{P}$ or $\mathcal{Q}$ by successive additions of finitely many points one point at a time. Every time a point is added one can apply the Lemma. It follows that

$$\underline{S}(f, \mathcal{P}) \leq \underline{S}(f, \mathcal{P} \cup \mathcal{Q}) \leq \overline{S}(f, \mathcal{P} \cup \mathcal{Q}) \leq \overline{S}(f, \mathcal{Q}),$$

which implies the desired estimate $\underline{S}(f, \mathcal{P}) \leq \overline{S}(f, \mathcal{Q})$. ∎

IV.6.2 *Functions with Bounded Derivatives*

Finally we prove Theorem 4.3 in the following slightly weaker form.

Theorem 6.2. *Suppose f is differentiable on the interval I and that its derivative $D(f)$ is bounded over the interval $[a, b] \subset I$. Then f is Riemann integrable over $[a, b]$.*

While this version of Theorem 4.3 does not cover *all* continuous functions, it is quite sufficient for most applications. Recall, for example, that all algebraic functions defined on an interval I satisfy the above hypothesis. Furthermore, all elementary transcendental functions that we have encountered, as well as any standard combinations of them, will also satisfy this hypothesis. Continuous functions that are not differentiable, or whose derivatives are not bounded on closed bounded intervals, are indeed quite rare. Furthermore, simple examples such as $f(x) = |x|$ can be handled, for example, by splitting up an interval $[-2, 3]$ into $[-2, 0] \cup [0, 3]$, so that the hypotheses of the Theorem are now satisfied on each subinterval. Also, this function is monotonic on each of these subintervals, so its integrability also follows from Theorem 4.2.

Proof. As we saw in the proof of the integrability of monotonic functions, a key step is to introduce appropriate upper and lower sums $\overline{S}$ and $\underline{S}$ corresponding to a partition $\mathcal{P}$, so that the difference $\overline{S} - \underline{S}$ can be readily estimated by $\|\mathcal{P}\|$. Since f is in particular continuous on $[a, b]$, and hence also on each subinterval $[t_j, t_{j+1}]$ of the partition $\mathcal{P}$, we can use the fact that f takes on both a maximum and a minimum on each such interval (Theorem II.4.6). Therefore, for each $j = 0, 1, ..., n - 1$ there are numbers m_j and $M_j \in [t_j, t_{j+1}]$ so that

$$f(m_j) \leq f(t) \leq f(M_j) \text{ for all } t \in [t_j, t_{j+1}].$$

We then define the *lower* sum $\underline{S}(f, \mathcal{P}) = \sum_{j=0}^{n-1} f(m_j)(t_{j+1} - t_j)$ and the *upper* sum $\overline{S}(f, \mathcal{P}) = \sum_{j=0}^{n-1} f(M_j)(t_{j+1} - t_j)$. Note that this naturally generalizes upper and lower sums for monotonic functions, in which case the maxima and minima occurred at appropriate endpoints of the intervals. Clearly one has, as in the monotonic case, that

$$\underline{S}(f, \mathcal{P}) \leq S(f, \mathcal{P}, \{t_j^*\}) \leq \overline{S}(f, \mathcal{P}).$$

for any choice of sampling points $\{t_j^*\}$. Furthermore, a simple modification of the proof of Lemma 6.1 yields the conclusion that $\underline{S}(f, \mathcal{P}) \leq \overline{S}(f, \mathcal{Q})$ for any two partitions $\mathcal{P}$, $\mathcal{Q}$ in this case as well. (See Problem 1 of Exercise IV.6.3.)

The crux of the matter is then an estimation of $\overline{S}(f, \mathcal{P}) - \underline{S}(f, \mathcal{P})$, as follows. Clearly one has

$$0 \leq \overline{S}(f, \mathcal{P}) - \underline{S}(f, \mathcal{P}) \leq \sum_{j=0}^{n-1} |f(M_j) - f(m_j)| (t_{j+1} - t_j). \qquad \text{(IV.9)}$$

Given a bound $|D(f)(x)| \leq K$ for all $x \in [a, b]$ for the derivative of f, the Mean Value Inequality (Theorem III.2.1) implies that

$$|f(M_j) - f(m_j)| \leq K |M_j - m_j| \leq K |t_{j+1} - t_j| \leq K \|\mathcal{P}\|. \qquad \text{(IV.10)}$$

Inserting (IV.10) into (IV.9) shows that

$$\overline{S}(f, \mathcal{P}) - \underline{S}(f, \mathcal{P}) \leq K \|\mathcal{P}\| \sum_{j=0}^{n-1} (t_{j+1} - t_j) = K \|\mathcal{P}\| (b - a).$$

One therefore has all the essential ingredients to complete the proof as in the case of monotonic functions. ∎

IV.6.3 *Exercises*

1. Suppose the partition $\mathcal{P}^{\#}$ is obtained from $\mathcal{P}$ by adding a single point. By suitably modifying the proof of Lemma 6.1 verify that $\underline{S}(f, \mathcal{P}) \leq \underline{S}(f, \mathcal{P}^{\#})$ and $\overline{S}(f, \mathcal{P}^{\#}) \leq \overline{S}(f, \mathcal{P})$ for any continuous function f.

2. Carefully complete the proof of Theorem 6.2 by using the estimates obtained in the text.

IV.7   Reversing the Chain Rule: Substitution

In Section III.2.3 we had already collected a few formulas for antiderivatives that were obtained by simply reversing basic rules for differentiation. The reader should review those formulas at this point.

IV.7.1 *Integrals that Fit the Chain Rule*

We now turn to the chain rule, one of the most important formulas for differentiation. Unfortunately, there is no simple way to invert the chain rule in all cases, that is, even if we know antiderivatives of f and g, in general there is no way to express $\int (f \circ g)(t)\, dt = \int f(g(t))\, dt$ in terms of $F = \int f$ and $G = \int g$. Instead, in order to invert the chain rule, one must start with a function that has a special structure. With $F = \int f$, recall that the chain rule states that

$$D(F \circ g)(t) = D(F)(g(t)) \cdot D(g)(t) = f(g(t)) \cdot g'(t).$$

To gain greater clarity, we introduce the variable u for the input into the function F, and we note that $D(F)(u) = f(u)$. The formula above can be read to state that $F \circ g$ is an antiderivative of the function $F'(g(t)) \cdot g'(t) = [(f \circ g) \cdot g'](t)$, i.e.,

$$\int f(g(t)) \cdot g'(t)\, dt = F(g(t)), \text{ where } F(u) = \int f(u)\, du. \qquad \text{(IV.11)}$$

There is a very useful notational device that helps to remember this relationship, and that provides a straightforward mechanism for implementation. Formally, the notation $g'(t) = \frac{dg}{dt}$ can be rewritten in the form $dg = g'(t)dt$, which states that the "differential" dg of the function g is given by multiplying the differential dt of the variable t by $g'(t)$. Therefore the integral on the right side in (IV.11) is transformed into the one on the left by simply substituting $u = g(t)$ and $du = g'(t)dt$, i.e.,

$$F(u) = \int f(u)\, du = \int f(g(t))\cdot g'(t)\, dt = F(g(t)), \text{ where } u = g(t).$$

$$(IV.12)$$

This "inverse" of the chain rule is also known as "integration by substitution", or simply as "substitution formula". The name refers to the fact that formal substitution of $u = g(t)$ transforms the expression on the left side of (IV.12) into the expression on the right side. It is important to remember that the substitution must occur not only in $F(u)$ and $f(u)$, but also in the term du, which accounts for the extra factor $g'(t)$. So, when looking for antiderivatives of functions that involve a composition $f \circ g$, the additional factor g' must be present in order to apply the formula above.

IV.7.2   *Examples*

1. The antiderivative $\int \cos(t^2)\, dt$, involves the composition of $\cos u$ with $u = g(t) = t^2$, but there is no factor $g'(t) = 2t$. So the above formula cannot be applied, at least not in an obvious way. On the other hand, $\cos(t^2)(2t)$ has the appropriate structure, and therefore

$$\int \cos(t^2)(2t)\, dt = \int \cos(g(t))\, g'(t)\, dt = \int \cos u \; du$$

$$= \sin u = \sin(g(t)) = \sin(t^2).$$

2. $\int (x^4 + 3)^5\, dx$ also does not fit the structure. In this case one could proceed differently: simply expand $(x^4+3)^5$ algebraically into a polynomial, whose antiderivative is then easy to find.

In contrast, $\int (x^4 + 3)^5 (4x^3)\, dx$ is of the form $\int f(g(x))\cdot g'(x)\, dx$, where $u = g(x) = x^4 + 3$, $f(u) = u^5$. It follows that

$$\int (x^4 + 3)^5 (4x^3)\, dx = \int u^5 du = \frac{u^6}{6} = \frac{1}{6}(x^4 + 3)^6.$$

Sometimes the given function differs from the appropriate structure only by a *constant* factor. This situation can easily be corrected because $\int c\, f(x)\, dx = c \int f(x)\, dx$ for $c \in \mathbb{R}$.

Example. $\int e^{kt}dt = \int f(g(t))\, dt$, with $f(u) = e^u$ and $u = g(t) = kt$. The missing factor $g'(t) = k$ is constant, so we note that if $k \neq 0$, then

$$\int e^{kt}\, dt = \int \frac{1}{k}(ke^{kt})\, dt = \frac{1}{k}\int e^{kt}k\, dt$$

$$= \frac{1}{k}\int e^u du = \frac{1}{k}e^u = \frac{1}{k}e^{kt}.$$

Similarly

$$\int e^{t^2} t \, dt = \frac{1}{2} \int e^{t^2} 2t \, dt = \frac{1}{2} \int e^{g(t)} g'(t) \, dt,$$

where $u = g(t) = t^2$. Hence

$$\int e^{t^2} t \, dt = \frac{1}{2} \int e^u du = \frac{1}{2} e^u = \frac{1}{2} e^{t^2}.$$

One also shows by substitution that if $F(x) = \int f(x) \, dx$, then for any constant $c \neq 0$ one has

$$\int f(cx) dx = \frac{1}{c} F(cx).$$

We emphasize once again that the technique used in the preceding examples only works for *constant* factors. If the "missing" factor explicitly involves the variable of integration, there is usually no procedure to correct the matter.

IV.7.3   *Changing Integrals by Substitution*

So far we have used the substitution formula to find antiderivatives of expressions that fit the left side of equation (IV.11), thereby reducing to the—hopefully—simpler antiderivative on the right side. But if presented with the problem of finding an explicit formula for an antiderivative $\int F(u) du$ in the case where there is no obvious known formula that gives the answer, we can start on the left side in (IV.12), introduce a substitution with an appropriate function $u = g(t)$ of our choice, and then work with the resulting integral on the right side of (IV.12). While this might appear to complicate matters, we could in fact be lucky and obtain an expression on the right side for which we can find an explicit antiderivative.

We illustrate the successful implementation of this method by working through an important example that came up earlier. Recall from Section 2.4 that the area of a quarter of a disc of radius R is given by

$$\int_0^R \sqrt{R^2 - t^2} \, dt.$$

In order to apply the Fundamental Theorem of Calculus to evaluate the definite integral exactly, one needs to find an antiderivative of $\sqrt{R^2 - t^2}$. This function, with $g(t) = R^2 - t^2$, does not fit into the left side of the substitution formula (IV.11); furthermore, the missing term $g'(t) = -2t$ is not constant, and therefore it cannot just be introduced out of nowhere as in some of the examples we just considered. So we try the alternative and view the given integral as the left side in the substitution formula (IV.12). In order to be consistent with the earlier notation we replace t by u, and we simplify by assuming $R = 1$ first. So we consider

$$\int \sqrt{1 - u^2} \, du.$$

The trigonometric identity $1 - \sin^2 t = \cos^2 t$ suggests the substitution $u = \sin t$ in order to eliminate the square root. We may restrict to $0 \leq u \leq 1$, and hence $0 \leq t \leq \pi/2$. On this interval $\cos t \geq 0$, so that this substitution gives

$$\sqrt{1 - u^2} = \sqrt{\cos^2 t} = \cos t.$$

Since $du = \cos t \, dt$, the substitution $u = \sin t$ leads to

$$\int \sqrt{1 - u^2} \, du = \int \cos t \cdot \cos t \, dt = \int \cos^2 t \, dt.$$

The antiderivative on the right side is still not immediately visible in explicit form. Still, we managed to get rid of the square root. Armed with trigonometric formulas, we can look for further modifications. One basic trigonometric identity states $\cos(2t) = \cos^2 t - \sin^2 t$. (This is a special case of the addition formula for the cosine function.) Add the identity $1 = \cos^2 t + \sin^2 t$ to the preceding equation to obtain $1 + \cos 2t = 2\cos^2 t$, that is

$$\cos^2 t = \frac{1}{2}(1 + \cos(2t)).$$

We have thus eliminated the exponent 2 on $\cos^2 t$, ending up with a much simpler function that we can easily handle. In fact, since $\int \cos(2t) dt = 1/2 \int \cos(2t) 2 dt = 1/2 \sin(2t)$, it readily follows that

$$\frac{1}{2} \int [1 + \cos(2t))] dt = \frac{1}{2}[t + \frac{1}{2}\sin(2t)].$$

After using the trigonometric formula $\sin(2t) = 2\sin t \cos t$, we end up with

$$\int \sqrt{1 - u^2} \, du = \frac{1}{2}(t + \sin t \cos t), \text{ where } u = \sin t.$$

In order to get the final answer in terms of u, we must reverse the substitution, that is, we must replace t on the right side by appropriate expressions in u. This is trivial for $\sin t = u$, and we had already seen that $\cos t = \sqrt{1 - u^2}$. For t itself we need to introduce the inverse of the sine function, i.e., $t = \arcsin u$, which is defined for $0 \leq u \leq 1$. (Recall Section

I.5.4.). After expressing the right side above in terms of u, we get the somewhat complicated and surprising formula

$$\int \sqrt{1 - u^2}\, du = F(u) = \frac{1}{2}(\arcsin u + u\sqrt{1 - u^2}).$$

The reader may wonder at this point whether what looks like "magic" indeed ends up with a correct result. However, one can verify the correctness of the result by carefully applying the various rules of differentiation from Sections II.6 and II.7. (See Problem 6 of Exercise IV.7.4 for more details.)

We can now evaluate the original definite integral exactly. Since $\arcsin 0 = 0$ and $\arcsin 1 = \frac{\pi}{2}$ (think $\sin \pi/2 = 1$), it follows that

$$\int_0^1 \sqrt{1 - u^2}\, du = F(1) - F(0) = \frac{1}{2}\frac{\pi}{2} - 0 = \frac{\pi}{4},$$

which is indeed the expected value.

Finally, in the case of a disc of arbitrary radius R, we reduce $\int \sqrt{R^2 - u^2}\, du$ on the interval $-R \le u \le R$ to the previous case by substituting $u = Rs$, $du = Rds$, resulting in

$$\int \sqrt{R^2 - u^2}\, du = \int \sqrt{R^2 - R^2 s^2}\, Rds = R^2 \int \sqrt{1 - s^2}\, ds$$

$$= R^2 \frac{1}{2}(\arcsin s + s\sqrt{1 - s^2}) \quad (\text{substitute } s = \frac{u}{R})$$

$$= R^2 \frac{1}{2}(\arcsin \frac{u}{R} + \frac{u}{R}\sqrt{1 - \frac{u^2}{R^2}}).$$

It then follows that

$$\int_0^R \sqrt{R^2 - u^2}\, du = R^2 \frac{\pi}{4},$$

so that the area of the full disc is indeed πR^2.

While this example turns out to be fairly complicated, it does illustrate how the substitution technique may transform an apparently intractable problem into one that can eventually be solved, although the solution may involve unexpected new functions, such as the inverse sine function in the case at hand.

IV.7.4   *Exercises*

1. Find the antiderivatives
   a) $\int x \sin(x^2)\, dx$;   b) $\int \sin x\ 2^{\cos x}\, dx$;   c) $\int \frac{x}{x^2+1}\, dx$;
   d) $\int (t^4 + 1)^{10}\, t^3 dt$.

2. Find $\int \frac{t}{\sqrt{t^2+1}} dt$.

3. Evaluate $\int_0^\pi \cos(\frac{1}{3}x)\, dx$.

4. a) Find the antiderivative $\int \sin x \, \cos^4 x \, dx$.
   b) Find the antiderivative $\int \sin^3 x \, \cos^4 x \, dx$.
   (Hint: $\sin^3 x = \sin x \, \sin^2 x = \sin x \, (1 - \cos^2 x)$; now use variations of a)).

5. Find $\int \sin^2 x \cos^3 x \, dx$. (Hint: Modify the techniques used in 4.)

6. Verify by differentiation that $F(x) = \frac{1}{2}(\arcsin x + x\sqrt{1-x^2})$ is an antiderivative of $f(x) = \sqrt{1-x^2}$ on the interval $(-1,1)$. (Hint: See Section II.6.5 for the derivative of $y = \arcsin x$; for the other part, use the product rule from Section II.7.1 and the Chain Rule.)

IV.8   Reversing the Product Rule: Integration by Parts

IV.8.1   *Partial Integration*

Recall the product rule for differentiation

$$(f \cdot g)' = f' \cdot g + f \cdot g'$$

from Section II.7.1. Translating this formula to antiderivatives results in the equation

$$f \cdot g = \int (f'g + fg')dt = \int f'g \, dt + \int fg' \, dt.$$

After solving for the second integral on the right one obtains

$$\int fg' \, dt = fg - \int f'g \, dt. \tag{IV.13}$$

This formula shows how finding the antiderivative of a product can be transformed into a *different* antiderivative problem if an antiderivative $g = \int g'$ of just one of the factors (g' in formula (IV.13)) can be found, i.e., if one "integrates just a part". This technique and the relevant formula (IV.13) is known as "partial integration", or also "integration by parts".

    Example. We consider $\int xe^x \, dx$. Since $\int e^x dx = e^x$ is known, we apply formula (IV.13) by choosing $g' = e^x$ and $f = x$, with $f' = 1$. One obtains

$$\int xe^x \, dx = \int x \, (e^x)' dx = xe^x - \int (x)' e^x dx$$

$$= xe^x - \int 1e^x dx = xe^x - e^x.$$

One can check the answer by careful application of the known differentiation rules, as follows.

$$[xe^x - e^x]' = (x)'e^x + x(e^x)' - (e^x)'$$
$$= 1e^x + xe^x - e^x$$
$$= xe^x.$$

Since we know the antiderivative $\int x = x^2/2$, the partial integration formula (IV.13) can also be applied by choosing $g' = x$ and $f = e^x$, with $g = \frac{x^2}{2}$ and $f' = e^x$. The result is

$$\int xe^x dx = \frac{x^2}{2}e^x - \int \frac{x^2}{2}e^x dx.$$

While this formula is indeed correct, it leads us to a more complicated antiderivative than we started with. This clearly is not useful. Choices need to be made, with an eye towards appropriate simplifications.

Example. A process similar to the preceding example, based on $\int \sin t = -\cos t$, leads to

$$\int t \sin t \, dt = t(-\cos t) - \int 1(-\cos t) \, dt$$
$$= -t \cos t + \int \cos t \, dt$$
$$= -t \cos t + \sin t.$$

Sometimes integration by parts needs to be repeated several times. For example,

$$\int x^2 e^x \, dx = x^2 e^x - \int 2xe^x \, dx$$
$$= x^2 e^x - 2 \int xe^x dx$$
$$= x^2 e^x - 2[xe^x - \int 1e^x dx]$$
$$= x^2 e^x - 2xe^x + 2e^x.$$

Again, the answer could be checked by differentiation. It is clear that repetition of this process allows us to determine explicitly $\int x^n e^x dx$ or $\int x^n \sin x \, dx$ for any fixed positive integer n. On the other hand, the reader should check that integration by parts is not helpful for finding the antiderivative $\int \sqrt{x}e^x \, dx$, no matter how the factors are chosen.

IV.8.2   *Some Other Examples*

Another twist arises with $\int \cos x \; e^x \, dx.$ Here the choice

$$g' = e^x \, , f = \cos x, \text{ with } g = e^x \text{ and } f' = -\sin x,$$

leads to

$$\int \cos x \, e^x dx = \cos x \; e^x - \int (-\sin x)e^x \, dx$$

$$= \cos x \; e^x + \int \sin x \; e^x \, dx.$$

Repeating the process with the choices $g' = e^x \, , \; f = \sin x,$ with $g = e^x$ and $f' = \cos x,$ results in

$$\int \sin x \, e^x dx = \sin x \, e^x - \int \cos x \, e^x \, dx.$$

Since we end up with the integral we started with, we seem to have gone around in a circle. But careful attention to the signs shows that

$$\int \cos x \; e^x \, dx = \cos x \; e^x + \sin x \; e^x - \int \cos x \, e^x dx,$$

so that

$$2 \int \cos x \, e^x dx = \cos x \, e^x + \sin x \; e^x.$$

It follows that

$$\int \cos x \, e^x dx = \frac{1}{2}e^x(\cos x + \sin x).$$

Sometimes the product is not visible right away. For example, $\int \ln x \, dx$ seems to involve only one function. Yet $\ln x = 1 \cdot \ln x,$ so that integration by parts with $g' = 1, g = x,$ and $f = \ln x, \; f' = 1/x,$ leads to

$$\int \ln x \, dx = x \ln x - \int x \cdot \frac{1}{x} \, dx$$

$$= x \ln x - \int 1 dx$$

$$= x \ln x - x.$$

IV.8.3 *Partial Integration of Differentials*

We had already seen the convenience of using differentials when we studied integration by substitution. Similarly, the formula for integration by parts is sometimes also written as a formula for differentials. In fact, this may provide a useful way to remember the formula. By using the equations $df = f'dt$, and $dg = g'dt$, the integration by parts formula (IV.13) takes the form

$$\int f\,dg = fg - \int g\,df.$$

This version of integration by parts provides an immediate solution for the antiderivative of $\ln x$ considered earlier, as follows. By the above,

$$\int \ln x\,dx = \ln x\ x - \int x\,d\ln x,$$

and one then proceeds with the remaining integral as before by using $d\ln x = dx/x$.

One must be careful when streamlining the symbolic notation. It is easy to overlook minor changes that can lead to major differences. For example, if in the substitution formula

$$\int (f \circ g)dg = \int (f \circ g)g'dt = \int f(u)du, \text{with } u = g(t),$$

the expression $(f \circ g)dg$ is erroneously contracted to $f\,dg$, one is led to $\int f\,dg$, which is the starting point for integration by parts.

IV.8.4 *Remarks on Integration Techniques*

Let us analyze the effect of applying different techniques to determine the antiderivative $\int \sin t \cos t\,dt$.

i) Integration by parts with $f' = \cos t, f = \sin t, g = \sin t, g' = \cos t$ gives

$$\int \sin t \cos t\,dt = \sin t\ \sin t - \int \sin t \cos t\,dt.$$

After rearranging one gets

$$2\int \sin t \cos t\,dt = \sin^2 t,$$

so that

$$\int \sin t \cos t\,dt = \frac{1}{2}\sin^2 t = F_1.$$

ii) Substitution with $u = g(t) = \cos t$, $g'(t) = -\sin t$ leads to

$$\int \cos t \, \sin t \, dt = -\int g(t) \, g'(t) \, dt$$

$$= -\int u \, du = -\frac{1}{2} u^2$$

$$= -\frac{1}{2} \cos^2 t = F_2.$$

iii) Using the trigonometric identity $\sin(2t) = 2 \sin t \, \cos t$ results in

$$\int \cos t \, \sin t \, dt = \frac{1}{2} \int \sin(2t) \, dt$$

$$= \frac{1}{2} \int (\sin u) \frac{1}{2} \, du \quad \text{(substitute } u = 2t, \ du = 2dt)$$

$$= -\frac{1}{4} \cos u = -\frac{1}{4} \cos(2t) = F_3.$$

Note that the three answers all look different, yet they are all correct, as is easily checked by differentiation. This apparent inconsistency is resolved by recalling that antiderivatives are only determined up to a constant. Let us therefore verify explicitly that any two of the solutions we obtained indeed differ only by a constant. First,

$$F_1 - F_2 = \frac{1}{2} \sin^2 t - \left(-\frac{1}{2} \cos^2 t\right)$$

$$= \frac{1}{2} (\sin^2 t + \cos^2 t) = \frac{1}{2} \cdot 1.$$

Next, by another trigonometric identity, $\cos(2t) = \cos^2 t - \sin^2 t$, so that (careful with the signs!)

$$F_1 - F_3 = \frac{1}{2} \sin^2 t - \left(-\frac{1}{4}(\cos^2 t - \sin^2 t)\right)$$

$$= \frac{1}{4} (\sin^2 t + \cos^2 t) = \frac{1}{4}.$$

It then follows that $F_2 - F_3 = F_1 - F_3 - (F_1 - F_2)$ is constant as well.

IV.8.5   *A Word of Caution*

The examples we discussed in the last two sections illustrate that explicit calculation of antiderivatives is not as straightforward a process as taking derivatives. Typically, it is not easy to recognize from the given problem which particular technique should be applied. Also, it is usually not clear at the outset whether a chosen technique will indeed lead to some progress, until one has worked out a fair amount of details. Consequently, finding explicit antiderivatives requires combining certain general principles with ingenuity, and it often involves trial and error. Traditional calculus courses included a detailed and lengthy exploration of many variations of these techniques in a variety of situations. Today's powerful symbolic computing software (such as *Maple* or *Mathematica*) can handle such computations much more efficiently and quickly, allowing the student to focus on understanding the essential ideas rather than getting lost in a multitude of special cases and techniques. Experience and practice are helpful, but there are limits even for the most experienced mathematician, since a particular problem may not have an explicit answer at all in terms of known functions. So, no matter how hard one tries to apply known techniques and tricks, it may all be of no use... .

IV.8.6 *Exercises*

1. Find explicit formulas for the antiderivatives

   a) $\int x \cos x \, dx$;   b) $\int x^2 \cos x \, dx$;   c) $\int x^3 \, e^x dx$;   d) $\int t \, 2^t \, dt$.

2. Find $\int \sin t \, e^t \, dt$.

3. Find $\int x^2 \ln x \, dx$. (Hint: Compare with $\int \ln x \, dx$ in the text.)

4. Find an antiderivative of $y = \arcsin x$ as follows.

   a) Apply integration by parts to $\int \arcsin x \cdot 1 \, dx$, with $g' = 1$. (See Section II.6.5 for the derivative of $y = \arcsin x$.)
   b) Use the substitution $u = 1 - x^2$ in the remaining integral.

IV.9 Higher Order Approximations, Part 2: Taylor's Theorem

In Section III.9 we had considered approximations of functions by so-called Taylor polynomials. The discussion culminated with some remarkable new representations for the exponential and trigonometric functions. The reader should briefly review the earlier discussion. At that time we accepted the main results based on intuitive principles and graphical evidence. We are now in a position to use integrals to formulate a precise formula for the error between a function and its Taylor polynomials. This result is known as *Taylor's Theorem with Remainder*. As an application, we then verify the convergence statements we had made in Section III.9, and we shall place these results in a broader context.

IV.9.1   *An Application of Integration by Parts*

Suppose f is an algebraic function, or more generally, a function whose derivatives of any order exist on an interval I. The derivatives of f are then continuous and hence bounded on any closed bounded interval contained in I. Therefore, by Theorem 6.2, any such derivative is then integrable. Fix a point $a \in I$. Since f is an antiderivative of $D(f)$, the Fundamental Theorem of Calculus implies that

$$f(x) - f(a) = \int_a^x D(f)(t)dt = \int_a^x f'(t)dt \quad \text{for any } x \in I. \quad \text{(IV.14)}$$

We now fix x and apply integration by parts to the integral on the right side, in the form $\int f'dg = f'g - \int g\,d(f')$, where we choose g so that $dg = 1dt$, that is, so that $g' = 1$. Hence $g(t) = t + c$, where we are free to choose the constant c. It follows that

$$\int f'(t)1dt = f'(t)(t + c) - \int f''(t)(t + c)dt.$$

The right side is an antiderivative of f' on the left. By the Fundamental Theorem of Calculus (Part 2) one therefore obtains

$$\int_a^x f'(t)dt = [f'(t)(t + c)]_a^x - \int_a^x f''(t)(t + c)dt$$

$$= f'(x)(x + c) - f'(a)(a + c) - \int_a^x f''(t)(t + c)dt.$$

In order to simplify this expression, we choose $c = -x$, so that the first term vanishes. (Remember: we treat x as constant; differentiation and integration are with respect to t!) We eliminate the minus signs in the preceding formula by interchanging the order in the resulting terms $(a - x)$ and $(t - x)$. We then obtain the formula

$$\int_a^x f'(t)dt = f'(a)(x - a) + \int_a^x f''(t)(x - t)dt. \quad \text{(IV.15)}$$

By combining formulas (IV.14) and (IV.15) and rearranging it follows that

$$f(x) = f(a) + \int_a^x f'(t)dt$$

$$= f(a) + f'(a)(x - a) + \int_a^x f''(t)(x - t)dt .$$

Recall that $y = f(a) + f'(a)(x - a) = P_1(x; a)$ is the equation of the tangent line to the graph of f at the point where $x = a$, that is, the Taylor polynomial of f of order 1 at $x = a$. So the integral on the right in the last equation measures exactly the error between the tangent line and the graph of the function at the point x. Notice that this error is controlled by the second derivative of f, that is by the convexity of the graph, as we had anticipated in Section III.9.

It now seems reasonable to repeat the integration by parts in the remaining integral, with the choice $dg = (x - t)dt$, and consequently $g(t) = \int (x - t)\,dt = -(x - t)^2/2$. After evaluating $\int f''dg = f''g - \int g\,df''$ between the bounds a and x, it follows that

$$\int_a^x f''(t)(x - t)dt = \frac{f''(a)}{2}(x - a)^2 + \int_a^x f'''(t)\frac{(x - t)^2}{2}dt.$$

The pattern is clearly visible. Because of the hypothesis on f, this process can be continued as many additional times as desired, resulting in the following theorem.

Theorem 9.1. *Suppose f is infinitely often differentiable on the interval I, and fix $a \in I$. Then for any positive integer n and $x \in I$ one has the representation*

$$f(x) = f(a) + f'(a)(x - a) + \frac{f''(a)}{2}(x - a)^2 + \dots + \frac{f^{(n)}(a)}{1 \cdot 2 \cdot 3 \cdot \dots \cdot n}(x - a)^n$$
$$+ \int_a^x f^{(n+1)}(t)\frac{(x - t)^n}{1 \cdot 2 \cdot 3 \cdot \dots \cdot n}dt.$$

Corollary 9.2. *Let $P_n(x; a)$ be the nth order Taylor polynomial of f centered at the point $a \in I$. Then for all $x \in I$ one has*

$$f(x) - P_n(x; a) = \int_a^x f^{(n+1)}(t)\frac{(x - t)^n}{n!}dt,$$

and hence

$$|f(x) - P_n(x; a)| \leq \max_{t \in [a,x]} \left| f^{(n+1)}(t) \right| \frac{|x - a|^{n+1}}{n!}.$$

Proof. The first part is clear. To prove the estimate, note that for t between a and x as in the definite integral, one has $|x - t| \leq |x - a|$. For such t one then has

$$\left| f^{(n+1)}(t)(x - t)^n \right| \leq \left(\max_{t \in [a,x]} \left| f^{(n+1)}(t) \right| \right) |x - a|^n ,$$

and the result follows by application of the standard estimate (see Section 4.2), which adds an additional factor $|x - a|$. ∎

If $|f(x) - P_n(x; a)| \to 0$ as $n \to \infty$ for all x in some interval I centered at a, one writes that

$$f(x) = \lim_{n \to \infty} P_n(x; a) = \sum_{k=0}^{\infty} \frac{f^{(k)}(a)}{k!}(x - a)^k \text{ for } x \in I, \qquad \text{(IV.16)}$$

where the meaning of the "infinite" sum identified by the symbol $\sum_{k=0}^{\infty}$ is given as the limit of the approximating partial sums P_n. This infinite sum on the right side of equation (IV.16) is called the *Taylor series* of f centered at a.

IV.9.2 *Taylor Series of Elementary Transcendental Functions*

We can now consider some examples.

The Exponential Function. The function $f(x) = e^x$ is particularly simple, since $f^{(n)}(x) = e^x$ for all n. We choose $a = 0$. From Corollary 9.2 one then obtains for fixed $x \in \mathbb{R}$ the estimate

$$\left| \int_0^x f^{(n+1)}(t) \frac{(x - t)^n}{n!} dt \right| \leq e^{|x|} \frac{|x|^{n+1}}{n!} .$$

The following lemma implies that for any fixed x the expression $e^{|x|} |x|^{n+1} /n! = (e^{|x|} |x|) |x|^n /n!$ has limit 0 as $n \to 0$. Therefore the Taylor polynomials

$$P_n(x; 0) = \sum_{k=0}^{n} \frac{1}{k!} x^k$$

converge to the function $f(x) = e^x$ at all points x, a result which we write in the form

$$e^x = \lim_{n \to \infty} \sum_{k=0}^{n} \frac{1}{k!} x^k = \sum_{k=0}^{\infty} \frac{1}{k!} x^k$$

$$= 1 + x + \frac{x^2}{2!} + \frac{x^3}{3!} + \frac{x^4}{4!} + \frac{x^5}{5!} + \dots .$$

The infinite sum that appears in the expression above is the *Taylor series of e^x* centered at 0. Note that it converges to e^x for all real numbers x.

Lemma 9.3. *For any fixed number $c > 0$ one has*

$$\lim_{n \to \infty} \frac{c^n}{n!} = 0.$$

Proof. The result may appear surprising, since numerical evidence shows that if c is quite large, say $c = 1000$, the numbers $1000^n/n!$ do at first grow quite rapidly as n increases. So one needs to look more carefully. Given c, we fix an integer $N > c$. Then $r_c = c/N < 1$, so that $\lim_{k \to \infty} r_c^k = 0$. (See Problem 3 of Exercise IV.9.5.) For $n > N$ one has

$$\frac{c^n}{n!} = \frac{c^N}{N!} \cdot \frac{c}{N+1} \cdot \ldots \cdot \frac{c}{n}$$

$$\leq \frac{c^N}{N!} \cdot \left(\frac{c}{N+1}\right)^{n-N}$$

$$\leq \frac{c^N}{N!} \cdot r_c^{n-N}.$$

Since $k = n - N \to \infty$ as $n \to \infty$, one has $r_c^{n-N} \to 0$ as $n \to \infty$, and the result follows. ∎

The essential message expressed by Lemma 9.3 is that factorials eventually grow much faster than powers, no matter how large the base.

We have thus verified the result that we had inferred in Section III.9 from the graphical evidence. In particular, we have verified the following representation for the base e of the natural exponential function:

$$e = e^1 = \lim_{n \to \infty} \sum_{k=0}^{n} \frac{1}{k!} = \sum_{k=0}^{\infty} \frac{1}{k!}.$$

Trigonometric Functions. For the function $f(x) = \sin x$, any derivative is $\pm \sin x$ or $\pm \cos x$, so that one always has $\left| f^{(n)}(t) \right| \leq 1$ for all $t \in \mathbb{R}$. With $a = 0$, it follows (use Lemma 9.3 again!) that

$$\left| \int_0^x f^{(n+1)}(t) \frac{(x-t)^n}{n!} dt \right| \leq 1 \cdot \frac{|x|^{n+1}}{n!} \to 0$$

as $n \to \infty$. Recalling the expression for the Taylor polynomials of the sine function from Section III.9. we therefore obtain from Corollary 9.2 that

$$\sin x = \lim_{m \to \infty} \sum_{k=0}^{m} (-1)^k \frac{1}{(2k+1)!} x^{2k+1} = \sum_{k=0}^{\infty} (-1)^k \frac{1}{(2k+1)!} x^{2k+1}$$

$$= x - \frac{x^3}{3!} + \frac{x^5}{5!} - \frac{x^7}{7!} + \frac{x^9}{9!} - \ldots$$

for any $x \in \mathbb{R}$.

Completely analogous arguments show that for $x \in \mathbb{R}$ one has

$$\cos x = \lim_{m \to \infty} \sum_{k=0}^{m} (-1)^k \frac{1}{(2k)!} x^{2k} = \sum_{k=0}^{\infty} (-1)^k \frac{1}{(2k)!} x^{2k}$$

$$= 1 - \frac{x^2}{2!} + \frac{x^4}{4!} - \frac{x^6}{6!} + \frac{x^8}{8!} - \dots .$$

We shall explore the relationship between exponential and trigonometric functions that is suggested by these formulas in the next section.

The Natural Logarithm. Finally, let us consider the function $f(x) = \ln x$ for $x > 0$. We choose the point $a = 1$, since we know the value $\ln 1 = 0$ precisely. Recall that $f'(x) = \frac{1}{x}$, $f''(x) = -1/x^2$, $f'''(x) = 1 \cdot 2/x^3, \dots$, $f^{(n)}(x) = (-1)^{n-1}(n-1)!/x^n$. Hence

$$f^{(n)}(1)/n! = (-1)^{n-1}/n.$$

It follows that the *nth* order Taylor polynomial centered at $a = 1$ equals

$$P_n(x; 1) = (x - 1) - \frac{1}{2}(x - 1)^2 + \frac{1}{3}(x - 1)^3 - \dots + (-1)^{n-1} \frac{1}{n}(x - 1)^n.$$

In order to estimate the difference

$$|\ln x - P_n(x; 1)| = \left| \int_1^x f^{(n+1)}(t) \frac{(x - t)^n}{n!} dt \right|$$

we first consider the case that $x \geq 1$. Then $t \geq 1$ as well, so that $\left| f^{(n+1)}(t) \right| = n!/t^{n+1} \leq n!$. The standard estimate then implies that

$$|\ln x - P_n(x; 1)| \leq \int_1^x \left| f^{(n+1)}(t) \right| \frac{(x - t)^n}{n!} dt$$

$$\leq \int_1^x n! \frac{(x - t)^n}{n!} dt = \int_1^x (x - t)^n dt$$

$$= \left[-\frac{1}{n + 1}(x - t)^{n+1} \right]_0^x = \frac{1}{n + 1}(x - 1)^{n+1}.$$

The latter expression goes to 0 as $n \to \infty$ as long as $0 \leq (x - 1) \leq 1$. We have thus shown that $\ln x = \lim_{n \to \infty} P_n(x; 1)$ for $1 \leq x \leq 2$. Note that this also includes the boundary value $x = 2$. Furthermore, it follows from the expression for P_n that there is no convergence if $|x - 1| > 1$. (See Problem 4 of Exercise IV.9.5.)

In the remaining case $0 < x < 1$ we must be more careful with the estimation. Note that since now $0 < x \leq t \leq 1$ in the integral, it follows that $x \leq x/t$, and therefore $1 - x/t \leq 1 - x$ for all $t \in [x, 1]$. Also $|x - t| = (t - x)$ and $1/t \leq 1/x$ for $x \leq t$. These remarks imply that for $0 < x < 1$ one has

$$\left| \int_1^x f^{(n+1)}(t) \frac{(x-t)^n}{n!} dt \right| \leq \int_x^1 \left| f^{(n+1)}(t) \frac{(x-t)^n}{n!} \right| dt$$

$$= \int_x^1 \frac{n!}{t^{n+1}} \frac{(t-x)^n}{n!} dt = \int_x^1 \frac{1}{t} \frac{(t-x)^n}{t^n} dt$$

$$\leq \int_x^1 \frac{1}{x} (1 - \frac{x}{t})^n dt \leq \frac{1}{x} \int_x^1 (1 - x)^n dt$$

$$= \frac{1}{x} (1-x)^n \int_x^1 dt = \frac{1}{x} (1-x)^{n+1}.$$

Since $0 \leq 1 - x < 1$, this last expression converges to 0 as $n \to \infty$. (See Problem 3 of Exercise IV.9.5.)

To summarize, we have shown that

$$\ln x = \lim_{n \to \infty} P_n(x; 1) = \lim_{n \to \infty} \sum_{k=1}^n (-1)^{k-1} \frac{(x-1)^k}{k}$$

$$= \sum_{k=1}^\infty (-1)^{k-1} \frac{(x-1)^k}{k} \quad \text{for all } x \text{ with } 0 < x \leq 2.$$

In particular, for $x = 2$ we get the remarkable representation

$$\ln 2 = 1 - \frac{1}{2} + \frac{1}{3} - \frac{1}{4} + \frac{1}{5} - \frac{1}{6} + \dots \ .$$

Finally, by replacing x with $1 + x$ one obtains the representation

$$\ln(1 + x) = \sum_{k=1}^\infty (-1)^{k-1} \frac{x^k}{k}.$$

This is the Taylor series of $\ln(1 + x)$ centered at $a = 0$. Note that it converges to $\ln(1 + x)$ for $-1 < x \leq 1$.

IV.9.3 *Power Series*

The Taylor series we obtained in the previous section are examples of so-called *power series* (with center a)

$$\sum_{k=0}^\infty c_k (x - a)^k.$$

These are to be interpreted as the limit of the partial sums

$$S_n(x) = \sum_{k=0}^{n} c_k (x - a)^k$$

as $n \to \infty$. The limit may or may not exist, depending on the particular coefficients $c_0, c_1, c_2, \ldots$ and on the value chosen for x. For example, we had seen that the Taylor series of e^x converges for all values x, while the Taylor series of $\ln x$ centered at 1 converges only for $0 < x \leq 2$. On the other hand, the power series $\sum k! \, x^k$ does not converge for any $x \neq 0$. (See Problem 6 of Exercise IV.9.5.) If the series does not converge at x we say that it diverges at x. The following result gives information about the set of points at which a given power series converges.

Theorem 9.4. *A power series* $S = \sum_{k=0}^{\infty} c_k (x-a)^k$ *either converges for all* x, *or else there exists a number* $R \geq 0$ *so that* S *converges if* $|x - a| < R$ *and diverges if* $|x - a| > R$.

The number R is called the radius of convergence of the power series. In the case where the power series converges for all x, one also says that $R = \infty$. The open interval $\{x : |x - a| < R\}$ is called the interval of convergence. A power series always converges at its center a, even when $R = 0$. Note that the theorem does not say anything about the endpoints x with $|x - a| = R$. Anything can happen here. For example, the Taylor series for $\ln x$ centered at $a = 1$ has radius of convergence $R = 1$, it converges at the boundary point 2, but not at the other boundary point 0 of the interval of convergence.

A power series $\sum_{k=0}^{\infty} c_k (x - a)^k$ with a positive radius of convergence R defines a function

$$F(x) = \sum_{k=0}^{\infty} c_k (x - a)^k$$

on the interval of convergence. Such a function can be viewed as a generalization of a polynomial, sort of a polynomial of infinite degree. In fact, the analogy goes much deeper. For example, one can apply the standard rules of calculus as if the sum were finite. More precisely, one has the following result.

Theorem 9.5. *The function*

$$F(x) = \sum_{k=0}^{\infty} c_k (x - a)^k$$

is differentiable (and hence, in particular, continuous) on the interval of convergence of radius R, and

$$D(F)(x) = \sum_{k=0}^{\infty} D(c_k(x-a)^k)$$

$$= \sum_{k=1}^{\infty} kc_k(x-a)^{k-1}$$

for $|x-a| < R$. An antiderivative of F can be obtained by taking antiderivatives of each summand, i.e.,

$$\int F(x)dx = \sum_{k=0}^{\infty} c_k \frac{1}{k+1}(x-a)^{k+1}$$

on the interval of convergence.

As the intention is to just give a very brief introduction into power series, we shall skip the proofs of the theorems stated here.

It is clear that Theorem 9.5 can be applied to the power series that represents the derivative $D(F)$, thereby showing that $D(F)$ is differentiable as well, and so on.

Example. Recall the Taylor series

$$\sin x = \sum_{k=0}^{\infty} (-1)^k \frac{1}{(2k+1)!} x^{2k+1}.$$

Taking the derivative according to Theorem 9.5 one obtains

$$(\sin x)' = \sum_{k=0}^{\infty} (-1)^k (2k+1) \frac{1}{(2k+1)!} x^{2k+1-1}$$

$$= \sum_{k=0}^{\infty} (-1)^k \frac{1}{(2k)!} x^{2k},$$

and we recognize the last power series to be the Taylor series of $\cos x$.

One of the simplest power series is the geometric series

$$\sum_{k=0}^{\infty} x^k = 1 + x + x^2 + x^3 + \dots .$$

It is well known that its partial sums satisfy

$$\sum_{k=0}^{n} x^k = \frac{1 - x^{n+1}}{1 - x} \quad \text{for } x \neq 1.$$

(See Problem 7 of Exercise IV.9.5.) Since $x^{n+1} \to 0$ as $n \to \infty$ whenever $|x| < 1$, while there is no limit if $|x| > 1$, it follows that the radius of convergence R is equal to 1, and that

$$\sum_{k=0}^{\infty} x^k = \lim_{n \to \infty} \sum_{k=0}^{n} x^k = \frac{1}{1-x} \text{ for } |x| < 1.$$

The geometric series is indeed the Taylor series of $f(x) = 1/(1-x)$ with center 0; a simple computation shows that $f^{(k)}(0) = k!$, so that $f^{(k)}(0)/k! = 1$ for $k = 0, 1, 2, \ldots$. This is a special case of the following general result.

Theorem 9.6. *If the power series $F(x) = \sum_{k=0}^{\infty} c_k(x-a)^k$ has positive radius of convergence, then*

$$F^{(n)}(a) = n!c_n, \text{ so that } c_n = \frac{F^{(n)}(a)}{n!}.$$

Hence the series agrees with the Taylor series of F centered at a.

This result is an immediate consequence of Theorem 9.5. Just compute successive derivatives of F by taking derivatives of each summand, and evaluate the result at $x = a$. (See Problem 8 of Exercise IV.9.5 for more details.)

The preceding result shows that if a function can be represented by a power series in a neighborhood of some point, then it can be done so in precisely one way, namely by its Taylor series.

IV.9.4   *Analytic Functions*

In the 17th and 18th centuries mathematicians viewed functions as algebraic expressions or as polynomials of infinite degree, i.e., as power series. They operated with these infinite sums in a purely formal way by using standard results for finite sums, without much concern about questions of convergence. Only much later in the 19th century was it recognized that the issue of convergence is quite delicate and cannot just be ignored. Even worse, it was discovered that the function concept is much broader than what was considered by the earlier mathematicians, and that it includes "awful" functions that are continuous on an interval but not differentiable at any point at all. In contrast, functions represented by convergent power series, i.e., by their Taylor series, are most "natural" and enjoy many good properties. But they still form quite a special class of functions. Even if a function f is infinitely often differentiable, it does not necessarily follow that its Taylor polynomials converge to the original function f, in other words, such a function is not necessarily representable by a power series.

Example. Define the function

$$G(x) = \begin{cases} e^{-\frac{1}{x}} & \text{for } x > 0 \\ 0 & \text{for } x \leq 0 \end{cases};$$

the graph of G is shown in Figure IV.23.

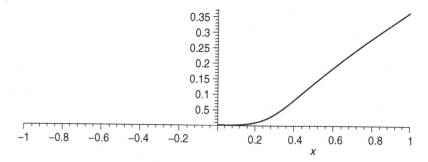

Fig. IV.23   The graph of the function $G(x)$.

It is clear that G is infinitely often differentiable at all points $x \neq 0$. More subtle arguments are required to show that G is also differentiable at 0, and that $G^{(n)}(0) = 0$ for all $n = 1, 2, \ldots$. For example, for the first derivative one needs to consider

$$\lim_{h \to 0} \frac{G(h) - G(0)}{h} = \lim_{h \to 0} \frac{G(h)}{h}.$$

Since the limit from the left $(h \to 0^-)$ is surely zero $(G(h) \equiv 0$ for $h < 0)$, it is enough to consider the limit from the right side, i.e., for $h > 0$, so that

$$\frac{G(h)}{h} = \frac{e^{-1/h}}{h} = \frac{1}{h} \frac{1}{e^{1/h}}.$$

This suggests to replace $t = 1/h$ and consider

$$\lim_{t \to \infty} \frac{t}{e^t}$$

instead. This latter limit exists and equals 0, since the exponential function grows much faster then any power function. (See Problem 11 of Exercise IV.9.5. for details.) It follows that G is differentiable at 0, with $G'(0) = 0$. Higher order derivatives are handled by similar arguments. The important fact that needs to be used is that

$$\lim_{t \to \infty} \frac{t^k}{e^t} = 0 \text{ for any positive integer } k.$$

Since all derivatives of G at 0 are equal to 0, it follows that all Taylor polynomials of G centered at 0 are identically equal to zero, so they do not approximate the values of G at any point $x > 0$.

Approximation by Taylor polynomials is thus a special property that is, in particular, useful for approximating the values of functions to any desired degree of accuracy. Functions that satisfy this property are called *real analytic,* or simply *analytic.* As we saw in the case of the logarithm, the Taylor polynomials centered at a certain point do not necessarily converge at every point of the domain of the function, so the relevant property is a *local* one. More precisely, we make the following definition.

Definition 9.7. *A function f is said to be **real analytic** on the interval I, if for each $a \in I$ the Taylor polynomials centered at a approximate f on some open interval centered at a, that is, there exists a positive number r_a, so that f is represented by its Taylor series*

$$f(x) = \sum_{n=0}^{\infty} \frac{f^{(n)}(a)}{n!}(x-a)^n$$

for all x with $|x - a| < r_a$.

Fortunately, most "natural" functions that one encounters in calculus enjoy this property. For example, appropriate variations of the arguments we used in Section 9.2 to study the Taylor series of e^x, $\sin x$, $\cos x$ centered at 0 show that the corresponding Taylor series centered at an arbitrary point a do indeed converge to the respective functions on the whole real axis.

Example. We show that the function $f(x) = 1/(1-x)$ is real analytic on its domain $\{x : x \neq 1\}$. We already considered the Taylor series centered at $a = 0$. More generally, if $a \neq 1$, the simple modification

$$\frac{1}{1-x} = \frac{1}{(1-a)-(x-a)} = \frac{1}{(1-a)}\frac{1}{1 - \frac{x-a}{1-a}}$$

$$= \frac{1}{(1-a)}\sum_{k=0}^{\infty}\left(\frac{x-a}{1-a}\right)^k \quad \left(\text{provided } \left|\frac{x-a}{1-a}\right| < 1\right)$$

$$= \sum_{k=0}^{\infty}\left(\frac{1}{1-a}\right)^{k+1}(x-a)^k \text{ for } |x-a| < |1-a| = r_a,$$

shows that $f(x) = 1/(1-x)$ can be represented by a power series, which must be its Taylor series (why?) in a neighborhood of a. We also note that the radius of convergence $r_a = |1 - a|$ is as large as it could possibly be, namely the distance from a to 1, the point where $f(x)$ has a singularity.

More generally, it is known that every algebraic function f in the class $\mathcal{A}$ is analytic on its domain. This latter result is significantly deeper than the results we have discussed in this section. It is usually proved by considering the extension of such functions to *complex* numbers and using fundamental results from "complex analysis". We shall give a brief introduction to complex numbers in the next section, where we shall discuss some remarkable formulas that connect the exponential function and the trigonometric functions.

IV.9.5 *Exercises*

1. Suppose $f(x) = x^3 - 2x^2 + 4x + 1$. Note that the explicit expression that defines f is the Taylor polynomial $P_3(x; 0)$.

   a) Find the Taylor polynomial $P_3(x; 1)$ of f centered at the point $a = 1$.
   b) Show that $f(x) = P_3(x; 1)$ for all x by using Taylor's Theorem to estimate $f(x) - P_3(x; 1)$.
   c) Verify $f(x) = P_3(x; 1)$ directly by algebra.
   d) Explain why $f(x) = P_3(x; a)$ for any point a.

2. Generalizing Problem 1, show that if f is a polynomial of degree n and $P_n(x; a)$ is its Taylor polynomial of degree n centered at the point a, then $f(x) = P_n(x; a)$ for all $x \in \mathbb{R}$.

3. Suppose $0 < r < 1$. Show that $r^n \to 0$ as $n \to 0$. (Hint: Set $\lambda = \inf\{r^n : n = 1, 2, 3, ...\}$; then $\lambda \geq 0$. Show that $\lambda > 0$ leads to a contradiction.)

4. According to the result proved in Section 9.2,

$$\ln x = \lim_{n \to \infty} \sum_{k=1}^{n} (-1)^{k-1} \frac{(x-1)^k}{k}$$

   for all x with $0 < x \leq 2$. Show that if $|x - 1| > 1$, in particular if $x > 2$, then the Taylor polynomials $P_n(x; 1)$ do not have any limit as $n \to \infty$, and hence this representation for $\ln x$ fails. (Hint: Show that if $x = 2 + d$, where $d > 0$, then $\frac{(x-1)^n}{n} = \frac{(1+d)^n}{n} > d$ for all $n = 1, 2, ...$.)

5. Find a positive integer N so that $\left| e - \sum_{j=0}^{N} \frac{1}{j!} \right| < 10^{-10}$, i.e., so the approximation of e given by $\sum_{j=0}^{N} \frac{1}{j!}$ is accurate to 10 digits past the

decimal point. (Hint: Let $f(x) = e^x$, and approximate $f(1)$ by Taylor polynomials centered at $a = 0$.)

6. Show that $\sum_{k=0}^{\infty} k! \, x^k$ does not converge for any $x \neq 0$. (Hint: Use the Lemma from Section 9.2, to the effect that $c^k/k! \to 0$, and hence the reciprocal $k! \, / \, c^k \to \infty$ for any $c > 0$.)

7. Let r be any number, and consider the finite geometric series $S_n(r) = \sum_{k=0}^{n} r^n$.

   a) Show that $rS_n(r) = S_n(r) - 1 + r^{n+1}$.

   b) Solve the equation in a) for $S_n(r)$ in the case $r \neq 1$ to obtain the summation formula for the geometric series.

8. Let $F(x) = \sum_{k=0}^{\infty} c_k(x - a)^k$ have positive radius of convergence R.

   a) Show that $F(a) = c_0$.

   b) Use the Theorem in the text to find the power series for $F'(x)$, and use the result to show $F'(a) = c_1$.

   c) Repeat b) to find $F''(x)$ from the series for $F'(x)$, and show that $F''(a) = 2c_2$.

   d) Show in general that

   $$F^{(n)}(x) = \sum_{k=n}^{\infty} c_k \, k \cdot (k-1) \cdot (k-2) \cdot \ldots \cdot (k-n+1) \, (x-a)^{k-n}$$

   on the interval of convergence and that $F^{(n)}(a) = c_n \, n(n-1)\ldots 2 \cdot 1$.

9. Find the Taylor series of

   $$f(x) = \frac{1}{(1-x)^2}$$

   centered at $a = 0$. (Hint: Differentiate the geometric series.)

10. The function $g(x) = \frac{e^x - 1}{x}$ is not defined at $x = 0$. Still, $g(x)$ is represented by a power series for $x \neq 0$. Find that power series!

11. a) Show that for any fixed integer k one has $\lim_{x \to \infty} \frac{e^x}{x^k} = \infty$. (Hint: $e^x \geq P_{k+1}(x)$ for $x > 0$. What happens with $\frac{P_{k+1}(x)}{x^k}$ as $x \to \infty$?)

    b) Explain why a) implies that $\lim_{x \to \infty} \frac{x^k}{e^x} = 0$ for fixed k.

IV.10   Excursion into Complex Numbers and the Euler Identity

As one compares the Taylor series of the exponential function and those of the basic trigonometric functions, one cannot help but be struck by the similarities. If it were not for the alternating signs, the exponential function $Ex) = e^x$ appears to be the sum of $\cos x$ and $\sin x$. The hidden connection becomes clearly visible if one expands the real numbers to the *complex* numbers, and if one allows these more general numbers as inputs for the exponential function. We shall therefore present a brief introduction into this important generalization of the real numbers that plays a fundamental role in analysis and in numerous applications.

IV.10.1   *Complex Numbers*

Complex numbers arise in the context of finding square roots of negative numbers. It looks like magic... just create a symbol i (for *imaginary*) to represent an object that satisfies the formal multiplication rule $i^2 = i \cdot i = -1$, so that i is a candidate for $\sqrt{-1}$. By formally applying addition and multiplication to real numbers and i, one is led to consider the set $\mathbb{C}$ of "complex numbers" defined by

$$\mathbb{C} = \{a + bi : a, b \in \mathbb{R}\}.$$

One then adds and multiplies such complex numbers according to the standard rules of arithmetic, keeping in mind to simplify $i^2 = -1$ wherever it occurs. For example,

$$(3 + 5i) + (2 - 3i) = 3 + 2 + (5i - 3i)$$
$$= 5 + (5 - 3)i = 5 + 2i,$$

or

$$(3 + 5i) \cdot (2 - 3i) = 3 \cdot 2 - 3 \cdot 3i + 5i \cdot 2 - 5 \cdot 3i^2$$
$$= 6 - 9i + 10i - 15(-1)$$
$$= 6 + 15 + (10 - 9)i$$
$$= 21 + i.$$

By following this process, one obtains the general rules

$$(a + bi) + (c + di) = (a + c) + (b + d)i$$

and

$$(a + bi) \cdot (c + di) = (ac - bd) + (ad + bc)i.$$

Clearly $a + bi = 0$ if and only if $a = b = 0$, and the formula $a = a + 0i$ shows that the real numbers are a subset of $\mathbb{C}$. One checks that all the familiar rules of algebra continue to hold in the set of complex numbers with the arithmetic operations defined as above. In particular, every non-zero complex number $z = a + bi$ has a multiplicative inverse z^{-1} given by

$$z^{-1} = \frac{a}{a^2 + b^2} - \frac{b}{a^2 + b^2} i.$$

(Note that $a^2 + b^2 > 0$ if $z \neq 0$, since at least one of the numbers a, b must be non-zero!). Just use the rule for multiplying two complex numbers to calculate

$$z \cdot z^{-1} = 1 + 0i = 1.$$

Much of the mystery surrounding the symbol i is removed by identifying a complex number $a + bi$ with the point in the plane whose Cartesian coordinates are (a, b). (See Figure IV.24.) In analogy to the real number *line*, this geometric representation of $\mathbb{C}$ is called the complex number *plane*.

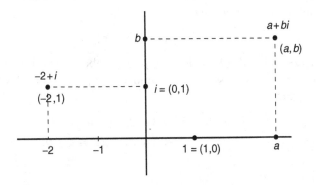

Fig. IV.24   The complex number plane.

Addition and multiplication of points in the plane are now defined according to the rules we had found earlier, i.e., in particular one has

$$(a, b) \cdot (c, d) = (ac - bd, ad + bc). \tag{IV.17}$$

Real numbers are identified with the points on the horizontal axis, i.e., $a \leftrightarrow (a, 0)$, and points $(0, b)$ on the vertical axis are identified with the *purely imaginary* numbers bi. In particular, the mysterious symbol i is now identified with the concrete point $(0, 1)$ in the plane, and according to the multiplication rule (IV.17) one indeed has

$$i \cdot i \leftrightarrow (0, 1) \cdot (0, 1) = (0 \cdot 0 - 1 \cdot 1, 0 \cdot 1 + 1 \cdot 0)$$
$$= (-1, 0) \leftrightarrow -1.$$

The coordinate plane is the favorite geometric model for the set $\mathbb{C}$ of complex numbers.

The absolute value $|z|$ of a complex number $z = a + ib$ is defined to be the (Euclidean) distance from z to 0, that is, $|z| = \sqrt{a^2 + b^2}$. This generalizes the absolute value $|a| = \sqrt{a^2}$ of a real number. While all *arithmetic* operations and rules extend from $\mathbb{R}$ to $\mathbb{C}$, it is not possible to extend the *order properties* of the real numbers to the complex plane. The relation $z > 0$ is meaningless for a complex number z. The only meaningful order relation in $\mathbb{C}$ applies to absolute values. In particular, one can show that the triangle inequality $|z + w| \le |z| + |w|$ holds for arbitrary complex numbers z and w.

IV.10.2   *The Exponential Function for Complex Numbers*

To consider functions of complex numbers is particularly simple for polynomials and rational functions, since only algebraic operations are involved. Just substitute $z = x + iy$ for x in the explicit expressions of such functions. For example, $p(x) = 3x^4 - 2x^3 + 4$ leads to

$$p(z) = 3z^4 - 2z^3 + 4$$

for any $z = x + iy$. Hence $p(z)$ can easily be evaluated for any x and y. However, for a function like $f(x) = 2^x$, the meaning of $f(z) = 2^z = 2^{x+iy}$ is not clear at all. Since for any $b > 0$ one has $b^x = e^{x \ln b}$, it is enough to concentrate on the natural exponential function $E(x) = e^x$. Here the Taylor series expansion provides a natural procedure for giving meaning to $E(z) = E(x + iy)$. In fact, since for any real number x the function $E(x)$ is the limit of the appropriate Taylor polynomials $P_n(x)$, one is led to consider

$$E(z) = \lim_{n \to \infty} P_n(z) = \lim_{n \to \infty} \sum_{k=0}^{n} \frac{1}{k!} z^k$$
$$= \sum_{k=0}^{\infty} \frac{1}{k!} z^k.$$

Given the representation of complex numbers as points in the plane, the concept of limit of a function of complex numbers has an intuitive geometric meaning. Many details need to be checked, but we can safely assume that the intuitive meaning can be given a precise interpretation that does indeed satisfy all the necessary properties that we are familiar with from the real numbers.

Given the central role of the functional equation for exponential functions, it is particularly important to verify that the power series representation for $E(z)$ still satisfies the functional equation for complex inputs, i.e., that one has

$$E(z + w) = E(z)E(w) \text{ for all } z, w \in \mathbb{C}.$$

In order to verify this, we recall the binomial theorem from algebra, which states that for any natural number n one has

$$(z + w)^n = \sum_{j=0}^{n} \binom{n}{j} z^j w^{n-j}, \tag{IV.18}$$

where the binomial coefficient is defined by

$$\binom{n}{j} = \frac{n!}{j!(n-j)!}.$$

By replacing $n - j = k$, we can rewrite equation (IV.18) in a more symmetric form

$$(z + w)^n = n! \sum_{\substack{j,k=0 \\ j+k=n}}^{n} \frac{z^j}{j!} \frac{w^k}{k!}.$$

By applying the formal rules of algebra and rearranging terms as needed, and assuming that all operations remain correct for the infinite sums, i.e., for the underlying limit statements, it follows that

$$E(z + w) = \sum_{n=0}^{\infty} \frac{1}{n!}(z + w)^n = \sum_{n=0}^{\infty} \sum_{\substack{j,k=0 \\ j+k=n}}^{n} \frac{z^j}{j!} \frac{w^k}{k!}$$

$$= \sum_{j=0}^{\infty} \sum_{k=0}^{\infty} \frac{1}{j!} z^j \frac{1}{k!} w^k = \left(\sum_{j=0}^{\infty} \frac{1}{j!} z^j \right) \left(\sum_{k=0}^{\infty} \frac{1}{k!} w^k \right)$$

$$= E(z)E(w).$$

We thus see that—subject to verification of the appropriate limit statements for complex numbers—the critical internal law of the exponential function is preserved as one extends the Taylor series for $E(x)$ to complex numbers. This confirms that the chosen definition of $E(z)$ for complex numbers retains the fundamental properties of the familiar exponential function for *real* numbers.

IV.10.3 *The Euler Identity*

We are now ready to look more in detail at the complex number $E(iy)$. Notice that

$$E(iy) = \sum_{k=0}^{\infty} \frac{1}{k!}(iy)^k = \sum_{k=0}^{\infty} \frac{1}{k!}(i)^k y^k.$$

The meaning of i^k depends on whether k is even or odd. For even $k = 2m$ one has

$$i^{2m} = (i^2)^m = (-1)^m,$$

while for odd $k = 2m + 1$ one has

$$i^{2m+1} = i^{2m}i = (i^2)^m i = (-1)^m i.$$

Therefore, after separating the terms where $k = 2m$ is even from those where $k = 2m+1$ is odd in the Taylor polynomial P_{2l+1}, and after factoring out the common factor i in the latter sum, one obtains

$$P_{2l+1}(iy) = \sum_{k=0}^{2l+1} \frac{1}{k!}(i)^k y^k$$

$$= \sum_{m=0}^{l} (-1)^m \frac{1}{(2m)!}y^{2m} + i \sum_{m=0}^{l} (-1)^m \frac{1}{(2m+1)!}y^{2m+1}.$$

As we take the limit $l \to \infty$, $P_{2l+1}(iy) \to E(iy)$, and hence

$$E(iy) = \lim_{l \to \infty} \sum_{m=0}^{l} (-1)^m \frac{1}{(2m)!}y^{2m} + i \lim_{l \to \infty} \sum_{m=0}^{l} (-1)^m \frac{1}{(2m+1)!}y^{2m+1}$$

$$= \sum_{m=0}^{\infty} (-1)^m \frac{1}{(2m)!}y^{2m} + i \sum_{m=0}^{\infty} (-1)^m \frac{1}{(2m+1)!}y^{2m+1}.$$

Here we recognize the two infinite sums as the Taylor series of $\cos y$ and $\sin y$, respectively. Hence

$$e^{iy} = E(iy) = \cos y + i \sin y.$$

This remarkable formula is known as the *Euler Identity*.[5] It ties together the basic elementary transcendental functions, and it reveals a deep connection that remains hidden as long as one considers only real numbers.

[5]The formula is named after its discoverer L. Euler (1707 - 1785), one of the most prolific mathematicians of all times.

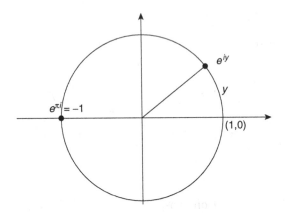

Fig. IV.25 e^{iy} is a point on the circle of radius 1.

While the formula may appear strange at first sight, it does give a concrete representation of the complex number e^{iy} as that point on the unit circle with coordinates $(\cos y, \sin y)$, where y is the length of the arc from $(1,0)$ to that point. (See Figure IV.25.).

Note that the function $f(y) = e^{iy}$ is periodic with period 2π. In particular,

$$e^{2\pi i} = e^0 = 1.$$

Since $e^{\pi i} = \cos \pi + i \sin \pi = -1$, one also obtains

$$e^{\pi i} + 1 = 0.$$

These remarkable formulas tie together some of the most important universal constants used in mathematics!

Finally, we can also consider the exponential function $e^z = e^{x+iy}$ with an arbitrary complex number as input. We already mentioned that the functional equation remains valid for complex numbers. Hence

$$e^{x+iy} = e^x e^{iy} = e^x (\cos y + i \sin y).$$

One evaluates $\left| e^{x+iy} \right| = e^x \neq 0$, so that the exponential function remains non-zero when extended to the complex plane. Furthermore, the equation

$$e^{z+2\pi i} = e^z e^{2\pi i} = e^z \cdot 1 = e^z \text{ for all } z \in \mathbb{C}$$

shows that the exponential function is periodic, with the purely imaginary period $2\pi i$. Of course this periodicity is invisible if one only studies the exponential function along the *real* number line.

Euler's Identity has revealed the advantages of enlarging one's point of view from the real numbers to an even larger number system. More generally, power series, which we briefly discussed in Section 9, are intimately connected to complex numbers, and many of their properties are revealed and better understood only in that context. In particular, it turns out that if a power series $\sum c_k(x-a)^k$ has a positive radius of convergence R then it converges for all complex numbers z with $|z-a| < R$ (the disc of convergence) and it diverges for all z with $|z-a| > R$. To illustrate this feature, let us consider the function

$$g(x) = \frac{1}{1+x^2},$$

which is a simple rational function without any singularities on the real axis. Its Taylor series expansion centered at 0 is obtained by using the geometric series in the form

$$g(x) = \frac{1}{1+x^2} = \frac{1}{1-(-x^2)}$$

$$= \sum_{k=0}^{\infty}(-x^2)^k = \sum_{k=0}^{\infty}(-1)^k x^{2k}.$$

This power series converges for all x with $x^2 < 1$, and it diverges if $x^2 > 1$. Hence its radius of convergence 1. As long as one stays within the real numbers, it is difficult to understand this result, given that g is well behaved on the real line. However, the reason becomes immediately obvious if one allows complex numbers: one must then include the singularities at the complex numbers i and $-i$, which are not visible in $\mathbb{R}$. Hence the complex disc of convergence cannot include these two points that have distance 1 from the center 0 of the power series. The radius of convergence must therefore be 1.

These amazing relationships are just the beginning of what is known as *Complex Analysis*—that is, calculus with complex numbers—a fascinating and central branch of mathematics that has a long history and that is rich in applications.

IV.10.4   *Exercises*

1. Evaluate the following complex numbers in the form $a + bi$, $a, b \in \mathbb{R}$.
   a) $(5+3i) + (2-2i)$;   b) $(3+2i)(4-i)$;   c) $(2+2i)^3$;   d) $(4-3i)^{-1}$.

2. If $w = a + bi \in \mathbb{C}$, introduce the *complex conjugate* $\overline{w}$ of w by $\overline{w} = a - bi$. Note that in terms of the representation of $w = (a, b)$ as a point in the plane, $\overline{w} = (a, -b)$ is just the reflection of w on the x−axis.

   a) Show that $w\overline{w}$ is a real number that satisfies $w\overline{w} \geq 0$, and that $w\overline{w} = 0$ if and only if $w = 0$.

   b) Show that if $w \neq 0$, then its reciprocal $\frac{1}{w} = w^{-1}$ is given by

   $$w^{-1} = \frac{\overline{w}}{w\overline{w}}.$$

3. Continuation of Problem 2. Define the *absolute value* $|w|$ of the complex number w by the formula $|w| = \sqrt{w\overline{w}}$.

   a) Show that if $w = a \in \mathbb{R}$, then $|w|$ as defined here is exactly the standard absolute value of the real number a.

   b) Show that $|w|$ equals the distance from 0 to the point w in the plane.

   c) Show that $\left|e^{it}\right| = 1$ for any real number t.

   d) Show that $|zw| = |z|\,|w|$ for all $z, w \in \mathbb{C}$.

4. Identify the points in the plane that correspond to the complex numbers $e^{i\pi/4}$, $e^{i\pi/2}$ and $e^{i3\pi/2}$.

5. a) Show that every complex number z can be written in the form $z = |z|\,e^{it}$ for some real number t, which may be chosen in the interval $[0, 2\pi)$.

   b) Give a geometric interpretation of the number t that appears in a).

   c) Show that if $z \neq 0$ and if t_1 and t_2 are any two real numbers so that $z = |z|\,e^{it_1} = |z|\,e^{it_2}$, then $t_2 - t_1 = 2k\pi$ for some integer k.

   Remark. The representation $z = |z|\,e^{it}$ is called the *polar representation* of the complex number z. Any number t that satisfies the equation $z = |z|\,e^{it}$ is called an *argument* (or polar angle) of z. Part c) shows that every $z \neq 0$ has exactly one argument t that lies in the interval $(-\pi, \pi]$, i.e., with $-\pi < t \leq \pi$.

Epilogue

The story that began with the ancient problem of finding tangents to familiar curves has led us to some powerful mathematical tools that have found wide applications. Along the way it has revealed to us some remarkable phenomena and relationships, such as:

The world of "limits" that forces us to consider a number concept that goes far beyond the familiar rational numbers...

Different sizes of infinity...

The "natural" base $e = 2.7182818...$ for exponential functions (how can this number be "natural"?)...

The amazing connection between the tangent problem and the area problem and its many applied versions...

The representation of simple algebraic and transcendental objects by computable infinite series...

And finally, a totally unexpected connection between basic exponential growth and fundamental periodic processes that becomes visible as we expand our horizon to the world of complex numbers.

While we have reached the end of this book, the story continues... . I hope that some of our readers have gotten sufficiently intrigued and inspired that they may wish to follow later chapters of the story through other sources. To help such readers, let us finish with a few suggestions. For example, the reader who would like to learn more about the precise technical language that has been developed to capture limits and other approximation processes, and who wants to learn about further topics in analysis involving functions of a single variable, could consult S.R. Lay, *Analysis, With an Introduction to Proof*, 5th ed., Pearson, 2014, R. G. Bartle and D. R. Sherbert, *Introduction to Real Analysis*, 4th ed., John Wiley, 2011, or W. Rudin, *Principles of Mathematical Analysis*, 3rd ed., McGraw-Hill,

1976. *Multidimensional* analysis is covered, for example, in G. B. Folland, *Advanced Calculus*, Prentice Hall 2002., although it would be helpful to first learn about the basics of Linear Algebra. Further applications of the methods of calculus can be found, for example, in G. F. Simmons, *Differential Equations with Applications and Historical Notes*, McGraw-Hill 1972. Finally, to learn more about the fascinating world of complex analysis, the reader could consult R. P. Boas, *An Invitation to Complex Analysis*, 2nd ed., Math. Assoc. of America, 2010, or J. P. D'Angelo, *An Introduction to Complex Analysis and Geometry*, Amer. Math. Society, 2011.

Index

Printed in the United States
By Bookmasters